The Urban Book Series

The Urban Book Series is a resource for urban studies and geography research worldwide. It provides a unique and innovative resource for the latest developments in the field, nurturing a comprehensive and encompassing publication venue for urban studies, urban geography, planning and regional development.

The series publishes peer-reviewed volumes related to urbanization, sustainability, urban environments, sustainable urbanism, governance, globalization, urban and sustainable development, spatial and area studies, urban management, transport systems, urban infrastructure, urban dynamics, green cities and urban landscapes. It also invites research which documents urbanization processes and urban dynamics on a national, regional and local level, welcoming case studies, as well as comparative and applied research.

The series will appeal to urbanists, geographers, planners, engineers, architects, policy makers, and to all of those interested in a wide-ranging overview of contemporary urban studies and innovations in the field. It accepts monographs, edited volumes and textbooks.

Indexed by Scopus.

Ransford A. Acheampong · Karen Lucas ·
Michael Poku-Boansi · Chinebuli Uzondu
Editors

Transport and Mobility Futures in Urban Africa

Springer

Editors
Ransford A. Acheampong
University of Manchester
Manchester, UK

Karen Lucas
University of Manchester
Manchester, UK

Michael Poku-Boansi
Kwame Nkrumah University of Science and Technology
Kumasi, Ghana

Chinebuli Uzondu
Federal University of Technology Owerri
Imo State, Nigeria

ISSN 2365-757X ISSN 2365-7588 (electronic)
The Urban Book Series
ISBN 978-3-031-17326-4 ISBN 978-3-031-17327-1 (eBook)
https://doi.org/10.1007/978-3-031-17327-1

This Springer imprint is published by the registered company Springer Nature Switzerland AG
The registered company address is: Gewerbestrasse 11, 6330 Cham, Switzerland

Contents

Editors and Contributors

About the Editors

Ransford A. Acheampong is a Lecturer and Presidential Academic Fellow in Transport and Urban Futures at the Department of Planning and Environmental Management, University of Manchester, UK. The first strand of his research is focused on the nexus between transportation and spatial structure and their integrated planning in cities. The second strand of his research addresses questions around the diffusion and socio-spatial impact of new and emerging transport technologies and mobility solutions, such as app-based mobility, autonomous vehicles, shared-mobility, and Mobility-as-a-Service.

Karen Lucas is a Professor of Human Geography at the University of Manchester where she also leads the Transport and Mobility hub for the Manchester Urban Institute. She has extensively researched issues of transport poverty in African cities and since 2007 has led the International Network for Transport and Accessibility in Low-Income Communities (INTALInC).

Michael Poku-Boansi is a Professor of Planning at the Department of Planning and the Vice Dean, School of Graduate Studies, Kwame Nkrumah University of Science and Technology, Kumasi, Ghana. Michael has over fifteen (15) years of transportation and land use planning, resettlement planning, climate change, and resilience planning experience spanning both academic and private sector appointments. Professor Poku-Boansi is a Fellow of the Rockefeller Foundation Professional Fellowship Programme and a Commonwealth Academic Fellow. He has authored/co-authored several peer-reviewed research articles published in reputable international journals and numerous project reports. Professor Michael Poku-Boansi is also a Fellow of the Ghana Institute of Planners (FGIP).

Chinebuli Uzondu is a Lecturer and Researcher at the Department of Transport Management Technology, Federal University of Technology Owerri, Nigeria. She has previously worked as a Post-doctoral researcher with the Human Factors and Safety Group at the Institute for Transport Studies, University of Leeds, United Kingdom, where she obtained her Ph.D. Her research seeks to understand road user behaviour through projects funded by various research organisations and the Nigerian and UK governments. This involves using both quantitative and qualitative research methods to inform the design of transport infrastructure and/or technology interventions and to evaluate the effectiveness of these interventions.

Contributors

Abraham Chris J. Department of Electrical & Electronic Engineering, Stellenbosch University, Stellenbosch, South Africa

Acheampong Ranford A. Department of Planning and Environmental Management, University of Manchester, Manchester, UK

Adams Charles Kwame Nkrumah University of Science and Technology, Kumasi, Ghana

Adanu Emmanuel Alabama Transportation Institute, Alabama, USA

Agergaard Jytte Department of Geosciences and Natural Resource Management, University of Copenhagen, Copenhagen, Denmark

Ambunda Robert University of Namibia, Windhoek, Namibia;
Stellenbosch University, Stellenbosch, South Africa

Anciaes Paulo Centre for Transport Studies, University College London, London, UK

Ankomah Marian Traffic Engineer/Road Engineer, Fekams Company Limited, Fekams House, Accra, Ghana

Asabere Stephen Boahen Department of Physical Geography, University of Goettingen, Goettingen, Germany

Asibey Michael Osei Department of Planning, Kwame Nkrumah University of Science and Technology, Kumasi, Ghana

Asimeng Emmanuel Theodore German Institute Of Development and Sustainability (IDOS), Bonn, Germany;
Institute of Urban and Regional Planning, Technische Universität Berlin, Berlin, Germany

Asuah Augustine Yaw Department of Planning and Environmental Management, University of Manchester, Manchester, UK

Behrens Roger University of Cape Town, Cape Town, South Africa

Booysen Marthinus J. Department of Electrical & Electronic Engineering, Stellenbosch University, Stellenbosch, South Africa;
Department of Industrial Engineering, Stellenbosch University, Stellenbosch, South Africa

Bradbury Annabel Ricardo Energy and Environment, Didcot, UK

Chabane Lila Research Centre in Applied Economics for Development, Bouzareah, Algeria

Dzisi Emmanuel Komla Junior Department of Civil Engineering, Kwame Nkrumah University of Science and Technology, Kumasi, Ghana

Epo Boniface Ngah University of Yaoundé II, Yaoundé, Cameroon

Etika Anderson Aja Department of Civil Engineering, Cross River University of Technology, Calabar, Nigeria

Jones Steven University of Alabama, Alabama, USA;
Alabama Transportation Institute, Alabama, USA

Lucas Karen Department of Geography, School of Environment, Education and Development, The University of Manchester, Manchester, United Kingdom

Lukomona Raymond Department of Physical Planning, Solwezi, Zambia

Møller-Jensen Maya Department of Geosciences and Natural Resource Management, University of Copenhagen, Copenhagen, Denmark

Nana Djomo Jules Médard University of Yaoundé II, Yaoundé, Cameroon

Nchito Wilma S. Department of Geography and Environmental Studies, University of Zambia, Lusaka, Zambia

Ndibatya Innocent Department of Electrical & Electronic Engineering, Stellenbosch University, Stellenbosch, South Africa;
Alabama Transportation Institute, University of Alabama, Tuscaloosa, AL, USA

Obeng Daniel Atuah Department of Civil Engineering, Kwame Nkrumah University of Science and Technology, Kumasi, Ghana

Poku-Boansi Michael Department of Planning, Kwame Nkrumah University of Science and Technology, Kumasi, Ghana

Rix Arnold J. Department of Electrical & Electronic Engineering, Stellenbosch University, Stellenbosch, South Africa

Ryseck Bianca University of Cape Town, Cape Town, South Africa

Srinivasan Sivaramakrishnan Department of Civil and Coastal Engineering, University of Florida, Gainesville, FL, USA

Uzondu Chinebuli Department of Transport Management Technology, Federal University of Technology Owerri, Owerri, Imo State, Nigeria

Wood Astrid School of Geography, Politics and Sociology, Newcastle University, Newcastle upon Tyne, UK

Yankson Eric Namibia University of Science and Technology, Windhoek, Namibia

Chapter 1
Introduction: Transport and Mobility Situations of African Cities

Ransford A. Acheampong, Karen Lucas, Michael Poku-Boansi, and Chinebuli Uzondu

Globally, the need to make cities inclusive, safe, resilient and sustainable is recognized. This recognition is reflected in the ambitious SDG target to: *"by 2030, provide access to safe, affordable, accessible and sustainable transport systems for all, improving road safety, notably by expanding public transport, with special attention to the needs of those in vulnerable situations, women, children, persons with disabilities and older persons"* (UN Department of Economic and Social Affairs 2020). Creating equitable, environmentally sustainable, safe, secure and efficient transportation systems remains a massive challenge in rapidly urbanizing areas in Africa. Across the continent, urban transport and mobility problems, including increasing levels of inequality of access to opportunities, rising levels of motorization, transport-related pollution and deteriorating urban environments, are becoming increasingly pronounced.

The prevailing challenges of mobility and accessibility in urban Africa are fundamentally rooted in the unfolding, largely unplanned and informal structural transformations of the continent's major towns and cities. Throughout the African continent,

R. A. Acheampong (✉)
Department of Planning and Environmental Management, University of Manchester, Manchester, UK
e-mail: ransfordantwi.acheampong@manchester.ac.uk

K. Lucas
Department of Geography, University of Manchester, Manchester, UK
e-mail: karen.lucas@manchester.ac.uk

M. Poku-Boansi
Department of Planning, Kwame Nkrumah University of Science and Technology, Kumasi, Ghana
e-mail: mpoku-boansi.cap@knust.edu.gh

C. Uzondu
Department of Transport Management Technology, Federal University of Technology, Owerri, Nigeria
e-mail: Chinebuli.uzondu@futo.edu.ng

R. A. Acheampong et al. (eds.), *Transport and Mobility Futures in Urban Africa*, The Urban Book Series, https://doi.org/10.1007/978-3-031-17327-1_1

historical population growth, urbanization and physical infrastructure development continue to drive rapid, unfettered outward expansion of major towns and cities. Consequently, the urban extent of major cities has increased significantly in the last few decades (Xu et al. 2019). For example, it is estimated that the urban extent of Lagos increased by four-fold (i.e. 297.38 to 826.84 km^2) between the mid-1980s and 2013, while that of Cairo tripled (i.e. 412.03 to 1363.96 km^2) between 1992 and 2013 (Angel et al. 2012). Within the Accra conurbation, it is estimated that urban expansion has doubled over the last three decades from 42.1 km^2 in 1985 to 89 km^2 in 2017 (Asabere et al. 2020).

The evolving urban spatial structures in the context of this rapid urbanization have created a growing mismatch between the demand for affordable mobility options and its supply, which together with the recent rapid growth in motorization, have contributed to urban land use and transportation systems that are socially, economically and environmentally unsustainable (Sietchiping et al. 2012; May and Marsden 2010).

The relatively affluent elites within African society increasingly resort to private car use dependency in response to the paucity and poor quality of public transport systems, which continue to be dominated by low-capacity vehicles and informal operators (*e.g. Trotros and Matutas*). Motorization levels are also on the increase (Green-Simms 2017; Adarkwa and Poku-Boansi 2011; Kenworthy 2011), and although in the early stages of motorization, most major African cities have already experienced the severely negative impacts of road traffic. Traffic congestion is a typical feature, especially in the central areas of Africa's major urban centres, worsening accessibility for non-car-based transport, in places such as Nairobi (Campbell et al. 2019), Dar es Salaam (Melbye et al. 2015), Accra (Møller-Jensen et al. 2012) and the Greater Kumasi conurbation (Acheampong and Asabere 2022). Aside from presenting major constraints to urban economic growth (Pirie 2014), urban traffic congestion is responsible for environmental pollution and poor air quality that have negative impacts on the health and well-being of Africa's growing urban population (see e.g., Rajé et al. 2018; Linden et al. 2012; Amegah and Agyei-Mensah 2017).

The benefits and disbenefits of the transport system are not distributed equally across different population groups. Most people living in material poverty within African cities live in informal and slum settlements, many of which are located a long way from the centralized economic and social activities of the city (Venter 2011). Even those, which are located more centrally, cover vast areas of land, most of which lack any formal pathways or transport infrastructures (Lucas et al. 2019). This causes a spatial dislocation between key destinations, such as employment centres, schools, colleges, hospitals and markets and home locations, often leading to long journey distances and travel times, and unaffordable travel costs for many of the poorest citizens (Lucas 2011).

At the same time, the most prevalent discourses and recent investments are focused largely on building new, often large-scale urban transportation infrastructures seen as crucial to the modernization and structural transformation of African cities. Investments have mainly prioritized infrastructures for motorized forms of transport in general and car-based transport in particular. In doing so, on-going interventions pay little or no attention to the everyday mobility needs and capabilities of existing populations or the well-being of future generations.

The challenge now and for the future is even greater for making urban transport humanly safe and environmentally sustainable, especially as urban development and transport planning continue to conceive and plan for motorization as a catalyst for economic and social development. Non-motorized forms of transport, such as cycling and walking, that could provide affordable and environmentally sustainable mobility for the majority of the urban population in Africa, are often positioned as outdated and symbols of poverty, and so are neglected in the planning and design of transport systems (Loo and Siiba 2019).

The vast majority of the urban population in Africa still rely on walking to reach their everyday destinations; the World Health Organization data on physical activity (WHO 2019) identifies that, as average across all African cities, 61% of men and 51% of women walk for over 55 min a day. The high reliance on walking in the current transport context within African cities leads to excessive human exposures to traffic-related risks. Indeed, road traffic crashes involving vulnerable road users is disproportionately higher in Africa than elsewhere globally, as the continent has the highest proportion of deaths at 44% (World Health Organization 2018) and the highest number of fatalities per 100,000 of children below 19 years (UNICEF 2015). This is because urban roads lack separate lanes for cyclists, footpaths and adequate crossings for pedestrians (Uzondu et al. 2018; Reynolds et al. 2009; Poku-Boansi et al. 2019).

Moreover, some intra-urban trips within African cities that are currently dominated by motorized modes, tend to cover relatively shorter distances, especially within inner-city locations. For example, motorized intra-urban work trips in the Kumasi metropolis, Ghana, was found to be relatively shorter, averaging about 4.5 km (Acheampong 2020). Such distances could be undertaken by non-motorized forms of transport such as bicycling. However, many people are unwilling to use non-motorized forms of transport out of choice due to the lack of safe infrastructure, coupled with the general negative perception and attitude that stems from the association of non-motorized transport use with poverty (Mokitimi and Vanderschuren 2017; Kim and de Jong 2011). Even where spaces (footpaths and cycle lanes) were originally made available for walking and cycling, they have been taken over by other uses such as for vehicle parking, rubbish dumping or they are occupied by street traders and vendors (TRL 2002; Acheampong and Siiba 2018).

Alongside the aforementioned long-standing issues and challenges of transport and mobility in urban Africa, on-going social and technological transformations, while presenting new opportunities, are also generating new complex challenges for the creation of sustainable mobility futures. The global agenda of decarbonizing urban transport raises several direct major implications and uncertainties for the sustainable futures of Africa's major cities (Collet et al. 2020). Developing eco-friendly transport systems will undoubtedly be critical to the sustainable futures of African cities. Cleaner transport technologies, such as electric powered vehicle engines, promise pathways to sustainable futures that may be relevant to Africa. However, harnessing these technologies to solve some of the pressing transport and mobility problems effectively without aggravating the already worsening problems of energy insecurity, poverty, social and economic exclusion and associated inequalities, remains a major challenge.

Furthermore, in the last few years, there has been a growing presence of digital platform mobility services, provided by Transportation Network Companies (TNCs) such as Uber and Bolt in major cities across Africa. The presence of TNCs and their ICT-mediated transport solutions on the continent has not only exposed major lapses in existing public transport regulatory regimes of city authorities, but also raises critical challenges for urban transport policy and governance in the years to come (see e.g., Agyemang 2020; Pollio 2019; Henama et al. 2017).

Furthermore, the emerging evidence suggests that in Africa, like the rest of the world, new digital platform/app-based mobility services, such as ride-hailing or ride-sourcing are already triggering new travel behaviours and presenting new safety and security challenges (Acheampong 2021; Acheampong et al. 2020; Dzisi et al. 2020; Giddy 2020; Huang et al. 2019). The larger mass of poor and vulnerable populations who are most affected by the prevailing mobility and accessibility challenges in urban Africa are largely excluded from benefiting from these new and emerging mobility solutions, due to limited access to ICTs, such as broadband internet, limited service coverage and affordability issues. Thus, ICT-mediated mobility solutions may be widening mobility inequalities, and so raising serious wider social and environmental sustainability implications for urban transport planning and operations within African cities.

As the contributions in this book will show, urban areas in Africa share many common challenges in creating inclusive and sustainable urban transport futures. It would, however, be impossible to address them effectively without a deeper understanding of the unique spatial, socio-cultural, political and economic contexts that generate and reproduce these challenges. It is, therefore, through empirically grounded, policy-relevant research, that we can have a better understanding of the challenges of urban transport and mobility and their interconnections with other complex urban problems in Africa. Through research, we can also identify the opportunities of today and critically engage with ways of creating desired transport and mobility futures in urban Africa.

The overarching goal of this book is to contribute to the aforementioned imperatives. To this end, this edited volume is written from an interdisciplinary perspective for an interdisciplinary and multi-sectoral audience exploring various aspects of transport and mobility in the African context. Our aim, therefore, with this book is to provide a rich collection of well-researched contributions addressing a wide range of transport and mobility issues from multiple countries and urban areas in

Africa. In doing so, we offer a series of insightful empirical studies that situate transport and mobility challenges in the unique context of individual countries and cities represented in this volume, while highlighting commonalities across the African continent. The theme of 'futures' is critical and central to the volume. Each chapter's contribution leverages a forward-looking perspective as a way of building on their empirical findings to engage explicitly with the normative imperatives that are critical to improving the transport and mobility situations of African cities now and in the future.

The contributions of this book are organized in Parts of five interrelated themes, following this introduction chapter. We will begin each section by providing a short introductory narrative that summarizes the focus of each chapter, and outlines the key concepts, issues and debates within the theme explored in that section.

A brief overview of the five Parts are presented here as follows: Part 1 addresses the theme of ***urban form, accessibility and travel demand***. Together, the contributions in this section provide insights into how rapid levels of historical urbanization have shaped emergent urban spatial structures, and the fundamental role of urban structural factors, such as the spatial configuration of land use and transport infrastructure in shaping mobility and accessibility outcomes in African cities.

The theme of ***transport poverty, equity and inclusion*** **is** addressed in Part 2. The contributions in this section address important questions including how we can conceptualize and measure the distributional impacts of transport and mobility interventions; the differential impacts of major road transport infrastructure development on communities; the everyday mobility experiences of women; and the interconnections between transport, accessibility to social amenities and livelihoods.

The contributions in Part 3 focus on ***public transport policy and governance***, covering important issues associated with the management of public transport facilities; governance of public transport reforms; and the politics of pricing urban transport services, based on case studies from multiple urban areas in Africa.

The primary thematic focus of Part 4 is ***non-motorized transport***. In this section, we examine the challenges of non-motorized transport (NMT), including walking and cycling in African cities. We learn through a number of case studies, the road safety challenges for NMT and vulnerable road users; the challenges and opportunities of getting more people to cycle and the steps being taken to make some urban areas in Africa walkable and bikeable.

Lastly in Part 5, we address the theme of ***ICTs, platforms and new technologies***. The contributions in this section explore the opportunities and challenges of new and emerging transport technologies and mobility services for sustainable transport in urban Africa. Topics covered in this section include the diffusion and emerging impacts of digital platform mobility services (ride-hailing or ride-sourcing), pathways and impacts of the transition from internal combustion engines to electric vehicles, and the prospects of leveraging ICTs to ensure safety and security among users of public transports.

References

Acheampong RA (2020) Spatial structure, intra-urban commuting patterns and travel mode choice: analyses of relationships in the Kumasi Metropolis. Ghana Cities 96:102432

Acheampong RA (2021) Societal impacts of smart, digital platform mobility services—an empirical study and policy implications of passenger safety and security in ride-hailing. Case Stud Transp Policy 9(1):302–314

Acheampong RA, Asabere SB (2022) Urban expansion and differential accessibility by car and public transport in the Greater Kumasi city-region, Ghana—A geospatial modelling approach. J Transp Geogr 98:103257

Acheampong RA, Siiba A (2018) Examining the determinants of utility bicycling using a socio-ecological framework: an exploratory study of the Tamale Metropolis in Northern Ghana. J Transp Geogr 69:1–10

Acheampong RA, Siiba A, Okyere DK, Tuffour JP (2020) Mobility-on-demand: an empirical study of internet-based ride-hailing adoption factors, travel characteristics and mode substitution effects. Transp Res Part c: Emerg Technol 115:102638

Adarkwa KK, Poku-Boansi M (2011) Rising vehicle ownership, road way challenges, and traffic congestion in Kumasi. In: Adarkwa KK (ed) The future of the tree: managing the growth and development of Kumasi. University Press, Kumasi, pp 128–152

Agyemang E (2020) Uber is here to stay: exploring the policy implications of the Uber-Local Taxis turf war in Accra, Ghana. Case Stud Transp Policy 8(1)

Amegah AK, Agyei-Mensah S (2017) Urban air pollution in Sub-Saharan Africa: time for action. Environ Pollut 220:738–743

Angel S, Blei AM, Civco DL, Parent J (2012) Atlas of urban expansion. Lincoln Institute of Land Policy, Cambridge, p 397

Asabere SB, Acheampong RA, Ashiagbor G, Beckers SC, Keck M, Erasmi S, Sauer D (2020) Urbanization, land use transformation and spatio-environmental impacts: analyses of trends and implications in major metropolitan regions of Ghana. Land Use Policy 96:104707

Campbell KB, Rising JA, Klopp JM, Mbilo JM (2019) Accessibility across transport modes and residential developments in Nairobi. J Transp Geogr 74:77–90

Collett KA, Byamukama M, Crozier C, McCulloch M (2020) Energy and transport in Africa and South Asia. Energy Econ Growth

Dzisi EK, Ackaah W, Aprimah BA, Adjei E (2020) Understanding demographics of ride-sourcing and the factors that underlie its use among young people. Sci Afr 7:e00288

Giddy JK (2020) Travel patterns of Uber users in South Africa. In: New Directions in South African tourism geographies. Springer, Cham, pp 113–127

Green-Simms LB (2017) Postcolonial automobility: car culture in West Africa. U of Minnesota Press

Henama US, Sifolo PPS (2017) Uber: The South Africa experience. Afr J Hosp, Tour Leis 6(2):1–10

Huang JY, Majid F, Daku M (2019) Estimating effects of Uber ride-sharing service on road traffic-related deaths in South Africa: a quasi-experimental study. J Epidemiol Community Health 73(3):263–271

Kenworthy J (2011) An international comparative perspective on fast-rising motorization and automobile dependence. Urban transport in the developing world: a handbook of policy and practice, pp 71–112

Kim P, de Jong R (2011) High time to change road investment patterns in http://www.unep.org/Transport/sharetheroad/PDF/StR_article_UrbanWorldDec2010.pdf Accessed 21 Dec 2020

Linden J, Boman J, Holmer B, Thorsson S, Eliasson I (2012) Intra-urban air pollution in a rapidly growing Sahelian city. Environ Int 40:51–62

Loo BP, Siiba A (2019) Active transport in Africa and beyond: towards a strategic framework. Transp Rev 39(2):181–203

Lucas K (2011) Making the connections between transport disadvantage and the social exclusion of low-income populations in the Tshwane Region of South Africa. J Transp Geogr 19(6):1320–1334

Lucas K, Tsoneva E, Abane A, Amoako-Sakyi R, Owusu S, Porter G, Jennings G, Behrens R, Zuidgeest M, Venter C, Asenime C, Oviedo D, Davila J, Mitullah W, Opiyo R, Basil P, Lwasa S, Kasaija P, Evans J (2019) Transport and social exclusion in five African cities. Online Volvo Research and Education Foundation (VREF). http://www.vref.se/download/18.45182a5f16a84e95fac8b2b5/1561370471910/Transport%20and%20Social%20Exclusion%20in%20five%20African%20Cities%20-%20June%202019.pdf

May A, Marsden S (2010) Urban transport and mobility. International transport forum, Germany, pp 3–5

Melbye DC, Møller-Jensen L, Andreasen MH, Kiduanga J, Busck AG (2015) Accessibility, congestion and travel delays in Dar es Salaam–A time–distance perspective. Habitat Int 46:178–186

Mokitimi MM, Vanderschuren M (2017) The significance of non-motorised transport interventions in South Africa—A rural and local municipality focus. Transp Res Procedia 25C:4802–4825

Møller-Jensen L, Kofie RY, Allotey AN (2012) Measuring accessibility and congestion in Accra. Nor Geogr Tidsskr-nor J Geogr 66(1):52–60

Pirie G (2014) Transport pressures in urban Africa: practices, policies, perspectives. In: Pieterse DE, Parnell S (eds) Africa's urban revolution: policy pressures, pp 133–147

Poku-Boansi M, Amoako C, Obeng Atuah D (2019) Urban travel patterns and safety among school children around Accra, Ghana. J Transp Health 15:100660. https://doi.org/10.1016/j.jth.2019.100660

Pollio A (2019) Forefronts of the sharing economy: uber in Cape Town. Int J Urban Reg Res 43(4):760–775

Rajé F, Tight M, Pope FD (2018) Traffic pollution: a search for solutions for a city like Nairobi. Cities 82:100–107

Reynolds CCO, Harris MA, Teschke K, Cripton PA, Winters M (2009) The impact of transportation infrastructure on bicycling injuries and crashes: a review of the literature. Environ Health 8(47):21

Sietchiping R, Permezel MJ, Ngomsi C (2012) Transport and mobility in sub-Saharan African cities: an overview of practices, lessons and options for improvements. Cities 29(3):183–189

Transport Research Laboratory (2002) Scoping study: urban mobility in three Cities. SSATP working paper no. 70

UN Department of Economic and Social Affairs (2020) Sustainable development goals. https://sdgs.un.org/goals

Uzondu CC, Jamson S, Lai F (2018) Exploratory study involving observation of traffic behaviour and conflicts in Nigeria using the traffic conflict Technique. Saf Sci 110(part a):273–284

Venter C (2011) Transport expenditure and affordability: the cost of being mobile. Dev South Afr 28(1):121–140

World Health Organisation (2019): Road traffic injuries report. https://www.who.int/news-room/fact-sheets/detail/road-traffic-injuries

World Health Organization (2018) Global status report on road safety 2018. World Health Organization, Luxembourg. Accessed from: https://www.who.int/violence_injury_prevention/road_safety_status/2018/en/

Xu G, Dong T, Cobbinah PB, Jiao L, Sumari NS, Chai B, Liu Y (2019) Urban expansion and form changes across African cities with a global outlook: spatiotemporal analysis of urban land densities. J Clean Prod 224:802–810

Ransford A. Acheampong is a Lecturer and Presidential Academic Fellow in Transport and Urban Futures at the Department of Planning and Environmental Management, University of Manchester, UK. The first strand of his research is focused on the nexus between transportation and spatial structure, and their integrated planning in cities. The second strand of his research addresses questions around the diffusion and socio-spatial impact of new and emerging transport technologies and mobility solutions, such as app-based mobility, autonomous vehicles, shared-mobility and Mobility-as-a-Service.

Karen Lucas is Professor of Human Geography at the University of Manchester where she also leads the Transport and Mobility hub for the Manchester Urban Institute. She has extensively researched issues of transport poverty in African cities and, since 2007 has led the International Network for Transport and Accessibility in Low Income Communities (INTALInC).

Michael Poku-Boansi is a Professor of Planning at the Department of Planning, and the Vice Dean, School of Graduate Studies, Kwame Nkrumah University of Science and Technology, Kumasi, Ghana. Michael has over fifteen (15) years' of transportation and land use planning, resettlement planning, climate change and resilience planning experience spanning both academic and private sector appointments. Professor Poku-Boansi is a Fellow of the Rockefeller Foundation Professional Fellowship Programme and a Commonwealth Academic Fellow. He has authored/co-authored several peer-reviewed research articles published in reputable international journals and numerous project reports. Prof. Michael Poku-Boansi is also a Fellow of the Ghana Institute of Planners (FGIP).

Chinebuli Uzondu is a lecturer and researcher at the department of Transport Management Technology, Federal University of Technology Owerri Nigeria. She has previously worked as a Post-doctoral researcher with the Human Factors and Safety Group at the Institute for Transport Studies, University of Leeds, United Kingdom where she obtained her Ph.D. Her research seeks to understand road user behaviour through projects funded by various research organisations and the Nigeria and UK governments. This involves using both quantitative and qualitative research methods, to inform the design of transport infrastructure and/or technology interventions, and to evaluate the effectiveness of these interventions.

Part I
Urban Form, Accessibility and Travel Demand

Chapter 2
Urban Form, Accessibility, and Travel Demand in African Cities: An Introduction

Ranford A. Acheampong

How are the various land use activities that constitute opportunities or valued destinations distributed? What transportation options exist to provide connectivity to opportunities? How do the existing land use and transportation systems combine to shape differential access to valued destinations? These are the fundamental questions addressed by the three case study chaps in the first sect of this book. In Chap. 3, Acheampong, Asabere, and Asuah examine differential access to public transport in the Accra metropolitan region of Ghana, in the context of rapid urban physical expansion. They extract built-up land from satellite images to quantify historical urban expansion trends, and demonstrate the close connection between the configuration of transportation infrastructure and urban expansion patterns in the region. Within the prevailing spatial form, the authors examine differential walking time access to the public transportation system and ultimately reveal significant first-and-last-mile accessibility deficits, which in turn, affect the overall levels of accessibility to opportunities such as jobs and social amenities. In Chap. 4, Chabane examines disparities in the public transportation system of Algiers, with a focus on bus network coverage and service quality. Chabane shows public transportation inequalities that are fundamentally rooted in the spatial configuration of the city's bus network, as well as poorer levels of service, especially in peri-urban municipalities. The chap highlights the detrimental effects of the prevailing disparities of access on the poorest populations in Algeria's capital and largest urban conurbation. In the fifth and final chap of Part I, Ankomah and Srinivasan present useful insights about travel demand patterns and the underlying determinants in Ghana. They model the effects of socio-demographic and locational factors on various aspects of travel demand, including

R. A. Acheampong (✉)
Department of Planning and Environmental Management, University of Manchester, Manchester, UK
e-mail: ransfordantwi.acheampong@manchester.ac.uk

R. A. Acheampong et al. (eds.), *Transport and Mobility Futures in Urban Africa*,
The Urban Book Series, https://doi.org/10.1007/978-3-031-17327-1_2

mode choice and trip frequencies, and provide the foundations for forecasting travel demand in the context of changing socio-economic characteristics of the population.

To set the wider context for the case studies presented in this sect, we offer a brief overview of the conceptual understandings of the relationship between urban land use and transportation systems. Most importantly, we briefly discuss the factors shaping the spatial forms of African cities and their transportation systems, and outline the fundamental role these systems play in setting the structural conditions for connectivity and potential access to urban opportunities for individuals and social groups.

2.1 Urban Form, Accessibility, and Travel Behaviour: Unpacking the Relationships

A city's land use and transportation systems are proximal determinants of its spatial form and structure. These two systems are dynamic and interact in complex ways to determine the location of activities and opportunities, as well as the overall emergent structure of the city. The 'land-use transport feedback cycle' conceptualizes the complex two-way dynamic link between land-use systems and transportation systems as follows (see Wegner 2004; Wegener and Fürst 2004): The distribution of land uses determines the location of activities. The spatial separation among the various land use activities in the city create the need for interaction, which manifests as travel by different mobility options, such as driving, walking, and cycling. The transportation system is, therefore, needed to facilitate this interaction, by providing connectivity between activity locations and valued destinations. Overtime, the land use and transportation systems co-evolve in response to changes within each system. For example, the development of transport infrastructure, such as highways generate accessibility benefits, which in turn attract new land-use activities. Similarly, the development of land usually is accompanied by the provision of some form of access such as local roads, cycling paths, and pedestrian walkways.

From the foregoing conceptualization, the land use system comprises all activity locations, including residential, employment, industry, and ancillary activities such as shopping, schools, and recreation. In any town or city, the emergent activity distributions are the result of interactions between complex land market processes and/or public sector interventions such as formal land use planning. A fundamental land market process that underpins the land use system is location choice decision in the property and employment markets. Various actors, including individuals, households, property developers, businesses, and the public sector make location decisions. Out of these location decisions ultimately emerge the land use patterns that reflect the distribution of various urban opportunities, generating the structural conditions shaping differential access to opportunities, travel behaviours, and observed mobility patterns (Batty 2013; Wegner 2004).

In the very basic sense, the transportation system consists of the physical infrastructure (e.g. roads, railways, walkways) and the various modes they support (e.g. car, bicycle, walking, etc.). The transportation system and the land use system together determine various aspects of mobility, including trip origins and destinations, individuals' trip scheduling behaviour, and the associated travel mode choices. Together, the two systems also determine the urban spatial form and the associated structural variables, such as land use density and diversity, destination accessibility, and distance to transit, that are known to influence travel behaviours (see e.g., Aditjandra et al. 2013; Gim 2013; Handy et al. 2005).

From the foregoing, one can recognize that the key mechanism linking the land use system and transportation system is accessibility (see e.g. Acheampong and Silva 2015; Hansen 1959). The exact meaning of the concept, however, depends on the context of usage. In the context of land use and transport interaction, the concept is often used to imply the reachability of land-use activities or valued destinations by means of a transport mode or combination of modes (Geurs and Van Wee 2004; Hansen 1959). It is also deployed in the economic sense to refer to restrictions to use specific transport modes due, for example, to their cost (Levinson and Wu 2020; El-Geneidy et al. 2016).

Moreover, several measures of accessibility have been developed and operationalized in different contexts (see Geurs and Van Wee 2004). The two most common measures are place-based and person-based accessibility. The former focuses on the individual with the aim to analyse and measure the activities that they can participate in at any given time. Person-based accessibility, therefore, takes into account the various limitations that affect individuals' activity participation, such as time budgets and availability and travel speed allowed by the transport system (Levinson and Wu 2020; Martens 2017). Location-based or place-based accessibility on the other hand, analyses accessibility between locations at multiple scales. For example, place-based accessibility can be analysed in terms of the number of activities, such as jobs, that can be reached within any time threshold between an origin and destination (see El-Geneidy et al. 2016).

2.2 Rapid Urbanization, Changing Urban Forms and Rising Inequalities of Access in African Cities

The fast-paced urbanization that has unfolded over the decades in Africa has had and will continue to have profound impact on the spatial forms of the continent's cities and their transportation systems. Today, more than 40% of the continent's over 1.2 billion population live in urban settlements, compared to just about 14% in the 1950s (United Nations 2014). By the first half of the twenty-first century, it is estimated that about 55% of the continent's population will be living in cities (United Nations 2014).

Within the context of rapid levels of historical urbanization, the spatial structures of major settlements have radically transformed. Rapid, outward, sprawling expansion typify urban settlements in Africa (Angel et al. 2012). As a result, the built-up land no longer conforms to administratively defined boundaries. Instead, major urban centres across the continent are emerging into large conurbations as individual settlements both in the historical origins of urban growth and formerly remote peri-urban areas expand and coalesce into contiguous urban-regions, blurring any existing administrative boundaries. At the same time as urban settlements are experiencing rapid population growth and expansion in physical size, investment in transportation systems, especially infrastructure for mass transit is lagging behind (Sietchiping et al. 2012; Falchetta et al. 2021).

The handful of studies that have so far investigated and measured accessibility suggest that the longstanding problem of unequal access to opportunities is worsening, as Africa increasingly becomes urban, with significant implications for the fight against extreme poverty, deprivation, and social exclusion (Lucas 2011). In Nairobi, Kenya, a recent study by Campbell et al. (2019) revealed a landscape of unequal access whereby accessibility is higher for car-based transport compared to walking and para-transit. Melbye et al. (2015) in their study of accessibility, congestion, and travel delays in Dar es Salaam, Tanzania, found overall levels of accessibility in the city to be poor. Residents faced longer travel times of about 60 min for journeys that covered relatively shorter distances of about 10–15 km from the city-centre. Studies in urban Ghana have also reached similar conclusions regarding accessibility. For example, Møller-Jensen et al. (2012) found that in Accra, travel times were substantially longer especially in morning and afternoon peak hours, partly because of inadequate road capacity across the city. Furthermore, a recent study by Acheampong and Asabere (2022) in which they assessed differential access by car and public transport in the Greater Kumasi city-region of Ghana, found significant deficits in terms of both access to public transport and access by public transport. They found that compared to public transport, car-based transport offered marginally better travel time access, but the overall levels of accessibility were poor irrespective of the travel mode, due to congestion.

What the foregoing evidence suggest is that the contours of accessibility are changing in tandem with the evolving spatial forms of Africa's major cities. Rapid urban expansion has meant that more population is increasingly living farther away from urban opportunities and services, such as jobs, schools, and hospitals. Within the dominantly mono-centric spatial structures that is typical of Africa's major towns and cities, everyday mobility and access to these opportunities is becoming a major challenge, especially for the larger mass of the urban population who do not have private modes of transport, but depend on public transport systems with limited coverage and/or walking.

The three chap in this part of the book are set within this context of rapid urbanization, emergent spatial forms and changing contours of accessibility in African cities. Each contribution utilizes various data sources and unique but complementary methodologies to provide in-depth analyses and understanding of accessibility and travel behaviour, accounting for the role of urban form and transport infrastructure in shaping differential accessibility levels, as well as the influence of socio-demographic and locational factors on travel behaviours.

References

Acheampong RA, Asabere SB (2022) Urban expansion and differential accessibility by car and public transport in the Greater Kumasi city-region, Ghana—A geospatial modelling approach. J Transp Geogr 98:103257

Acheampong RA, Silva EA (2015) Land use–transport interaction modeling: a review of the literature and future research directions. J Transp Land Use 8(3):11–38

Aditjandra PT, Mulley C, Nelson JD (2013) The influence of neighbourhood design on travel behaviour: empirical evidence from North East England. Transp Policy 26:54–65

Angel S, Blei AM, Civco DL, Parent J (2012) Atlas of urban expansion. Lincoln Institute of Land Policy, USA, pp 397

Batty M (2013) The new science of cities. MIT Press

Campbell KB, Rising JA, Klopp JM, Mbilo JM (2019) Accessibility across transport modes and residential developments in Nairobi. J Transp Geogr 74:77–90

El-Geneidy A, Levinson D, Diab E, Boisjoly G, Verbich D, Loong C (2016) The cost of equity: asessing transit accessibility and social disparity using total travel cost. Transp Res Part a: Policy Pract 91:302–316

Falchetta G, Noussan M, Hammad AT (2021) Comparing paratransit in seven major African cities: an accessibility and network analysis. J Transp Geogr 94:103131

Geurs KT, Van Wee B (2004) Accessibility evaluation of land-use and transport strategies: review and research directions. J Transp Geogr 12(2):127–140

Gim THT (2013) The relationships between land use measures and travel behavior: a meta-analytic approach. Transp Plan Technol 36(5):413–434

Handy S, Cao X, Mokhtarian P (2005) Correlation or causality between the built environment and travel behavior? Evidence from Northern California. Transp Res Part D: Transp Environ 10(6):427–444

Hansen WG (1959) How accessibility shapes land use. J Am Inst Plann 25(2):73–76

Levinson D, Wu H (2020) Towards a general theory of access. J Transp Land Use 13(1):129–158

Lucas K (2011) Making the connections between transport disadvantage and the social exclusion of low income populations in the Tshwane Region of South Africa. J Transp Geogr 19(6):1320–1334

Martens K (2016) Transport justice: designing fair transportation systems. Routledge

Melbye DC, Møller-Jensen L, Andreasen MH, Kiduanga J, Busck AG (2015) Accessibility, congestion and travel delays in Dar es Salaam–a time–distance perspective. Habitat Int 46:178–186

Møller-Jensen L, Kofie RY, Allotey AN (2012) Measuring accessibility and congestion in Accra. Nor Geogr Tidsskr-nor J Geogr 66(1):52–60

Sietchiping R, Permezel MJ, Ngomsi C (2012) Transport and mobility in sub-Saharan African cities: an overview of practices, lessons and options for improvements. Cities 29(3):183–189

United Nations (2014) World urbanization prospects: the 2014 revision highlights. Department of economic and social affairs. Population Division, United Nations, 32

Wegener M (2004) Overview of land-use transport models. Handb Transp Geogr Spat Syst 5:127–146
Wegener M, Fürst F (2004) Land-use transport interaction: State of the art. Available at SSRN 1434678

Ransford A. Acheampong is a Lecturer and Presidential Academic Fellow in Transport and Urban Futures at the Department of Planning and Environmental Management, University of Manchester, UK. The first strand of his research is focused on the nexus between transportation and spatial structure, and their integrated planning in cities. The second strand of his research addresses questions around the diffusion and socio-spatial impact of new and emerging transport technologies and mobility solutions, such as app-based mobility, autonomous vehicles, shared-mobility and Mobility-as-a-Service.

Chapter 3
Urban Form and Access to Public Transport in Accra, Ghana

Ransford A. Acheampong, Stephen Boahen Asabere, and Augustine Yaw Asuah

Abstract Major cities in Africa have experienced fast-paced expansion in their physical size as a result of rapid levels of urbanization. The emergent urban forms have fundamental influence on the spatial distribution of activities and the levels of access that the growing urban population can have to opportunities, such as jobs and social amenities. A key determinant of connectivity to opportunities and valued destinations in any urban context is access *to* public transport. This chapter analyses public transport access in the Accra city-region, a contiguous conurbation comprising Accra, Ghana's capital city and the surrounding peri-urban settlements. In order to show the prevailing urban form and understand the role of accessibility in shaping urban expansion patterns, the relationship between the region's observable built-up land as of 2021 and the existing primary road infrastructure is examined and quantified. Within this urban structural context, the analysis also examines differential walking time access to public transport service routes for both paratransit and Bus Rapid Transit (BRT), as well as public transport stops across the city-region. Ultimately, a huge accessibility deficit with respect to the public transport system is revealed. The chapter highlights the need for an integrated, accessibility-oriented paradigm to land use and transport planning that prioritizes urban growth management as well as public transport and non-motorized transport infrastructure development. This is vital in creating healthy, inclusive and environmentally sustainable urban futures.

Keywords Urbanization · Urban form · Accessibility · Public transport · Accra city-region

R. A. Acheampong (✉) · A. Y. Asuah
Department of Planning and Environmental Management, University of Manchester, Manchester, UK
e-mail: ransfordantwi.acheampong@manchester.ac.uk

A. Y. Asuah
e-mail: augustine.asuah@postgrad.manchester.ac.uk

S. B. Asabere
Department of Physical Geography, University of Goettingen, Goettingen, Germany
e-mail: stephenboahen.asabere@uni-goettingen.de

R. A. Acheampong et al. (eds.), *Transport and Mobility Futures in Urban Africa*,
The Urban Book Series, https://doi.org/10.1007/978-3-031-17327-1_3

3.1 Introduction

Africa is increasingly becoming urban as a result of rapid levels of historical population growth and rural–urban migration that are generating larger concentrations of population and economic activities across the continent. Although only 14% of the continent's population lived in cities in the 1950s, the urban population had increased to 40% in 2015, and it is projected that by the first half of the twenty-first century, about 55% of the continent's population will be living in cities (UN/DESA 2014). One of the easily observable impacts of the unfolding urbanization process is the fast pace of expansion in the physical size of human settlements, as a consequence of land conversion for housing, employment, and infrastructure and other activities. Rapid peri-urbanization, sprawling and fragmented patterns of spatial development typify urban growth across the continent (Angel et al. 2012).

Closely linked to the on-going process of rapid urban expansion are the challenges of meeting the everyday transport and mobility needs of the growing urban population. Although most cities in Africa are at early stages of motorization, overall levels of motorization and car use are fast rising in recent years (Mitullah 2017; Sietchiping et al. 2012). Rising motorization levels stem from increasing car ownership, and the use of other new car-based modes of transport such as ride-hailing. Motorization and the associated impacts also result partly from the very nature of the public transport system in these contexts. In most African cities, non-scheduled minibus (also known as paratransit) and taxis are the common forms of public transport. Given their limited capacities, more vehicles are required to meet travel demand. Consequently, major African cities are characterized by congested and often polluted central areas, which negatively affect accessibility to opportunities, including jobs and social services (see e.g. Campbell et al. 2019; Melbye et al. 2015; Møller-Jensen et al. 2012).

While paratransit serves the daily mobility needs of the larger mass of the urban population, they do not always do so efficiently. As operators attempt to balance revenue generation and the associated costs, paratransit tends to respond to demand in high population concentrations, especially during peak travel times (Saddier et al. 2017; Cervero and Golub 2007). This implies that in low-density suburban and peri-urban areas, coverage tends to be poor, especially during off-peak demand periods. The way paratransit operates also means that in a typical African city, the overall public transport service coverage, and by extension accessibility *to* and/or *by* the available public transport modes can vary significantly across space and time. As a consequence, the prevailing levels of inequality of access could further worsen for suburban and peri-urban residents, as well as for poor and low-income populations living in congested urban core areas. Indeed, the growing challenges of accessibility in urban Africa are now recognized as constituting some of the key dimensions and expressions of urban poverty (Lucas 2011).

Access plays an important role in urban development by shaping location decisions and emergent patterns of development (Levinson and Wu 2020; Acheampong and Silva 2015). The development of transport infrastructure, such as roads, provides geographical accessibility between individual settlements and within towns and cities. This accessibility benefit drives urban physical development by attracting more population and economic activities. Where urban containment measures are weak or non-existent, the geographical accessibility benefits provided by transport infrastructure have been found to promote rapid outward expansion often characterized as fragmented, leap-frogging or sprawling patterns (see e.g. Akubia and Bruns 2019; Olvera et al. 2003; Barredo and Demicheli 2003). The emergent forms of cities, in turn, interplays with factors such as socio-economic characteristics of individuals and households and available mobility options to determine the levels of access to valued destinations realized by individuals and households (Martens 2017).

Accessibility can be interpreted in two ways: accessibility *to* public transport, which refers to how easy it is for people to access public transport services, and accessibility *by* public transport—the ease with which individuals can reach opportunities *by* public transport (Cui et al. 2022). Regarding the latter, accessibility is fundamentally determined by the spatial distribution of transport infrastructure, such as public transport service routes and stops. Using these critical infrastructures of the transportation system as valued destinations, one can assess the ease with which individuals or different groups living across a geographical area can reach them in terms of walking time or distance. The resulting measures can, therefore, reveal prevailing first-and-last-mile accessibility deficits in relation to the existing public transport system and even whole 'transit desert' in parts of the city.

As accessibility implies reachability to valued destinations or opportunities, it is also important to define the subject and unit of analysis. At one level, the subject could be individuals or households, in which case person-based accessibility measures are employed to measure the activities that they can participate in at a given time (Geurs and Van Wee 2004). Person-based accessibility analysis and evaluation require a lot of individualized data that is difficult and expensive to collect, especially in global south cities. Another measure is place-based, which involves analyses of accessibility between locations at multiple scales, such as, the travel time it takes to reach a defined activity zone (e.g. census tract or neighbourhood) from another. Thus, in contexts where detailed socio-economic information about the population at the individual or household levels are lacking to allow for person-based accessibility analysis, the place-based alternative could be computed using built-up land data and available public transport infrastructure data (see e.g. Acheampong and Asabere 2022). Built-up land gridded at any resolution can reflect population and land-use distribution at that spatial resolution for a given period, and provide the frame to measure place-based accessibility levels.

This chapter presents an analysis of differential access to public transport in Accra, Ghana's capital and the surrounding areas that together constitute the functional Accra city-region. A place-based accessibility assessment across the conurbation is presented by employing the built-up land extracted from satellite images and existing public transport infrastructure. The analysis covers two main interrelated aspects. Firstly, the relationship between urban physical expansion and road transport infrastructure development is examined and quantified to show the broad emergent pattern and the role of accessibility in shaping the observed patterns. Secondly, the analysis focuses on access to public transport by computing walking time to public transport service routes and stops at a 100 m^2 spatial resolution of built-up land. Ultimately, the prevailing accessibility deficits across the Accra conurbation are revealed and quantified.

3.2 Quantifying Urban Expansion and Public Transport Access: Overview of Data and Approach

The geographical extent of the analysis presented in this chapter is the Accra City-Region (Fig. 3.1), also known as the Greater Accra Metropolitan Area (GAMA). The city-region, which we refer to simply as Accra in the rest of this chapter, is a changing functional geography currently comprising about twenty-six local government administrative units (i.e. Metropolitan, Municipal, and District and Assemblies). As the analysis will later show, the geographical extent of the city-region reflects the unfolding patterns of urban growth within Accra and the surrounding peri-urban districts. The area, therefore, comprises the city of Accra, Ghana's capital, and the surrounding built-up areas that forms a contiguous conurbation.

One of the main data inputs for the accessibility analysis is satellite-based Landsat multispectral images with ~30 m spatial resolution. The images covered two-time steps, 1985 and 2021, spanning a period of 36 years. As the analysis will later show, Accra has experienced substantial growth and expansion in physical size over this

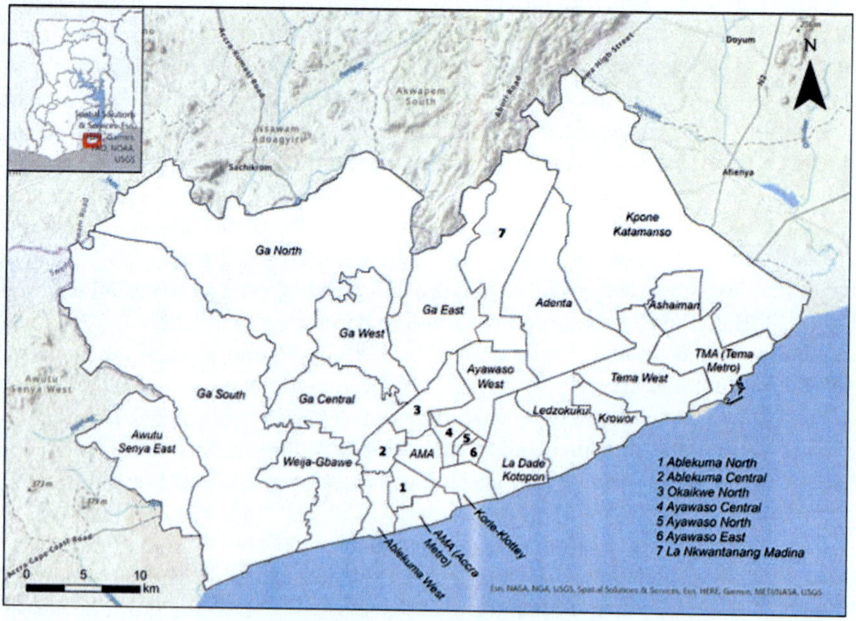

Fig. 3.1 Extent of the Accra city-region showing the 26 administrative units within the conurbation

period. The images were selected based on availability and quality considerations such as the absence or minimal presence of clouds. To account for the differences in images and sensor acquisition dates, images were processed to the top of atmosphere reflectance. For each image, a combination of different spectral bands including false and true colour composites were generated in order to identify and systematically sample points for classification. Our sampled points were based on two main categories, i.e., built and non-built areas. The built areas were defined by private and public buildings, concrete surfaces, rooftops, roads, and other artificial surfaces, whereas the non-built were mainly areas that did not fulfil the built criteria including water surfaces and vegetated areas. A total of 1200 and 1800 sampled points were obtained for the 1985 and 2021 image years, respectively. At each of these locations, spectral band values were extracted as input to train a support vector machine classifier in R with the "caret" package, whereby 70% of the sampled data was used for training and the remaining for validating, yielding >90% overall accuracies. Using a fuzzy logic approach in assigning a probabilistic value ranging 0–1 to each pixel, the images were classified whereby a probability threshold of 0.7 was used to achieve a binary output of built-up and non-built-up areas, comparable to the previous classification (see Asabere et al. 2020).

In addition to the built-up land data extracted from the satellite images, available data on Accra's public transport infrastructure were obtained. Two main datasets were relevant for the analysis, namely primary road networks, which are also the public transport service routes for paratransit and BRT, and public transport stops. The stops and routes data were obtained in the form of general transit feeds (GTFS) made publicly available in OpenStreetMap. These datasets were generated through the AccraMobile3 project, which was launched in 2017 by the Department of Transport of the Accra Metropolitan Assembly to map all paratransit (i.e. Trotro) network, and stops in the study area (see Saddier et al. 2016). A total of 4,171 public transport stops were identified in the data. A heatmap overview showing the distribution and density of the stops as well as the public transport service routes in the study area is presented in Fig. 3.2.

Employing the built-up and public transport infrastructure datasets, two key pieces of interrelated analyses were conducted using geospatial modelling techniques in a GIS and linear regression modelling. In the first aspect of the analysis, a buffer extent of 6 km, comprising discrete buffers of 100 m was overlaid on the satellite-based built-up data for 2021, to quantify the amount of built-up land observed at varying distances from a primary road. The 6 km buffer extent covered nearly all the built-up area of the conurbation. Piecewise regression modelling was then employed to examine the uniform relationship between primary transport links and urban expansion, by accounting for the varying quantities of the latter as the distance increases in discrete continuous buffers from the former. The piecewise regression model is expressed as Eq. 3.1, and the regression coefficient, r^2, is used to assess the strength of the relationship. The advantage of using piecewise regression to model this relationship is that it captures the different segments of the linear relationship in different regions

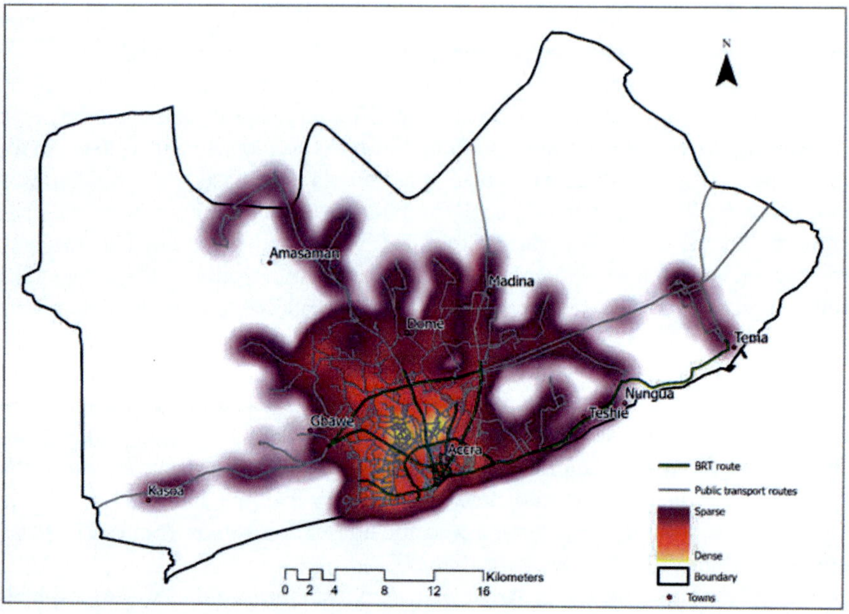

Fig. 3.2 Heat map showing the distribution of public transport stops and service routes, and density of stops

between the dependent (built-up area) and explanatory (distance to primary road) variables.

$$Built - up\left[km^2\right] = b0 + b1 * Distance[km] + b2(Distance[km] - K) * d \tag{3.1}$$

where $b0$, $b1$, $b2$ are coefficients of the model; K is break-point value (or knot value); and d is the knot dummy variable which is either 0 or 1 based on K.

The second aspect of the analysis involved computing access in terms of walking time to key infrastructures of the public transport, which are the service routes for both paratransit and BRT and the ancillary stops on the network. To this end, a 100 m^2 lattice was overlaid on the study area, and intersected with the built-up areas to provide the spatial unit of analysis for a location-based accessibility analysis. Access was measured by computing in a GIS a Euclidian proximity from the centroid of each 100 m^2 location of built-up land to the nearest main public transport service routes and public transport stations in the study area. Euclidian, instead of networked-based distances were computed because of the lack of detailed road and pathway data for the study area. The resulting distances were converted to walking time by assuming a walking time of 4.8 km hr^{-1}.

3.3 Urban Physical Expansion and its Relationship with Transport Infrastructure in Accra

Over the last three-and-half decades, urban physical expansion in the Accra conurbation has unfolded at a rapid, sprawling pace (Fig. 3.3). In the mid-1980s, 105 km^2 of built-up land was observed in the study area. In the present day, this has increased significantly to an estimated 622 km^2. This represents nearly a six-fold increase between the observed periods (i.e. 1985–2021), or 493% increase in the physical size of the built-up area, at an annual average growth rate of 13.69%. As of 2021, it is estimated that about 39.67% of the area within the conurbation boundary was built. As can be seen from Fig. 3.3, much of the emergent built-up area in recent years has occurred in peri-urban districts outside of coastal towns that constituted the origins of historical growth in the wider conurbation. The observed built-up land expansion is due primarily to the conversion of hitherto undeveloped peri-urban land to residential dwellings and ancillary amenities.

Moreover, a strong relationship can be deduced between the existing transport infrastructure (i.e. primary road network) and the urban expansion observed in 2021. About 96.65% of the total built-up area in 2021 (i.e. 601.79 km^2) was captured in the 6 km buffer employed in this analysis, meaning that the buffer extent sufficiently covers nearly all the built-up area. The piecewise regression analysis illustrates that there is a strong negative relationship ($r^2 = 0.94$) between the distance from transport

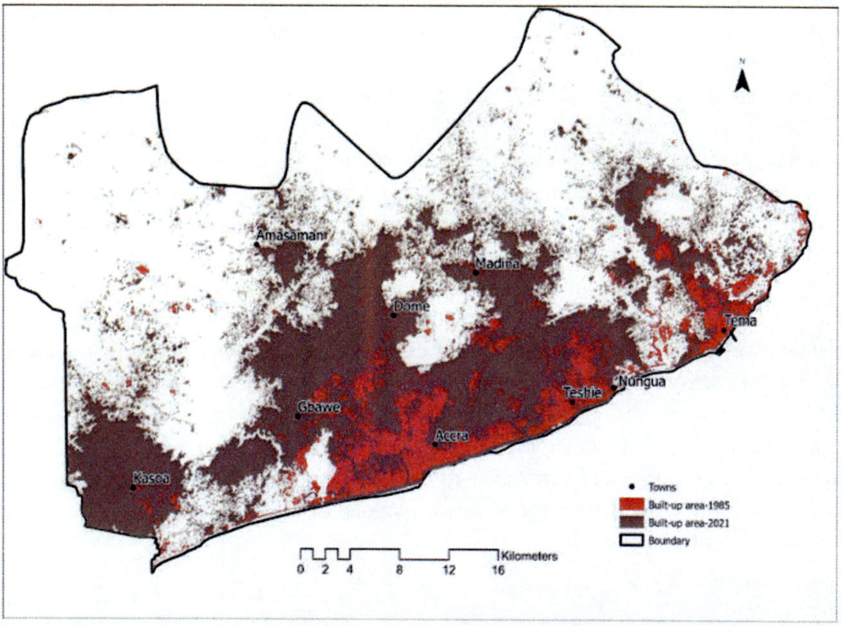

Fig. 3.3 Historical urban expansion in the Accra city-region observed in 1985 and 2021

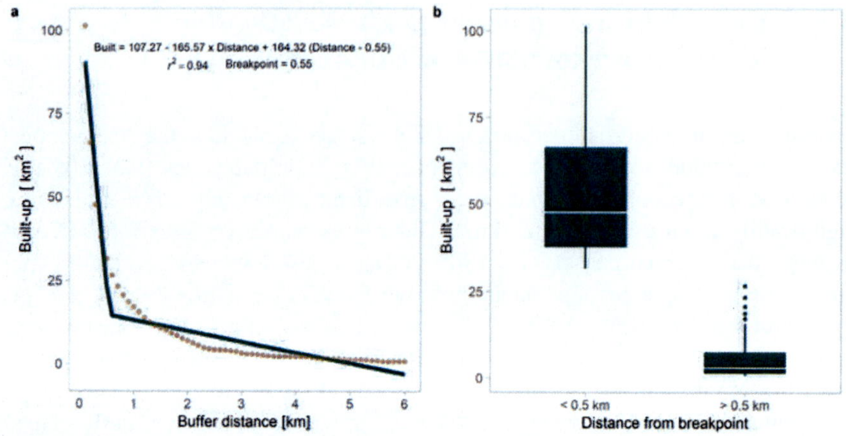

Fig. 3.4 Modelled relationship between distance from a primary road and amount of urban physical development

infrastructure (i.e. primary road network) and the amount of urban physical development. The regression analysis further reveals that there is a threshold of (<0.5 km) where there is noticeable increase in the concentration of the built-up area in relation to a primary road infrastructure (Fig. 3.4). For example, it was observed that about 47.36% of all built-up land in the conurbation is within half-a-kilometre distance to a primary road. At a distance of 1 km of a primary road, about 65.06% of the built-up land was observed. The amount of built-up land, however, decreases with increasing distance from a primary road network (Fig. 3.4). From the buffer analysis, it is estimated that only about 8.36% of the observable built-up land is located 4 km or more from a primary road.

These findings reveal an urban growth pattern whereby land development, as reflected in the observed built-up land, occurs linearly along major transport links, before spreading outward into the surrounding areas. Through this process of linear growth and expansion, hitherto smaller settlements in peri-urban areas coalesce into the main built-up area, which has over time, formed the contiguous conurbation observed today. Most importantly, the observed relationship also affirms the fundamental co-evolutionary nexus that exists between accessibility provided by transport links and emergent patterns of urban physical development. Together, the built-up land and transportation systems form the structural configuration determining the urban extent, and the distance separation between activities and opportunities. The emergent urban structural form also fundamentally shapes the need for mobility as well as levels of accessibility to various valued destinations that are realized by households and individuals.

3.4 Differential Access to Public Transport in Accra

Differential access to public transport in the wider city-region has been examined in this section focusing on three critical infrastructures of the transportation system as the valued points of interest. These are the public transport service routes (i.e. primary routes plied by para-transit); the recently designated BRT service route (which is not exclusively used by buses, but also accommodates paratransit), and public transport stops used by both paratransit and BRT. It is worth mentioning that in Accra, as is similar with other cities where paratransit operates, passengers do not only board vehicles at stations, but often do so anywhere along the service routes. Indeed, this flexibility is seen as one of the main advantages of the paratransit or Trotro system in Accra. In recognition of this, access in terms of walking time to both public transport service routes and stops is measured. The goal is to reveal the extent to which each of these transport infrastructures are within reach from any 100 m^2 location in the study area. Figure 3.5 shows the cumulative walking time to these three points of interest while Figs. 3.6 and 3.7 maps the differential accessibility levels to the service routes and stops, respectively.

Regarding access to public transport service routes, the analysis shows that an estimated 273.98 km^2 of the total built-up area (i.e. 44%) of the conurbation is within 5 min walking time (Fig. 3.5b). If the walking time thresholds are set at 10 and 15 min, the area reachable increases to 59% and 69%, respectively (see Figs. 3.5b and 3.6). Within a threshold of 5 min walking time, the total area with access to public transport stops is relatively smaller at 37% (232 km^2) compared to that of the available public transport access routes (see Figs. 3.5c and 3.7). Moreover, within 10 min and 15 min walking time, 53% and 63% of the study area could reach a public transport station. This means that the available stops cover 7% less of the area covered by public transport service routes for a time threshold of 5 min, and 6% less for a time threshold of 10 and 15 min, respectively. Furthermore, the analysis reveals that the

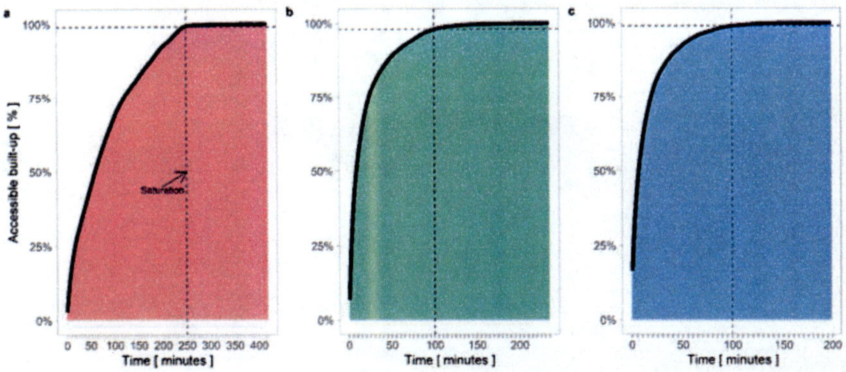

Fig. 3.5 Cumulative walking time to: **a** BRT service route **b** all public transport service routes and **c** public transport stops

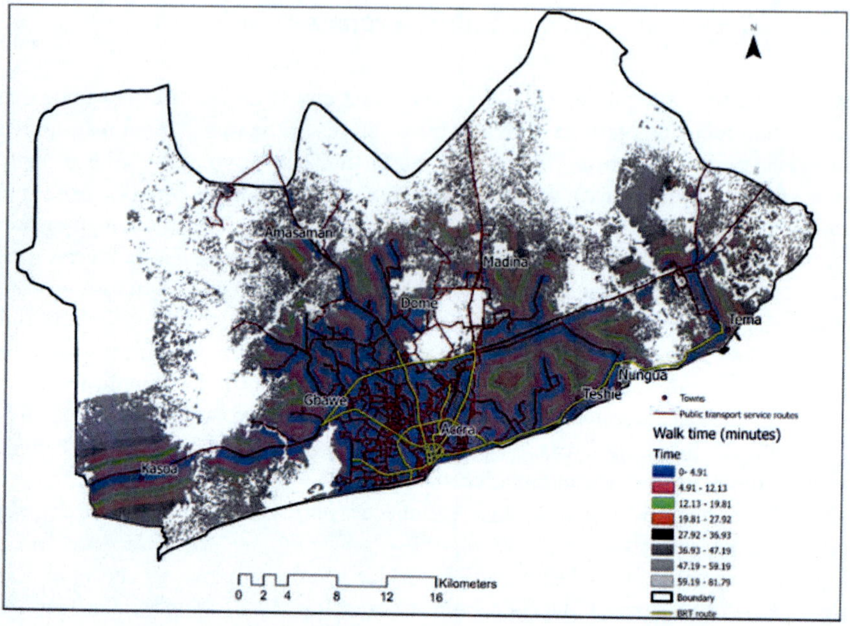

Fig. 3.6 Walking time to public transport service routes in the Accra city-region

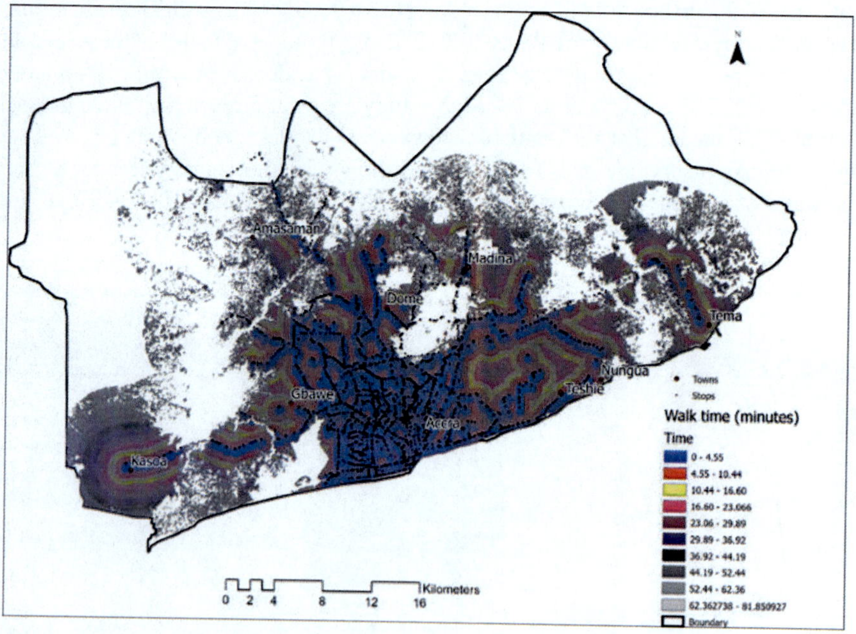

Fig. 3.7 Walking time to public transport stops in the Accra city-region

BRT service route offers the least coverage in terms of physical access or walking time. Within a 5-minute walking time threshold, only 11% of the study area can potentially reach the BRT (see Fig. 3.6a). Even for walking time thresholds of 10 and 15 min, only 18 and 23% of the study area could potentially reach a BRT service route. Given that the BRT service does not operate the flexible stopping that typifies paratransit, but only allows passengers to board and alight at designated stops, the actual coverage could be significantly lower than the estimates presented here, based on the service routes alone.

The 'saturation' lines shown in Fig. 3.5 indicate the time thresholds at which no noticeable change in accessible built-up land is observed in terms of walking access to BRT (5a), all public transport service routes (5b), and public transport stops (5c). What this means is that cumulative walking access stabilizes at these time thresholds. An estimated 99%, 98%, and 99% of the total built-up area within the study area is within 250 min and 100 min walking time to a BRT line and public transport service routes and stops, respectively. Thus, they essentially represent the maximum walking time to these infrastructures for almost all areas within the Accra conurbation.

The accessibility analysis presented here yields two major revelations. Firstly, the findings show that a huge deficit exists in the study area regarding access to the public transport system. This means that first-and-last-mile accessibility challenges exist for the *majority* of the study area, and by extension, the populations that live in these areas. Overall, the findings suggest that the reachability to the public transport system is influenced by the linear growth pattern along major transport links. Even so, only a small percentage of the study area that is within very close proximity to public transport service routes and/or stops can have better walking access to public transport.

Secondly, despite that overall coverage in terms of potential reachability by walking is poor for both stations and public transport service routes, the latter offers more coverage than the former. Thus, relying on the available stops alone would mean that a significant proportion of the potentially served areas would be excluded. By providing flexible boarding and alighting along the entire service route, paratransit (Trotro) provides more access opportunities to those who would otherwise not have been able to do so. Thus, to some extent, the way paratransit operates is shaped by the distribution of public transport stations in the study area.

3.5 Discussion and Policy Implications

This chapter has quantified historical urban expansion in the Accra conurbation and examined the relationship between access provided by primary road transport infrastructure and urban physical development. Different levels of access that could be realized to key infrastructures of the public transportation system from any location in the city-region have also been assessed and quantified. The results show that over the last three-and-a-half decades, rapid pace of outward physical expansion from the main built-up area of Accra to the outlying peri-urban settlements has unfolded. The results show a strong negative association between the amount of urban physical development and distance to a primary road network. In other words, the amount of physical development decreases as distance increases from a primary road. These findings suggest a broad pattern of largely unplanned outward, linear sprawling expansion facilitated by the development of primary road infrastructure linking the main built-up area and historical origins of urban development to hitherto relatively smaller suburban and peri-urban towns, forming the contiguous urban conurbation that is observable today. The co-evolutionary relationship between accessibility and urban physical development as well as the overall pattern of urban physical expansion observed in the Accra city-region has also been observed in a recent study in the Greater Kumasi city-region of Ghana, suggesting that this could be typical for most towns and cities across Ghana (see Acheampong and Asabere 2022).

Within the prevailing urban form, the analysis also examined walking time access to public transport service routes for both paratransit and BRT and public transport stops. The results reveal that huge accessibility deficits exist for each of these valued destinations of the existing public transport system. What these findings suggest is that for the vast majority of the region's built-up area and by extension the populations that live in those areas, the prevailing urban form, and public transport infrastructure configurations present significant accessibility challenges. While the established core areas of the region are better served by the existing public transport infrastructure in terms of first-and-last-mile reachability, the analysis suggests that access worsens in all directions towards the outlying sub-urban and peri-urban areas. As the analysis has shown, the density of the primary road networks determines the geographical coverage of public transport service routes and stops, which in turn, shape the contours of overall public transport access in the study area. One possible explanation is that, as Accra has urbanized and expanded in physical size rapidly over the years, and public transport infrastructure development has not kept pace, especially in the newly emerging sub-urban and peri-urban areas. Indeed, as Møller-Jensen et al. (2012) observed, the road infrastructure and capacity is inadequate in almost all areas of Accra. Given that more and more people are living in the region's peri-urban towns and commuting to the centre, traffic conditions also tend to be particularly worse during peak-hour periods to and from the central areas (Agyemang 2017; Agyapong and Ojo 2018).

Although constituting a fundamental determinant of the possibilities that the public transport system could offer individuals across the region, the differential levels of accessibility presented in this chapter is only a part of the larger picture. Beyond physical access, economic access in terms of affordability of public transport to individuals and households of different socio-economic means is also important. While the analysis presented in this chapter does not explicitly cover the latter, issues of public transport affordability have come to the fore in recent years in urban areas in Ghana, including Accra. Among the pool of travel alternatives, public transport has been seen as the most affordable to the larger mass of Accra's urban poor and low-income households. However, in recent years, fuel price increases have often been accompanied by rises in public transport fares and overall rising cost of living in Accra. In addition, the common practice whereby paratransit operators increase fares during peak travel periods to levels significantly higher than what they would usually charge or are permitted to charge (see e.g. Abane 2011; Poku-Boansi and Adarkwa 2013; Stasik 2015), further worsens the accessibility challenges faced by public transport users.

The prevailing accessibility challenges are also compounded by severe congestion especially in central locations of the city-region, which results in substantially longer journey times (Møller-Jensen et al. 2012). The foregoing implies that deficits not only exist in terms of reachability to public transport service routes and stops, as evidenced by this chapter, but also in terms of overall reachability to valued destinations such as employment, shopping and social services, by public transport accounting for affordability and overall journey times.

If the existing public transport system is to meet the mobility needs of the urban population now and in the future, then the evidence presented in this chapter suggests the need for policy action in a number of areas. Firstly, there is the need for integrated land use and transport planning that has improving accessibility levels as the overriding priority. Historically, the planning system in Ghana has been weak in managing urban growth. As the analysis has shown, unfettered and unplanned growth leads to unsustainable outward urban expansion that pushes the limited public transport infrastructure and urban opportunities farther away from the resulting suburban and peri-urban populations. Addressing the prevailing accessibility challenges and averting the possibility of them worsening should begin with addressing the problem of uncontrolled urban expansion.

Beyond effective growth management and integrated urban and transport planning that is responsive to the emerging existing challenges alluded to in this chapter, accessibility-oriented policy and strategies must not focus narrowly on transport supply measures that leads largely to road infrastructure building. While road building may play an important role in increasing the geographical coverage of public transport to the mostly peri-urban areas experiencing the accessibility deficits revealed in this study, it should constitute part of a portfolio of interventions. We know that in most African cities, including Accra, most of the urban population depend on non-motorized transport, in particular walking. Yet, non-motorized forms of transport receive limited attention in transport infrastructure provisioning measures that heavily prioritize motorized modes in general and car-based transport in particular. For healthy, inclusive and environmentally sustainable outcomes, urban transport intervention should prioritize walkability as a key component of accessibility improvement measures. Moreover, despite the overall longer travel journey times experienced by commuters, intra-urban journeys in urban areas in Ghana and elsewhere tend to cover relatively shorter distances that could easily be covered by cycling. Thus, another strategy worth considering in improving overall accessibility levels is investing in the infrastructure required to support and promote cycling.

3.6 Conclusion

This chapter has shown that regarding access to public transport, huge deficits exist in the Accra city-region. The prevailing accessibility challenges stem from decades of rapid outward urban expansion and the limited geographical coverage of the existing public transport service routes and stops. Improving access to and by public transport is critical to creating healthy, inclusive, and environmentally sustainable urban futures. In view of this, an integrated, accessibility-oriented land use and transport planning response is urgently needed. Policy responses now and in the future should design and implement a portfolio of measures to increase the supply of public transport infrastructure with the aim of improving geographical coverage. Most importantly, investment in sustainable, active forms of transport, including walking and cycling ought to be prioritized.

References

Abane AM (2011) Travel behaviour in Ghana: empirical observations from four metropolitan areas. J Transp Geogr 19(2):313–322

Acheampong RA, Asabere SB (2022) Urban expansion and differential accessibility by car and public transport in the Greater Kumasi city-region, Ghana—a geospatial modelling approach. J Transp Geogr 98:103257

Acheampong RA, Silva EA (2015) Land use–transport interaction modeling: a review of the literature and future research directions. J Transp Land Use 8(3):11–38

Agyapong F, Ojo TK (2018) Managing traffic congestion in the Accra central market, Ghana. J Urban Manag 7(2):85–96

Agyemang E (2017) Mode choice for long distance trips: evidence from the greater accra metropolitan area of Ghana. J Transp Geograph 64(C):150–157

Akubia JE, Bruns A (2019) Unravelling the frontiers of urban growth: spatio-temporal dynamics of land-use change and urban expansion in Greater Accra Metropolitan Area, Ghana. Land 8(9):131

Angel S, Blei AM, Civco DL, Parent J (2012) Atlas of urban expansion. Lincoln Institute of Land Policy, Cambridge, MA, p 397

Asabere SB, Acheampong RA, Ashiagbor G, Beckers SC, Keck M, Erasmi S, Sauer D (2020) Urbanization, land use transformation and spatio-environmental impacts: analyses of trends and implications in major metropolitan regions of Ghana. Land Use Policy 96:104707

Barredo JI, Demicheli L (2003) Urban sustainability in developing countries' megacities: modelling and predicting future urban growth in Lagos. Cities 20(5):297–310

Campbell KB, Rising JA, Klopp JM, Mbilo JM (2019) Accessibility across transport modes and residential developments in Nairobi. J Transp Geogr 74:77–90

Cervero R, Golub A (2007) Informal transport: a global perspective. Transp Policy 14(6):445–457

Cui B, DeWeese J, Wu H, King DA, Levinson D, El-Geneidy A (2022) All ridership is local: accessibility, competition, and stop-level determinants of daily bus boardings in Portland, Oregon. J Transp Geograph 99:103294

Geurs KT, Van Wee B (2004) Accessibility evaluation of land-use and transport strategies: review and research directions. J Transp Geogr 12(2):127–140

Levinson DM, Wu H (2020) Towards a general theory of access. J Transp Land Use

Lucas K (2011) Making the connections between transport disadvantage and the social exclusion of low income populations in the Tshwane Region of South Africa. J Transp Geogr 19(6):1320–1334

Martens K (2017) Transport justice: designing fair transportation systems. Routledge

Melbye DC, Møller-Jensen L, Andreasen MH, Kiduanga J, Busck AG (2015) Accessibility, congestion and travel delays in Dar es Salaam–a time–distance perspective. Habitat Int 46:178–186
Møller-Jensen L, Kofie RY, Allotey AN (2012) Measuring accessibility and congestion in Accra. Norsk Geografisk Tidsskrift-Norwegian J Geograph 66(1):52–60
Mitullah WV (2017) Non-motorized transport integration into urban transport planning in Africa. In: Vanderschuren MJWA, Khayesi M (eds) Routledge, New York, NY, USA
Olvera LD, Plat D, Pochet P (2003) Transportation conditions and access to services in a context of urban sprawl and deregulation. The case of Dar es Salaam. Transp Policy 10(4):287–298
Poku-Boansi M, Adarkwa KK (2013) The determinants of demand for public transport services in Kumasi, Ghana. J Sci Technol (ghana) 33(3):60–72
Saddier S, Patterson Z, Johnson A, Chan M (2016) Mapping the Jitney network with smartphones in Accra, Ghana: the AccraMobile experiment. Transp Res Rec 2581(1):113–122
Saddier S, Patterson Z, Johnson A, Wiseman N (2017) Fickle or flexible? Assessing paratransit reliability with smartphones in Accra, Ghana. Transp Res Record 2650(1):9–17
Stasik M (2015) Vernacular neoliberalism: how private entrepreneurship runs public transport in Ghana. Sociologus, 177–200
Sietchiping R, Permezel MJ, Ngomsi C (2012) Transport and mobility in sub-Saharan African cities: an overview of practices, lessons and options for improvements. Cities 29(3):183–189
UN/DESA (2014) world urbanization prospects: the 2014 revision, highlights. United Nations (9789211515176)

Ransford A. Acheampong is a Lecturer and Presidential Academic Fellow in Transport and Urban Futures at the Department of Planning and Environmental Management, University of Manchester, UK. The first strand of his research is focused on the nexus between transportation and spatial structure, and their integrated planning in cities. The second strand of his research addresses questions around the diffusion and socio-spatial impact of new and emerging transport technologies and mobility solutions, such as app-based mobility, autonomous vehicles, shared-mobility and Mobility-as-a-Service.

Stephen Boahen Asabere is a Postdoctoral Researcher at the Department of Physical Geography University of Göttingen, Germany. The interdisciplinary nature of his research combines remote sensing/GIS, landscape ecology, and soil biogeochemistry. His research examines how global change processes, such as urbanisation, affect natural and managed ecosystems at different spatial scales using interdisciplinary approaches. Currently, he has a grant from the German Research Foundation (DFG) to study the spatiotemporal effects of urbanization on soil carbon dynamics in tropical agroecosystems.

Augustine Yaw Asuah is a Ph.D. Researcher at the University of Manchester UK, where his research is on understanding the relationship between spatial structure and inequality of access in sub-Saharan African cities. Prior to this, he worked as a research associate in a Fondecyt Project; aimed at developing new methodologies and frameworks for evaluating the sustainability levels of urban areas in medium cities of Southern Chile. He was a researcher affiliated to the sustainable urban studies lab in the South of Chile. He holds a Joint-MSc in Regional Development Planning and Management from the Technical University of Dortmund, Germany, and University Austral de Chile, Chile.

Chapter 4
Disparities in Public Transport Coverage and Bus Service Quality in Algiers

Lila Chabane

Abstract Urban public transport services offer mobility opportunities to people who do not have a private car or other means of travel. This crucial public service can, therefore, only fulfil its mission if it is equitably available throughout the city and for all populations, and if it functions efficiently. This research aims to test the hypothesis that the problem of inequalities in access to public transport services in Algiers is mainly due to a bad spatial configuration of the bus network and an unbalanced distribution of the public transport offer. A methodological approach that combines data analysis by classification and geospatial analysis is used by the author. The analyses showed that there is an inequality in the spatial distribution of the supply of the public transport network combined with a poorer performance of the public transport system in the peripheral municipalities, where there is a lack of basic infrastructure, which increases the risk of exclusion of the poorest populations. The Algiers Urban Transport Authority should take passengers' expectations into account in the decision-making process, as well as in the establishment of transport plans and in the definition of quality indicator requirements for bus service operators eligible for the activity. In addition, information and communication technologies can be used to optimize the efficiency of the public transport system by making travel information readily available and accessible to users.

Keywords Algiers · Bus network performance · Spatial inequality

4.1 Introduction

The transport sector encounters two kinds of problems in Algeria, particularly in Algiers, the capital city. The capital city has sprawled and car ownership/use has increased, leading to congestion problems in central areas. Indeed, most of North African cities suffer from traffic congestion that impacts the citizens' quality of life and

L. Chabane (✉)
Research Centre in Applied Economics for Development, Bouzareah, Algeria
e-mail: chlila7@gmail.com; l.chabane@cread.dz

R. A. Acheampong et al. (eds.), *Transport and Mobility Futures in Urban Africa*, The Urban Book Series, https://doi.org/10.1007/978-3-031-17327-1_4

their access to job opportunities and health services (UITP 2016a). The rapid motorization of households has exacerbated the problem of congested urban areas affecting life quality of citizens in terms of air quality, commuting time, safety and accessibility (El-Geneidy et al 2013), due to differences between the rapid evolution of urbanization and increasing travel needs and slowness of institutional actors to respond to the growing demand for transport (Godard 2013).

Opening up public transport sector to private investment in Algeria in order to resolve public sector difficulties, led to quantitative improvement of supply at the expense of qualitative aspect such as service availability and ensuring equitable access (Matouk 1992; ISTED 1999; Baouni 2003; Chabane 2014).The current supply of public transport is insufficient and does not meet the everyday travel demand of Algiers' residents. In Algiers there are only 0.6 buses per 1000 population.[1] Indeed, the bus transport sector is in its majority organized on very small-business basis (operators tend to have only one bus).Buses do not have a dedicated corridor, which would allow for priority passage, and the overall levels of service provided by the public transport system is of poor quality (Chabane 2014).The low quality of public transport service has led to rising motorization levels (UITP 2016b). Also, the existing road network capacity is exceeded by daily traffic volumes leading to traffic congestion. The average daily traffic on the Algiers south ring road for example, increased from 33,000 vehicles per hour and per direction (v/h/d)in 1997 to 115,930 v/h/d in 2002 and to 250,000 v/h/d currently, with an average traffic speed of 15 km/h in urban areas (Kerbachi et al. 2017, Bakour et al 2018).

Moreover, more than two-thirds of the population in Algiers is affected by critical situations, including low incomes, remoteness from home and work, longer journey times, limited captive public transport users, and inadequate transport services in the periphery. Better spatial coverage by public transport would allow better access to urban amenities for people who do not have another motorized mode of travel (Jehanno 2008).Empirical research has shown that difficulties of some population segments in travelling for financial reasons or because of poor quality of service and quality of a public transport network supply may lead to social exclusion and marginalization (Orfeuil 2004; Le Breton 2002). Thus, good access to a city's resources is facilitated by the existence of a better supply of public transport networks, which in turn, contributes to social cohesion and reduction of spatial disparities in living standards of the population (Caubel 2007).

The empirical research has shown that "the geographic structure of the network exerts a clear influence on various aspects of supply localization and allocation of demand to that supply" (Arnold et al. 2000). Also, any modification of a transport system in terms of infrastructure, service, frequencies or timetables has spatial repercussions (Chapelon 1996).Our case study explores the hypothesis that urban transit service inequalities in Algiers is mainly due to a poor spatial configuration of the bus network and to an unbalanced distribution of transport service on existing public transportation lines. The paper aims to address the problem of supply quality typology by combining indicators of spatial distribution of bus network structure and

[1] Calculated by the author on the basis of public transport supply and total population in Algiers.

by using Cluster Analysis and spatial analysis. In this paper, we present results of the first comprehensive survey we carried out on the entire public transport network in Algiers. The results of our research could enrich knowledge on the spatial differentiation of public transport supply and its relationship with transport inequality and its social consequences. Also, the results of this study may inform public policies to guide future investments programmed in city master plans in order to ensure balance of territories and promote public transport.

The chapter is organized as follows: Sect. 4.2 briefly describes an exhaustive survey of the bus network service in Algiers and methodology. Section 4.3 summarizes the main results obtained with analysis of the designed survey; the last section brings some conclusions of this study.

4.2 Data and Methodology

To analyse service quality in transit service by bus, the 57 municipalities making up the urban transport perimeter were taken into account. GIS is used in this research for mapping and spatial analytics of the collected data. Three types of data are used: the performance data of the public bus transport network, the spatial accessibility to this network and the aggregated socio-economic data on households broken down by municipality.

We carried out transportation surveys in "the weekday peak time", during the mornings (07:00–09:00) and evenings (15:30–18:00), because it is a period when a large number of people use the buses to go to work or home, apart from students and trainees. Traffic conditions are also very difficult due to traffic jams in various places, mainly on the eastern and southern access roads and the roads leading to these axes.

The majority of trips for public transport are made by buses (84.13%). To better analyse public transport coverage, to evaluate bus network performance and to show under-served areas, we have used a comprehensive survey of the bus transport network in Algiers. This survey took place in urban stations, using a questionnaire and by taking the bus on all network trips to investigate mainly the bus line functioning, operation characteristics (speed, starting number, fare, range of work), bus road and stopping points. The results of this survey enabled us first to map bus networks, to identify the operation and offer the quality of each bus line and finally to detect territorial imbalances availability in both qualitative and quantitative terms. To carry out our analysis, we have used indicators of spatial distribution of supply and connectivity of territories. Some of these indicators are commonly used in a study of territorial performance of a public transport network (Stathopoulos 1994; Bavoux et al. 2005).

In order to obtain an accurate picture of the quality of public bus transport service in the municipalities served, we have chosen the hierarchical bottom-up classification as the statistical method of data analysis. Hierarchical ascending classification uses mathematical partitioning algorithms or hierarchies, to discover a dendrogram or classification tree for hierarchical ascending classification (Kaufman and Rousseeuw 1990; Duda et al. 2001).An ascending hierarchical classification, with similarity index: correlation coefficient of Pearson, carried out on the variables (factorial axes) makes it possible to define six classes of bus network quality. Statistical software SPSS (version 18.0) was used for the hierarchical ascending classification application and MapInfo (version 8.0) for spatial representation of results.

4.3 Results

4.3.1 General Structure of the Bus Network

Public transport network by bus contains 190 lines. We identify the bus line operations according to the service supply and geographical coverage. Classification of bus line network according to the type of service provided is as follows: 32 lines are local, they serve the centre, 62 lines are radial (connecting city centre to periphery) and 96 lines are peripheral, these are lines that serve the periphery zone without going through the centre. As shown in Fig. 4.1, half of the bus lines serve peripheral zone of the agglomeration.

The predominance of peripheral lines is due to the large size of the area and the existence of many short bus routes, which increases the number of load cuts on the bus route network and the loss of time for users. The network of bus transport is organized around 43 stations located in the centre and periphery of Algiers city (Fig. 4.1). The high number of bus stops characterizes some lines, and the bus network in Algiers has

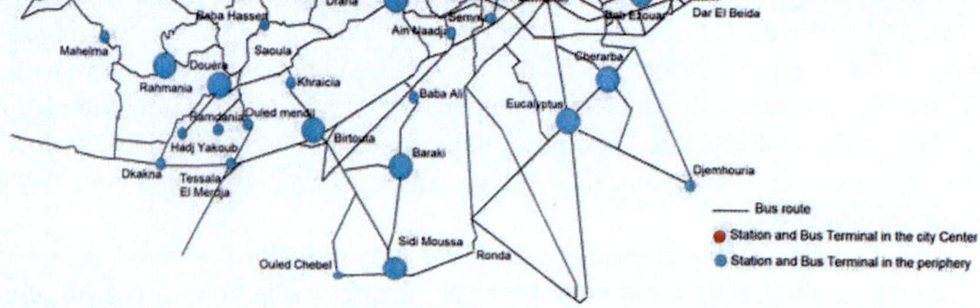

Fig. 4.1. Bus routes and spatial distribution of stations and terminals in Algiers. Realized by the author

1992 stops. An estimated 68% of those stopping points are not marked, and those that are either marked by a bus shelter or poles do not give any information on the lines that pass through this stopping point, the network map, schedules, etc.62% of non-identifiable stopping points are on peripheral lines, that encourages the inappropriate stops of the carrier on user request, in dangerous conditions and are of non-respect of traffic conditions and road safety.

4.3.2 *Characteristics of the Bus Line Network Supply*

To analyse the quality of service, we look in this section for key performance indicators: speed, amplitude, frequency, rotation and fare.

(i) ***Bus speed***

The bus speed is the result of the ratio between the length of the line and the travel time between the two ends of the line, the average speed of the bus line network is 13.4 km/h, and the average speed on the network according to the type of line is as follows:

- Radial line 13,5 km /h
- Local line 9,5 km/h
- Peripheral line 15,6 km/h

These speeds are those of rush hour. The lowest speeds on local lines are due to very poor traffic conditions characterized by traffic jams mainly in the centre, the commercial speed is higher in the periphery because of the provision of road infrastructures such as hoppers, but this speed remains at a low level compared to the bus line networks in the developed countries.

Classification of radial lines according to geographic division by sector enabled us to calculate the average speed during morning and evening rush hours for different corridors of the bus line network as follows:

- Average speed on the West axis: 12,5 km/h
- Average speed on the East axis: 17,7 km/h
- Average speed on the South axis: 12,2 km/h

These results reflect the inconvenience caused by the general circulation of bus efficiency and the importance of traffic. The West and South roads are negatively impacted by poor bus performance. There is no big difference in speeds on the roads entering the city's hyper centre, in fact for some lines of 5.5 and 8.5 km, the speed reaches 4.3 and 6.8 km/h, respectively.

The past evolution of commercial speed can be apprehended from RSTA (1976), SOFRETU (1982) and the author's surveys (2008) data. The average speed was 16.4 km/h in 1982, thus marking a decrease of 18% in 2008.Results show that between

Table 4.1 Evolution of bus commercial speed by type of line

	1982 (km/h)	2008 (km/h)	Change (%)
Radial line	18,5	13,5	– 27
Local line	12,7	9,5	– 25
Peripheral line	21,4	15,6	– 27

Source calculated by the author according to RATP, SOFRETU and SNTF 1982, author survey 2008 data

the two periods the speed of circulation has dropped significantly on different types of lines (Table 4.1).

(ii) **Amplitude of the service**

The 'amplitude' of a bus service refers to the hours per day in which it is in operation. The amplitude of bus service between the first and last run during a day in Algiers varies between 10 and 18 h and 30 min. A few bus services provided by the public company ETUSA in the centre starts at 5:30 am and ends at midnight. This service is provided by the public company ETUSA with a double crew. The average of the amplitude of the service on the entire network is of 12 h and 41 min.

First departures vary between 04h30 and 06h30 for the majority of bus lines, and the service stops between 18h00 and 19h00 for 68% of the lines, which is a very early schedule for those who work during the night and limit the displacements for leisure or visit (Fig. 4.2).

(iii) **Service frequency**

The frequency of passage by-line is an important element of service quality, which characterizes the service offered to users. There is no passage timetable to be respected, but the operators regulate their passage according to the request and order of the bus arrival station.

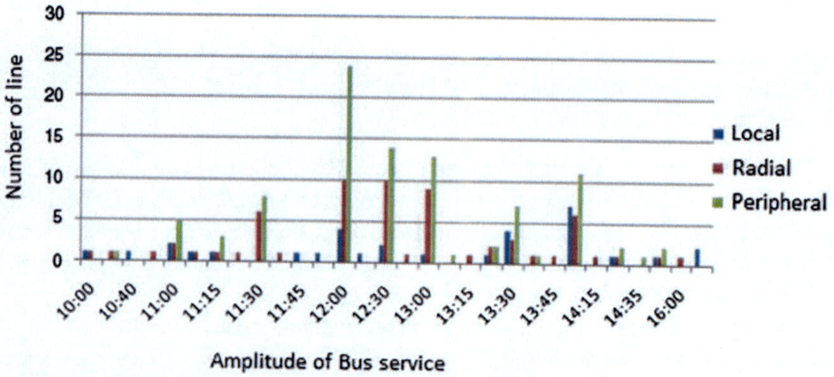

Fig. 4.2 Amplitude of Bus line service in Algiers

The frequency has been calculated in terms of bus departure number in the quarter of an hour, which we have multiplied by 4 to find the departure number in one hour; on an average, we have 8 vehicles/hour on all the line times. This average is the same for radial and peripheral lines, with only 4 vehicles/hour in the local lines. In this case, the average waiting time to get on a bus is 15 min (supposing that operation was reliable). These results are due to the fact that the bus is subject to the vagaries of general road traffic, and therefore, the user is penalized in terms of waiting time at bus stops and stations.

(iv) **Number of rotation**

The number of rotations is another indicator of the service quality, defined as the number of round trips made by carrier on the bus line during a working day. The number of rotation varies according to a period (summer, winter, day of the week, weekend), the average number of rotation on the bus line network is of 9/working day and 6/weekend day.

The average number of rotation is 10/day on peripheral lines, 9/day on local lines and 7/day on radial lines.

(v) **Affordable fare but with several service disruption**

Bus fare varies between 10 and 30 DA[2] for peripheral lines, 10 and 35DA for radial lines and between 10 and 20DA for the local lines. Ticket price per unit is not excessive, but the fact of taking a lot of connections to get to the final destination increases the total cost of bus travel mainly for those who live on the outskirts and are not served by radial bus lines.

Pricing is theoretically regulated, but there is a consensus among private operators on the ticket price. As far as the ETUSA tariff is concerned it is single on lines that it operates, nevertheless, given the bus size that it operates and the importance of the staff employed. The tariffs that ETUSA practices are higher than those of private operators on certain lines and sections of bus lines (example: trip fare place of May 1st–Ben Aknoun 10DA on private operator bus and 15DA on ETUSA bus) (Fig. 4.3).

The difference in bus travel costs depending on the spatial location of the starting point of the trip is due to the existence or non-existence of a bus line that serves a given municipality directly to the hyper-centre. For example, to make a trip from Ouled-Chebel municipality based in the south of Algiers city to downtown, the traveller must take two connections. The first connection is to BirTouta, and to reach this point you have to pay 20DA; the 2nd connection is at the Birkhadem station, where the cost of this trip is 15 DA and to reach downtown from Birkhadem you have to pay 20 DA. Finally, the total cost of the journey OuledChebel-Hyper-centre is 55 DA.

[2] DA: Algerian Dinars, current exchange rate in euro: 1euro = 133 DA.

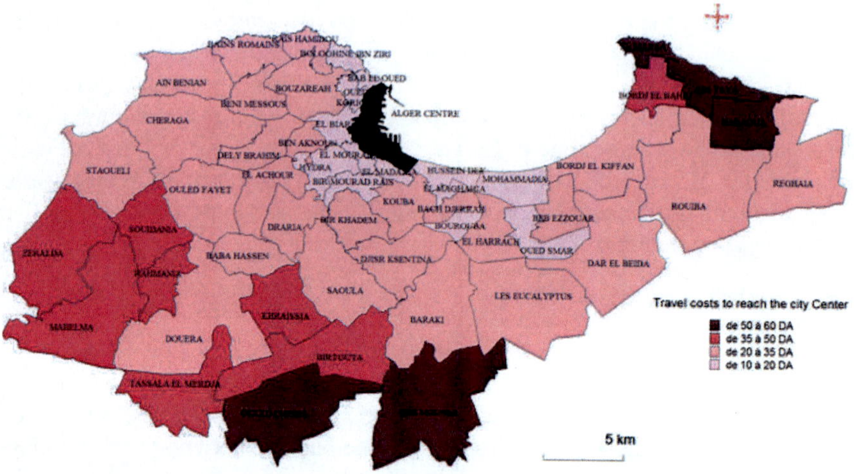

Fig. 4.3 Spatial distribution of cost travelling by bus to reach downtown

4.3.3 Typology of Municipalities by Quality of Public Transport

To classify municipalities according to the quality of public transport, we used indicators defined as follows:

- NL: Number of bus lines serving a municipality, a municipality is served by a bus line if at least one stop of this line is located on the territory of this municipality.
- CL 30mn: number of municipalities accessible within 30 minutes from at least one stop in the municipality concerned.
- CDA: number of municipalities directly accessible from the municipality concerned.

Some indicators, which could have been foreseen, were not retained, such as bus frequency, service amplitude, fares, traffic speed or the number of urban stations and stops.

In fact, following the analysis of the survey data, bus frequency appears to be a difficult indicator to use in the evaluation of the spatial distribution of public transport supply by bus.

It is not possible to judge the average frequency of several lines serving a municipality, because a municipality may be served by a line with a very high frequency and others with very low frequency or even none at all when the municipality is not served by a bus line. In terms of supply, the frequency of bus passage is above all directly dependent on the number of lines serving a municipality, which is one of the indicators used. The number of lines serving a municipality makes it possible to quantify the supply of bus transport on the territory of a municipality, and this indicator is somewhat aggregated as long as the municipality is considered to be served by the bus itself. If there is only one stop of the bus line that is located on the territory of this municipality, but, combined with the other selected indicators, it can give us additional information on the quality of the spatial distribution of the bus transport offer.

In addition, the level of fares can be an asset or a disincentive to the use of public transport. The price per kilometre of a single ticket is the same, regardless of the bus line concerned in the same municipality. It is the fact of taking several connections that makes the cost of transport by bus vary. It is, however, difficult to support all destinations from all stops within a municipality and to simulate all possible trips. Pricing is, therefore, not a discriminating indicator between bus lines.

In other cases, commercial speed by definition requires measuring the travel time between two given geographical points, which may include one or more connections. It is, therefore, an indicator that is impossible to measure between all the stops of all the bus lines in the province of Algiers.

On the other hand, the number of municipalities accessible within 30 min from at least one stop in the municipality concerned was easier to quantify in our survey. The number of municipalities that can be reached within a maximum of 30 min from at least one stop in the municipality in question provides information on the quality of the bus user's accessibility to the city's territory by bus.

Finally, the number of urban stations and stop points is an indicator of the geographical coverage because of its aggregative nature; it gives information about opportunities offered by a "stop point" in terms of accessibility because of its place in the network and its connections with other points. Also, it has another disadvantage inherent in its interpretation. It is assumed that the population within a radius of 300 m around the stopping point or station is served by a bus line. In this practice, differences between stations do not matter. Qualitative and quantitative parameters that each station offers are not taken into account. We considered that it is more judicious to take into consideration the number of municipalities that were directly accessible from a reference municipality. The number of municipalities directly accessible from a reference municipality is an indicator that allows us to perceive a degree of territorial accessibility that is offered to a public bus user, and to discriminate municipalities about the spatial offer of public transit by bus.

By analysing the map (Fig. 4.4) that we have produced on the spatial distribution of the bus lines on the territory of the province of Algiers, it emerges that 31 communes have less than 9 lines (average number of lines per commune), 84% of these lines are located on the eastern or southern outskirts of the central city. In addition, these lines only serve very close municipalities, so the bus user has to make at least one connection to make a long-distance trip.

Also, less than 40% of households are equipped with a private vehicle, which further deteriorates the travelling conditions of people dependent on a motorized means of transport, given the very low incomes of households residing in these municipalities; the average household income which does not exceed 16,430 DZD/Month is lower than the average for the whole province evaluated at 26,495 DZD/Month.

As regards the municipalities that are well served by the bus network (at least 20 lines), there are 6 municipalities in the central city and one municipality in the eastern

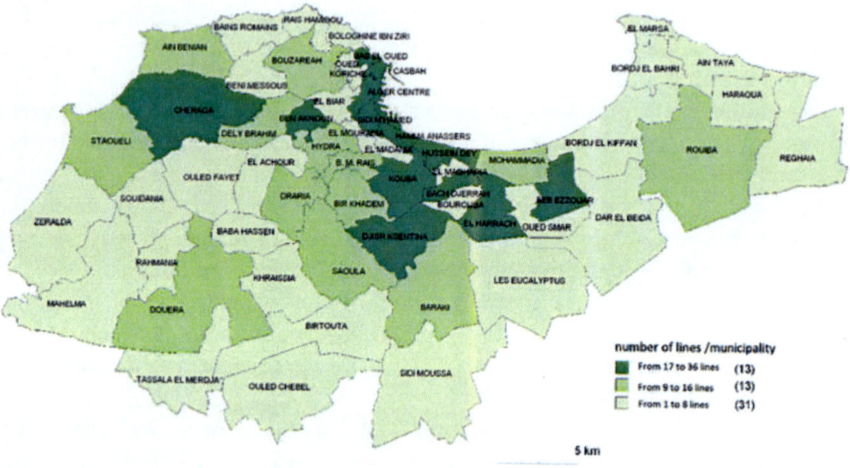

Fig. 4.4 Number of bus lines serving municipalities. Realized by the author

suburbs of Algiers (El Harrach), and another municipality in the western periphery of Algiers (Chéraga).

For the municipalities of the central city, many lines converge towards them in addition to the local lines, as regards the municipality of El Harrach which is at the Eastern border of the central city and thus on the course of these radial lines. The municipality of Chéraga beneficiary of two urban stations is a point of departure and destination of several lines which serve the Western periphery of Algiers and the central city.

The result of the classification as presented in Fig. 4.5, gives us 6 categories of services, from the weakest to the strongest. Municipalities in the category of weakest services are unfortunately very poorly classified in terms of household income, illiteracy rate, labour market, activity rate and housing conditions.

4.4 Discussion

Through analysis of the current public transport offer; based on the results of the first comprehensive survey of the bus network in Algiers, a statistical and cartographic analysis was carried out on the quantitative and qualitative offer of the bus transport system. We tried to make a diagnosis of the existing urban transport network. Algiers has an important public transport network but the narrowness of the road network, its saturation, added to that a sinuous road, have made the current public transport network by bus not very effective in terms of frequency of passage and time of course.

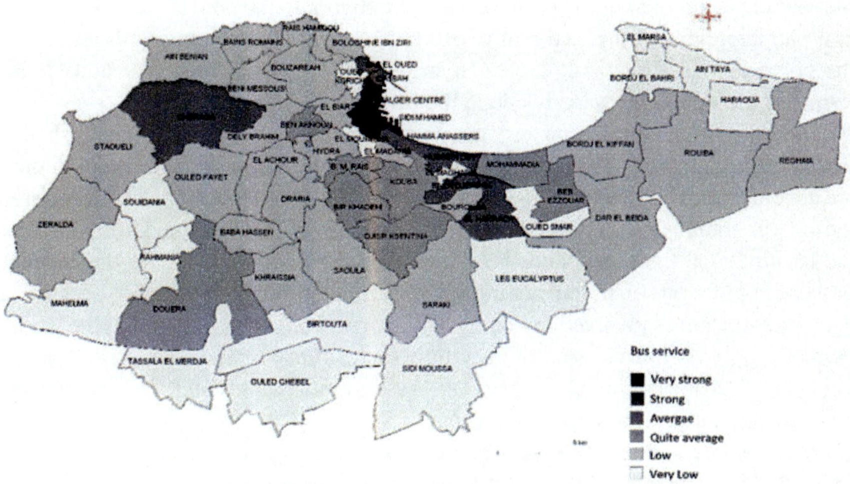

Fig. 4.5 Classification of municipalities according to the quality of bus line network offered. Realized by the author

In this case study, we used three indicators to represent different levels of the quality of the public bus transport service network in Algiers in terms of the operational characteristics of the service provided and the infrastructure. The results show that between 1982 and 2008 traffic speed dropped significantly on all lines. These results reflect the inconvenience caused by the general bus traffic and the amount of traffic, as well as the negative impact on the performance of the bus.

In the classes of the best classified services, we find only 2 municipalities which are located in the hyper-centre of the city of Algiers, whereas as regards the 'very weak' services we find 16 municipalities of which the majority are located at the extreme eastern and southern periphery of the central city. Thus, in total, 33.1% of the Algerian territory is poorly served by public transport.

It so happens that the municipalities least served by the bus are for some who only crossed by one bus line. The travel costs of the users residing in these municipalities are conditioned by the bus service that passes through the stop located on the territory of that municipality and which may not stop in case of overload. These municipalities have the lowest levels of trip emissions (less than 9600 trips/day of emissions) and the highest number of walking trips in terms of emissions of these municipalities. Households living in these municipalities spend an average of 11% of their income on transport; this share is estimated at 15% for households earning the minimum wage (12,000 DA/month), which accounts for 11% of all households living in these municipalities. The average share of transport expenditure in the income of households residing in these municipalities varies between 8 and 17.6%, significant proportions of which are already low.

The analysis of the results of the comprehensive survey reveals territorial inequalities at the operational level of the public transport system. There is an inequality in the spatial distribution of the public transport network to the detriment of the peripheral municipalities of the city centre, in addition to the lack of basic infrastructure; the poorest populations risk exclusion when it is known that 76% of households living on the minimum wage reside in the periphery of the city centre.

The correlation between poverty and unemployment has been well established in various research studies (Xue and Zhong 2003; Osinubi 2005). Thus, these communes are devoid of employment, and the job offer (calculated as a percentage of the offer in the agglomeration) varies between 0.1% (Rahmania) and 2.07% (Oued Koraichi). For the 16 municipalities as a whole, the rate is 14.83%, which increases the dependence of these populations on motorized transport, particularly public transport, because of the longer distances involved in finding a job or returning to work. 78% of the labour market supply is in services (administration, trade, transport and other services), and the lack of territorial competitiveness of these municipalities and their modest tax revenues do not favour their development.

These municipalities are economically weakened and deprived of jobs. In these municipalities, the employment offer does not exceed 2.07% of the total offer of the conurbation, which increases the dependence of these populations on motorized transport, and in particular public transport, due to the increase in the distances travelled to find work or to reach the workplace.

These are socially disadvantaged municipalities with the highest illiteracy rates and a large stock of precarious housing. As a result, among the communes underserved by public transport, some communes have been classified as poor communes according to the wealth ratio in the UNDP poverty map (UNDP 2001). In addition to economic and social inequalities, an ineffective transport policy exacerbates the social exclusion of certain groups, with some people without jobs or training perpetuating their lack of qualifications by refraining them from earning a living (Orfeuil 2004; Lucas 2011; Le Breton 2002). This situation pushes the most disadvantaged people into immobility and isolation, "locking people into restricted territories, a locking that then becomes an obstacle to inclusion" (Le Breton 2002).

To achieve the Sustainable Development Goals and the climate agenda for African cities, it is recommended that policies be designed to encourage a shift towards cities with relatively high residential densities, multifunctional neighbourhoods and a multi-modal transport system combining walking, cycling and public transport (African Union Commission 2015). Also, information and communication technologies can be used to reduce non-renewable energy consumption by optimizing travel and the efficiency of public transport through greater visibility and legibility of its network.

4.5 Conclusion

Public transport remains the first means of transport for users in Algiers, due to the low level of household income. Unfortunately, the lack of public transport is manifested by a lesser quality offering. Algiers residents are strongly requesting for public transport, but unfortunately, the offer still does not satiate this important request. The analysis of spatial equity and efficiency of a transport system, taking into account the characteristics of the network, has shown disparities between network supply and demand for travel, in terms of public transport network quality.

Referring to residents' movements in Algiers to the spatial configuration of transport lines reveals a gap between mobility practices and the structuring of public transport network, those favour links serving downtown, and people wishing to travel between peripheries who must pass through the centre. One of the reasons for inability of public passenger transport by bus to meet the travel demand of Algiers' inhabitants is an unbalanced spatial structure of bus transport network, and quantitative supply combined with a supply quality which is below standards.

In addition to economic and social inequalities, a policy of inefficient transport will exacerbate social exclusion of some groups, as people without a job or training thus perpetuate their lack of qualifications refrain them from making a living. This situation is pushing the most deprived people to immobility and isolation (Lucas 2011).

The preponderant role of public transport in opportunities offered to people to access services is recognized which also improves their living conditions daily by facilitating access to health, education and employment. This situation aggravates the social exclusion of the inhabitants who already suffer from a deficit of accessibility. Insufficiency of means of transport would constitute an obstacle to access to services.

The main objective of the transport policy is not only to transport users but to improve the overall welfare of inhabitants. Social and spatial justice in transport must be considered a priority when defining transports policies and planning investments (Martens 2017). The transport organizing authorities must establish transport plans and define quality indicators which reflect benefit for all and on the whole territory, so that spatial distribution of supply meets the real needs of the population without creating discrimination. Also, public transport operators will have to meet service quality requirements predefined by the authority to organize to be eligible for operation and benefit from subsidies.

References

African Union Commission (2015) Agenda 2063, the Africa we want. https://au.int/sites/default/files/documents/36204-doc-agenda2063_popular_version_en.pdf

Arnold P, Charlier J, Thomas I (2000) Transport geography at the turn of the century. Belgeo 1-2-3-4, 219–226

Bakour M, Baouni T, Thévenin T (2018) La dépendance automobile à Alger: entre efficacité du système automobile et précarité du système de transport (Automobile dependency in Algiers: between efficiency of automobile system and precariousness of transport system), RTS-Recherche Transports Sécurité, IFSTTAR, pp 26

Baouni T (2003) Croissanceurbaine et Transport collectif à Alger (Urban Growth and Public Transport in Algiers). Transports 420:227–233

Bavoux JJ, Beaucire F, Chapelon L, Zembri P (2005) Géographie des transports (Geography of Transport), Armand COLIN éd. Paris, p 231

Caubel D (2007) Réduire les disparités d'accès à la ville ? Une réponse concrète, mais imparfaite, par les transports collectifs (Reduce disparities of access to the city? A concrete but imperfect response by public transport). Les Cahiers Scientifiques Du Transport 51:9–36

Chabane L (2014) Le secteur privé des transports urbains de voyageurs, quelles logiques de fonctionnement ? Une enquête auprès des opérateurs privés à Alger (The privatesector of urbanpassenger transport, whichlogics of operation? A survey of private operators in Algiers). Les Cahiers Du Cread 109:89–119

Chapelon L (1996) Modélisation multi-échelles des réseaux de transport: vers une plus grande precision de l'accessibilité. Mappemonde 3:28–36

Duda RO, Hart PE, Stork DG (2001) Pattern classification. John Wiley and Sons, New York, p 654

El-Geneidy A, Diab E, Jacques C, Mathez A (2013) Sustainable urban mobility in the Middle East and North Africa. Thematic study prepared for Global Report on Human Settlements, p 48. :https://unhabitat.org/sites/default/files/2013/06/GRHS.2013.Regional.Middle.East_.and_.North_.Africa.pdf

Godard X (2013) Comparisons of urban transport sustainability: lessons from West and North Africa Research. Transp Econ 40:96–103

Jeahanno A (2008) Stratégie de mobilité durable dans les villes des pays en développement, Certu, Lyon, p 199

ISTED (1999) Eléments de stratégie du développement des transports urbainsdans les pays endéveloppement (Elements of strategy for urban transport development in developing countries)-Bilan et propositions Groupe Transport Urbain Paris, p 50

Kaufman L, Rousseeuw PJ (1990) Finding groups in data. John Wiley and Sons, New York, p 342

Kerbachi R, Chikhi S, Boughedaoui M (2017) Development of real exhaust emission from passenger cars in Algeria by using on-board measurement. Energy Procedia 136:388–393

Le Blanc F (1995) SIG pour transports (GIS for Transport). Techniques Et Politiques D'équipement 127:36–39

Le Breton E (2002) La mobilitéquotidiennedans la vie précaire, Note de recherche pour l'Institut pour la villeenmouvement (Daily mobility in precarious life, Research Note for the Institute for the City in Movement)–PSA Peugeot, Citroën et Abeille Aide et Entraide, p 46

Lucas K (2011) Making the connections between transport disadvantage and the social exclusion of low income populations in the Tshwane Region of South Africa. J Transp Geogr 19:1320–1334

Martens K (2017) Transport justice: designing fair transportation systems. Routledge, pp 256

Matouk A (1992) Les transports urbainsenAlgérie (Urban Transport in Algeria), Actes de conférence : villes et développement, Montréal, p 26

Orfeuil J-P (2004) Transports, Pauvreté, exclusions (Transport, Poverty, Exclusions), éditions de l'aube, p 183

Osinubi T-S (2005) Macroeconometric analysis of growth, unemployment and poverty in Nigeria. Pak Econ Soc Rev XLIII(2):249–269

Stathopoulos N (1994) Distribution territoriale de l'offre et couverture spatiale du réseauferroviaireparisien: uneméthodologied'évaluation de la performance des réseaux de transport urbain (Territorial distribution of supply and spatial coverage of Parisian rail network: a methodology for evaluating the performance of urban transport networks). Les Cahiers Scientifiques Du Transport 29:3–16

UITP MENA Centre for Transport Excellence (2016a) MENA transport report 2016. UITP MENA CTE, Dubai

UITP MENA CTE (2016b) Mobility in cities database–lessons learned from the cities of the MENA region, Dubai

UNDP (2001) Map of poverty in Algeria. http://www.arabstates.undp.org/content/rbas/en/home/library/Sustainable_development/map-of-poverty-in-algeria.html

Xue J, Zhong W (2003) Unemployment, poverty and income disparity in Urban China. Asian Econ J 17(4):383–405

Lila Chabane is a senior researcher at Center for Research in Applied Economics for Development (CREAD) in Algeria. She hold a Ph.D. in geography at Paris Diderot university (France), a Postgraduate Diploma in Transport Economy from ENTPE at Lyon (France) and Statistical Engineer from National School of Statistics and Applied Economics at Algiers. Currently, she is responsible of a research team on urban dynamics; his research is broadly focused on the city problems in particular people mobility, urban systems, transport policy and urbanism. Also, she is Head of research project on governance and financing of public transport in the city of Algiers.

Chapter 5
An Analysis of the Determinants of Travel Demand Patterns in Ghana

Marian Ankomah and Sivaramakrishnan Srinivasan

Abstract Understanding the travel behaviour of road users is important in the development of transportation plans. This chapter present models for household vehicle fleet composition and person-level trip frequencies of both working and non-working adults in Ghana. The study utilizes data from the 2012 Ghana Transport Indicator Database Survey. A Multinomial logit model for vehicle fleet ownership (cars, bicycles, and motorcycles) captured the effects of socio-economic and locational factors. The negative binomial regression models for trip frequencies by mode (walk, bicycle, motorcycle, car, taxi, and bus) highlight important differences in the impacts of socio-economic and location factors on trip frequencies of workers and non-workers. The models collectively demonstrate that factors such age, gender, household composition, income and residential location all have statistically strong impacts on choices about vehicle ownership and trip frequencies by mode. To a large extent, these results appear intuitively reasonable based on common sense expectations and socio-cultural underpinnings of Ghana. The models also highlight the heterogeneity in the vehicle ownership and trip frequency choices across the different parts of the country. Overall, the models from this study can be used to forecast the impacts of changes in socio-economic characteristics of the population on both vehicle ownership and travel demand (trip frequencies) for different parts of Ghana.

Keywords Trip frequencies · Vehicle ownership · Demand forecasting · Developing countries

M. Ankomah
Traffic Engineer/Road Engineer, Fekams Company Limited, Fekams House, Tabora No. 6, Accra, Ghana
e-mail: moankomah@ufl.edu

S. Srinivasan (✉)
Department of Civil and Coastal Engineering, University of Florida, 365 Weil Hall, 1949 Stadium Road, Gainesville, FL 32611, USA
e-mail: siva@ce.ufl.edu

R. A. Acheampong et al. (eds.), *Transport and Mobility Futures in Urban Africa*,
The Urban Book Series, https://doi.org/10.1007/978-3-031-17327-1_5

5.1 Background and Objectives

Travel demand forecasting plays a significant role in the development of transportation plans and in the evaluation of transportation infrastructure. There are several factors that contribute to changes in travel demand. In the developed world (for example, United States and several European countries), the critical influence of socio-economic factors (population growth, age/gender distribution, household structure, income, vehicle ownership, etc.) on travel demand patterns is well established (see for example, Martin et al. (2016) for a detailed discussion in the context of United States). Further, the advancement of information and communication technologies (ICTs) has led to new mobility options and opportunities to undertake activities virtually leading to significant changes in travel demand.

These socio-economic and technological factors are also relevant in the context of developing countries. For example, Li et al. (2018) studied the impacts of socio-economic factors on travel demand in the context of China and Bansal et al. (2018) studied the impacts in the context of India. In the context of the Middle-East and North Africa, Masoumi (2018) identifies the influence of social factors such as family size, gender and income in contributing to vehicle ownership and travel demand. Based on discussions from a workshop conducted at the 12th International Conference on Travel Behavior Research (IATBR) in 2009, Srinivasan et al. (2012) report that the *rapid* expansion of the economies of several developing countries is leading to personal affluence, urbanization, and motorization at, potentially, a greater speed than what was experienced by many of the currently developed nations. The associated rapid increases in travel demand along with low/poor supply of infrastructure is leading to unique transportation planning challenges in developing nations.

To develop comprehensive plans and policies to address the growing transportation needs using sustainable solutions, there is a critical need to develop improved travel-demand forecasting models. Such models are commonplace in the United States and in Europe. These models typically comprise of two main components: models for long-term choices such as residential location and vehicle ownership and models for short-term choices of daily trip frequency by purpose, destination, mode and route (popularly called the "four-step model system"). Over the last three decades, there have also been several enhancements to the demand forecasting models of the developed world driven by the need to capture transportation choices better, the availability of data to support the development of models and the enhancements in statistical techniques and software to estimate and apply advanced models. The development of "activity-based models" as an alternative to four-step models is a key consequence of the new needs and capabilities. The features of an activity-based model also make it more suitable to model the travel demand patterns in developing countries (Srinivasan et al. 2012); however, the data needed are not yet readily available to make activity-based models a practical reality.

The demand forecasting models in developing nations are often rudimentary. The state of practice is dictated, in the author's opinion, by both the lack of perceived need for models (possibly related to lower importance of transportation planning compared to other aspects of nation building) and the lack of data. However, it can be argued that there is not only the critical need for data-driven/model-based policy making for shaping the future transportation infrastructure, but also that the improved capabilities in the fields of data collection and analysis can indeed support such an endeavor.

The focus of this paper is on developing travel demand models for Ghana. Automobile ownership in Ghana is increasing by about 40,000 vehicles per year (Organisation Internationale des Constructeurs d'Automobiles 2019). A significant proportion (60%) of the vehicles owned in the country are in Accra, the capital city and business hub of Ghana. The vehicle ownership level in Accra is expected to rise to over 1 million by 2023 (Fiagborlo 2017). The dramatic growth in vehicle population is driven by population growth, urbanization, and a sense of pride in vehicle ownership. Owning a vehicle is considered an "achievement", and individuals who usually patronize public transport are from poor or low-income households. Therefore, households who own vehicles are rarely seen using public transport as it is perceived to diminish their social status (Moody 2019; Gatersleben, 2011). The increased travel demand along with the preference for automobiles increases the pressure on the road infrastructure. In fact, according to Adogla-Bessa (2016), it is estimated that about 15% of road users driving personal cars in Accra use 85% of the road space. Therefore, the introduction of Bus Rapid Transport in Accra in 2016 aimed to shift the dependency on small occupancy vehicles to some form of public transport. Acheampong (2020) also highlights "rapid historical urbanization" of African cities as a critical trend that would shape the future commute patterns of its workers.

Given the rapid changes in vehicle ownership patterns, urbanization, and socioeconomic changes, data-driven planning efforts are needed to quantitatively evaluate alternate scenarios and policies to produce an efficient transportation system that is truly reflective of the growing needs of its population. However, studies on the development of comprehensive travel forecasting models (four-step models or activity-based models) for the entire country of Ghana appear to be scarce. There are, however, efforts aimed at understanding certain elements of travel behaviour such as mode choice patterns often in the context of urban centres in the country. For example, Abane (2011) conducted a study about the attitudes and travel behaviour of Ghanaians from four metropolitan areas—Accra, Tamale, Kumasi and Sekondi-Takoradi. This study showed that mode choice is significantly impacted by the affordability and availability of those modes. Agyemang (2017) identified perceptions of convenience (that can be related to affordability and availability) as a strong predictor for choosing a 'trotro' (minibus) over a car. In comparing the mode choice of employees in the formal sector in the two largest cities in Ghana, that is Accra and Kumasi, Abane (2011) and, Amoh-Gyimah and Aidoo (2013) found that high-income workers are less likely to use public transport. Furthermore, findings from Amoh-Gyimah and Aidoo (2013) showed that workers with households of family size greater than two are less likely to choose non-motorized transport. From their study, the distance from home significantly affects mode choice as workers are less likely to travel to work by public transport for distances greater than 5 km. Most recently, Acheampong

(2020) developed models to examine the impacts of land use and spatial patterns on commuting patterns in the Kumasi metropolis in Ghana, using data from over a thousand workers, models for car ownership and commute mode choice. The results indicate that suburbanization and poor public transportation systems are key drivers of increased car ownership and car-based commuting for the segment of the population that can afford cars.

The discussions thus far highlight both Ghana's need for travel-demand models to support its transportation planning programs and the limited availability of such models in practice. The objective of this chapter is to demonstrate the development of models that determine the travel demand patterns for Ghana using the latest set of data available and, subsequently, highlight the needs and opportunities for enhanced data collection and approaches for building better models in the future. Data from the 2012 Transport Indicator Database surveys are used to develop models for household vehicle fleet ownership (long-term choice) and daily trip frequencies by mode (short-term choice) for both workers and non-workers. Both the models for household vehicle fleet ownership and trip frequencies using data from Ghana are new empirical contributions to the literature. The experience from assembling and analysing these data is used to make recommendations about future data needs and opportunities.

The rest of this chapter is organized as follows. Section 5.2 presents an overview of the data, Sect. 5.3 describes the modelling methodology and the empirical model results are presented and discussed in Sect. 5.4. In each of Sects. 5.2, 5.3 and 5.4, the first subsection focuses on the household-level vehicle fleet ownership while the second subsection focuses on trip frequencies by mode. Section 5.5 presents the summary and conclusions.

5.2 Data

This study used data from the 2012 Ghana Transport Indicator Database Survey which was conducted by the Ghana Statistical Service. This first-ever nationwide household-based transport survey was conducted over a 3-month period between September 2012 and December 2012. The survey was administered to over 23,000 persons from about 6000 households across the entire country. Further details about the survey and data assembly are available from Ankomah (2019). The dataset also provides information on the region from Ghana to which the households belong. Figure 5.1a identifies the regions in the country. It is useful to note that these are the pre-2019 regional boundaries as used in the travel survey. Figure 5.1b shows the extent of urbanization in each of the regions. The Great Accra (capital Accra) and Ashanti (capital Kumasi) are the most urbanized regions in Ghana.

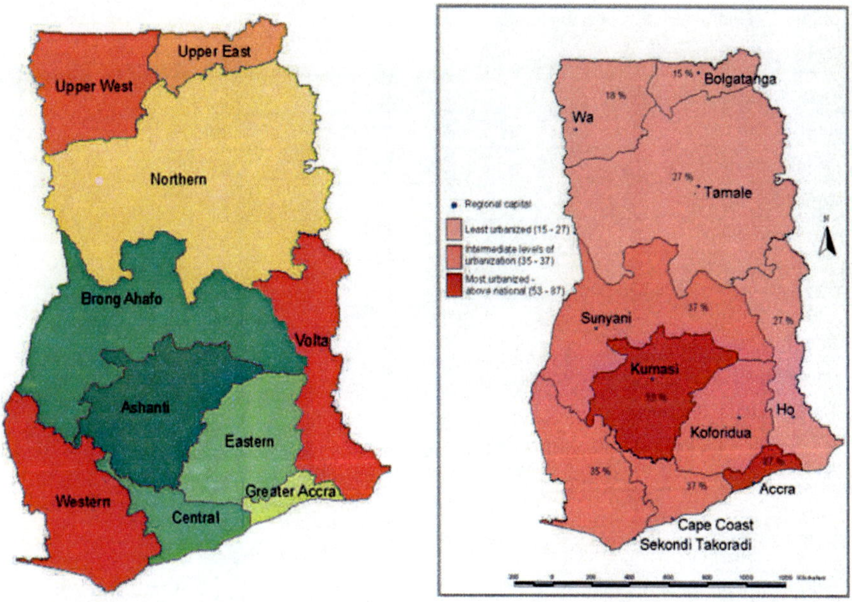

(a) (b)

Fig. 5.1 Regions of Ghana

5.2.1 Data for Modelling Household Vehicle Fleet Ownership

As vehicle ownership is modelled at the household level, the descriptive statistics for the 5921 households from the cleaned sample (a small proportion of households were excluded because of missing and/or inconsistent data) used for modelling are presented in Table 5.1. The household size ranges from 1 to over 6 persons with an average size of 3.99 persons per household. The households were also classified based on the gender and age (child/adult) of the members. The highest level of education was determined by examining the education levels of all persons in the household. An examination of household income indicates that about 70% of the households surveyed belong to the two lowest levels of income. About 57% of the households were in rural areas and the remaining 43% were in urban areas. Other descriptors of residential locations such as city centre, in a town, in the suburb, near a major road, and near a transit terminal are also available. These aggregate indicators available from the survey serve as the best proxies for various detailed descriptors of land-use (such as density, diversity, and design) around the residence which can be constructed if detailed GIS maps were available.

The survey provided information on the availability of each of the cars, motorcycles, bicycles, and other vehicles for each household. The overall household vehicle fleet was determined based on all these variables. Six categories emerged: (1) no

Table 5.1 Characteristics of households

Socio-economic variables	Percent frequency
Household size structure	
Single Person HH	20.30%
Two Persons HH	13.30%
Three Persons HH	15.10%
Four Persons HH	15.60%
Five Persons HH	13.30%
Six or More Persons HH	22.40%
Other HH size structure	
Households with a single adult and at least one child	13.00%
Married couple household	70.20%
Households with an equal proportion of adult male to adult female	38.20%
Predominantly female households	35.10%
Predominantly male households	26.80%
Highest Education	
Less than middle school graduate	18.80%
Middle school graduate/BECE	30.90%
Some High School	2.70%
Vocational Training	2.30%
SSCE certificate/A level certificate	13.40%
Training College/Polytechnic	4.30%
Bachelors	3.70%
Unknown	23.90%
Household Monthly Income	
Income less than 200.01 ghc	43.40%
Income between 200.01 ghc and 600 ghc	31.90%
Income between 600.01 ghc and 1000 ghc	10.50%
Income between 1000.01 ghc and 1200 ghc	2.90%
Income greater than 1200 ghc	4.30%
Unknown	6.90%
Residential Location	
HH in urban area	42.70%
HH in rural area	57.30%
HH lives in city centre	5.10%
HH lives in town	35.90%
HH lives in the suburb	32.00%

(continued)

Table 5.1 (continued)

Socio-economic variables	Percent frequency
HH lives along a major road	8.60%
HH lives on the farm	6.80%
HH lives near trans terminal	0.20%
Other Location	8.10%
Unknown	3.2
Regional Location	
Western Region	9.60%
Central Region	8.50%
Greater Accra Region	14.30%
Volta Region	9.50%
Eastern Region	11.40%
Ashanti Region	18.80%
Northern Region	10.40%
Upper East Region	4.80%
Upper West Region	3.00%

vehicles, (2) only car, (3) only motorcycles, (4) single bicycle, (5) multiple bicycles, and (6) mixed fleet (one or more of cars, bicycles, motorcycles, and other vehicles). Table 5.2 presents the fraction of household belonging to each household fleet category. The results are presented by urban and rural locations and in the overall. Almost 70% of the households have no vehicles. 6.1% of households in urban areas and 1.2% of households in rural areas have only cars and 4.9% of households in urban areas and 2% of households in rural areas have only motorcycles. These clearly indicate that urban households are more likely to have motorized vehicles even though the overall level of motorization is still very low. In contrast, ownership of bicycles (one or more) is higher in rural areas when compared to urban areas. About 4.4% of households own a mixed fleet of vehicles. The model is estimated using all data while including various locational indicators to reflect the effect of urbanization and other factors.

Table 5.2 Household vehicle fleet ownership

	Urban (%)	Rural (%)	Overall (%)
Zero Vehicles	71.7	68.3	69.8
Only Car	6.1	1.2	3.3
Only Motorcycle	4.9	2.0	3.3
Single bicycle	10.3	17.2	14.2
Multiple bicycles	2.8	6.8	5.1
Mixed fleet	4.2	4.5	4.4

5.2.2 Data for Modelling Trip Frequencies by Mode

Unlike in the case of household vehicle fleet modelling, the analyses for trip frequencies are conducted at the person level. Among the 22,784 total persons in the cleaned dataset 12,734 were adults and these persons were used for modelling the frequency of the trips by mode (i.e., travel demand of children are not modelled in this study). The persons were further stratified into 9648 workers and 2849 non-workers and separate models were estimated in each case.

The survey queried each respondent about the total number of trips by car, bus, taxi, motorcycle, and bicycle *in the last seven days from the day the respondent completed the survey*. In contrast, data on *daily* (one day) trips by foot (walking) were collected. Among the different modes, the trips by foot, bus, and taxi are most common while trips by car, motorcycle, and bicycle were generally the least. It is useful to note that walking, taxi, and bus are generally available modes to all persons while trips by car, motorcycle, and bicycle by a person are feasible only when the household owns the corresponding vehicle. It is evident from the vehicle fleet analysis presented previously that the proportions of all households owning such vehicles are relatively low.

The distribution of frequency of trips for the common modes (Foot, bike, and taxi) is presented in Tables 5.3 (for workers) and 4 (for non-workers). The cell values represent the percentage of persons making the corresponding level of the trip. The results are also stratified based on urban and rural residential locations of the workers/non workers. Workers and non-workers in rural areas are more likely to not make any trips (i.e., 0 trips) by bus and taxi when compared to corresponding persons in urban areas. In contrast, workers and non-workers in urban areas are more likely to not make any trips (i.e., 0 trips) by foot when compared to corresponding persons in rural areas. It is also evident that workers are more likely to make trips (less likely to make 0 trips) compared to non-workers in both rural and urban areas. Finally, while over 90% of all people make foot trips *daily* (irrespective of location or employment status), more than 50% of people do not make any trips by either taxi or bus *even over a week* (Table 5.4).

The survey data includes socio-economic characteristics of the individuals of a household (such as age, gender, ethnic group) in addition to variables that show the detailed characteristics of the individual's educational background (such as highest educational level, current grade, and means of transport to and from school). Also, the data provides socio-economic characteristics of the household characteristics such as size and income, and some location variables such as region and district where the household exists. Tables 5.5 and 5.6 present these characteristics for workers and non-workers, respectively. The first major column presents these descriptions for all workers (9648 persons) and (2849 persons) non-workers. The subsequent columns are for the subset of workers and non-workers who have car, bicycle, or motorcycle available in their households. In modelling the trip frequencies by car, bicycle, and motorcycle, these appropriate subsets of data were used. Trip generation by foot, taxi, and bus were modelled using all data.

Table 5.3 Distribution of trips by taxi, bus, and foot for workers

# Trips	Taxi		Bus		Foot	
	Urban	Rural	Urban	Rural	Urban	Rural
0	49.00	65.19	52.94	66.17	3.48	1.94
1	1.67	1.42	1.26	1.06	0.08	0.16
2	12.28	12.14	10.36	12.90	16.85	16.12
3	1.44	0.77	1.36	0.81	0.45	0.65
4	9.24	8.16	8.58	7.40	19.70	20.45
5	0.96	0.47	0.58	0.57	0.63	0.95
6	5.34	4.39	5.33	4.26	13.16	15.38
7	0.89	0.11	0.75	0.13	0.13	0.11
8	3.32	1.94	3.22	1.65	9.28	10.23
9	0.35	0.02	0.15	0.04	0.00	0.04
10	4.76	2.61	4.70	2.10	11.23	12.63
10+	10.74	2.79	10.76	2.91	25.02	21.33

Table 5.4 Distribution of trips by taxi, bus, and foot for non-workers

# Trips	Taxi		Bus		Foot	
	Urban	Rural	Urban	Rural	Urban	Rural
0	59.59	77.59	64.45	77.69	7.07	6.12
1	2.06	0.72	0.95	0.85	0.63	0.07
2	10.94	8.79	9.79	9.39	19.48	22.59
3	0.87	0.59	1.34	0.65	0.24	0.20
4	7.05	4.50	6.16	4.37	19.40	18.75
5	0.55	0.39	0.71	0.39	0.79	0.59
6	4.60	3.00	4.19	2.28	11.63	13.09
7	0.40	0.26	0.71	0.07	0.39	0.20
8	3.09	1.24	2.45	0.85	6.91	8.33
9	0.40	0.00	0.08	0.00	0.08	0.07
10	3.25	1.37	3.55	1.76	11.70	11.33
10+	7.21	1.56	5.61	1.70	21.68	18.68

Broadly, the workers are likely to be older (25–44 years has highest frequency) compared to non-workers (18–24 years has the highest frequency). Workers are also likely to have a higher level of education than non-workers and are more likely to be married. There were no significant gender differences or locational differences between workers and non-workers. The subset of workers and non-workers from households with cars are also more likely to be from higher income households compared to workers and non-workers in general.

Table 5.5 Characteristics of working Adults in Ghana

Variable	*Proportion of households by trip mode*			
	All modes	Car	Bicycle	Motorcycle
Sample	9648	467	2748	840
Gender				
Male	46.8	50.7	51	53.1
Female	53.2	49.3	49	46.9
Age Categories				
18–24	13.9	14.1	14.4	13.6
25–44	56.1	60.4	55.9	62.7
45–64	25	24	24.7	20.7
>= 65	5	1.5	4.9	3
Highest Education Level				
Less than middle school	19.1	8.4	15,5	13.1
Middle school graduate/BECE	26	23.3	15.8	19.2
Some High School	2	2.8	1.5	2
Vocational Training	2	3.6	1.2	1.7
SSCE certificate/A level certificate	8.6	14.8	6.6	10.5
Training College/Polytechnic	2.7	7.9	1.7	4.3
Bachelors	2.5	18.6	1.3	4
Other	0.7	2.8	0,5	1.1
HH member currently in school	5.2	13.3	5.9	5.6
Employment Status				
Employee	1.3	0.9	1.2	1.5
Employer	0	0	0,5	0.1
Self Employed	52.2	53.3	56.8	55.4
Unpaid Family Worker	0.1	0	0.1	0
Apprentice	45.8	45	41.3	42
Other Employment Status	0.5	0.9	0.5	1
Household Monthly Income				
Income less than 200.01 ghc	38.1	5.4	43.3	32
Income between 200.01 ghc and 600 ghc	34.3	26.3	33.8	36.4
Income between 600.01 ghc and 1000 ghc	12.2	19.7	10.3	15.2
Income between 1000.01 ghc and 1200 ghc	3.6	10.5	3.3	6.2
Income greater than 1200 ghc	6.1	32.5	6.5	7.4
Marital Status				
Married	68.4	67.7	76.7	81
Divorced/Separated	6	3.2	2.7	2

(continued)

Table 5.5 (continued)

Variable	*Proportion of households by trip mode*			
	All modes	Car	Bicycle	Motorcycle
Widowed	5.2	2.4	3.5	1.7
Never married	20.4	26.8	17.1	15.4
Ethnicity				
Akan	47	67.7	23	25.7
Ga/Dangme	6.4	7.3	2.5	1.9
Ewe	13.2	12.6	9	8.2
Mole Dagbani	19.8	8.1	39.3	36.8
Other ethnicity	12.7	3	25.4	26.2
Religion				
Christian	72.1	88.7	52.1	53.3
Islam	17.7	7.9	32.8	37.3
Traditional	5.2	0.2	11.2	6
Other religion	5.1	3.2	3.9	3.5
Residential location				
Household member in urban area	41.7	77.1	30.5	48.5
Household member in rural area	58.3	22.9	69.5	51.5
Household lives in city centre	4.7	8.1	2.3	1.9
Household lives in town	36	28.1	27.7	35.8
Household lives in a suburban area	32.4	53.5	26.7	28.6
Household lives along a major road	8.4	6.4	13.1	10.5
Household lives on the farm	8.4	1.3	14.8	9.8
Household lives near transport terminal	0.2	0	0	0
Other household location	8.6	1.9	14.6	12.7
Regional census				
Western	10	4.3	3.8	3.8
Central	6.3	4.3	1.2	1.2
Greater Accra	12.8	32.1	6.3	6.1
Volta	8.1	2.1	8.4	9
Eastern	11	13.9	4.1	3.3
Ashanti	20	28.9	9.7	10.5
Brong Ahafo	10.8	7.3	16.2	13.5
Northern	11.3	5.8	27	31.4
Upper east	6.6	0.4	15.9	11.8
Upper west	3.1	0.9	7.5	9.4

(continued)

Table 5.5 (continued)

Variable	*Proportion of households by trip mode*			
	All modes	Car	Bicycle	Motorcycle
Sekondi Takoradi (Western Region)	1.4	1.1	0.1	0.2
Gomoa East (Central Region)	0.6	1.3	0	0
Accra Metropolitan Area (Greater Accra Region)	4.8	13.7	1.6	2.5
Ho Municipal (Volta Region)	0.2	0	0.2	0.2
Kwaku North (Eastern Region)	0.8	0.4	1.3	1.5
Kumasi Metropolitan Area (Ashanti Region)	7	20.8	1.5	5.1
Techiman (Brong Ahafo Region)	0.9	0.9	1.2	1.4
Tamale Metro (Northern Region)	1.7	4.5	4	9.8
Bolgatanga Municipal (Upper East Region)	0.9	0	1.8	1.4
Wa Municipal (Upper West Region)	0.3	0.9	0.5	1.4
Vehicle Ownership				
Bicycle	28.48	17.1	100	57.6
Motorcycle	8.71	10.28	17.61	100
Car	4.84	100	2.91	5.71
Bus	0.44	3	0.18	0.8
Truck	0.39	1.93	0.55	1.55

5.3 Modelling Methodolgy

Vehicle ownership (a long-term choice) is an important predictor of daily (short-term) travel demand and such decisions are made at the household level. Therefore, it is necessary to understand the factors that determine household vehicle fleet ownership. The substantive focus of vehicle-ownership studies in the developed world is on automobile (car) ownership (Train and Lohrer 1982; Potoglou and Kanaroglou 2008; Potoglou and Susilo 2008). However, car ownership levels are still very low in Ghana, but households do own other vehicles such as bicycles and motorcycles. So, our models consider the vehicle fleet mix (cars, motorcycles, and bicycles) owned by a household instead of simply focusing on cars. The approach for modelling household vehicle fleet mix is presented in Sect. 3.1.

Trip generation is the first step in the traditional demand forecasting procedure which determined the frequency of trips. Models routinely used by transportation planning agencies are stratified by trip purpose (work, shopping, leisure, school, etc.). However, the models developed in this study are stratified by the travel mode. This was necessitated by the nature of available data. The need for collecting data on demand patterns by purpose (work, shopping, recreation, etc.) is identified as an important future need. The approach for modelling person-level trip frequencies by mode is presented in Sect. 3.2.

Table 5.6 Characteristics of non-working adults in Ghana

Variable	*Proportion of households by trip mode*			
	All modes	Car	Bicycle	Motorcycle
Sample	2849	148	763	229
Gender				
Male	46.6	60.1	55	51.5
Female	53.4	39.9	45	48.5
Age Categories				
18–24	56.2	39.2	66.7	65.9
25–44	9	4.7	12.1	13.1
45–64	13.2	30.4	9.6	9.6
> = 65	21.6	25.7	11.7	11.4
Highest Education Level				
Less than middle school	15	6.8	14.2	12.7
Middle school graduate/BECE	20.2	20.3	12.7	10.5
Some High School	1.5	2	0.7	0.4
Vocational Training	1.8	6.8	1.2	1.3
SSCE certificate/A level certificate	9	13.5	5.8	8.7
Training College/Polytechnic	2.6	10.8	2.2	3.9
Bachelors	2.1	13,5	1.2	1.7
Other	0.4	0.7	0.1	0.9
HH member currently in school	22.7	23.6	31.6	31.4
Household Monthly Income				
Income less than 200.01 ghc	39.4	6.1	40.1	28.4
Income between 200.01 ghc and 600 ghc	30.4	20.3	34.3	37.1
Income between 600.01 ghc and 1000 ghc	12.1	23	9.4	14
Income between 1000.01 ghc and 1200 ghc	4.1	15.5	3.8	7
Income greater than 1200 ghc	6.9	26.4	8.5	10
Marital Status				
Married	35.1	48.6	35.3	36.2
Divorced/Separated	4.4	2.7	2	2.2
Widowed	12.7	7.4	6.3	6.6
Never married	47.8	41.2	56.5	55
Ethnicity				
Akan	43.9	60.8	20.4	22.7
Ga/Dangme	8.3	5.4	4.2	5.2
Ewe	14.2	14.2	8.7	7.4
Mole Dagbani	19.5	4.1	38.9	36.7

(continued)

Table 5.6 (continued)

Variable	*Proportion of households by trip mode*			
	All modes	Car	Bicycle	Motorcycle
Other ethnicity	13.6	5.4	26.7	27.5
Religion				
Christian	72.6	95.3	52.7	49.3
Islam	17.9	2	33.2	40.6
Traditional	5.2	0.7	11	6.1
Other religion	4.4	2	3.1	3.9
Residential location				
Household member in urban area	45.1	83.8	33.9	50.7
Household member in rural area	54.9	16.2	66.1	49.3
Household lives in city centre	5.5	10.8	2.4	0.9
Household lives in town	34.9	22.3	25.6	28.4
Household lives in a suburban area	31.1	55.4	26.9	31.4
Household lives along a major road	9.8	10.1	14.7	14
Household lives on the farm	8	0	13.8	10.9
Household lives near transport terminal	0.2	0	0.3	0
Other household location	8.8	0.7	15.7	13.5
Regional census				
Western	8.2	0.7	2.5	2.2
Central	6.5	1.4	0.9	0.4
Greater Accra	14.4	41.9	8.3	8.7
Volta	9.7	0.7	8.8	7.9
Eastern	11.9	25	5.4	2.6
Ashanti	19.3	20.9	8.3	11.8
Brong Ahafo	9	6.1	14.2	15.7
Northern	11.5	0.7	28.7	30.6
Upper east	6.7	1.4	17	10
Upper west	2.9	1.4	6	10
Sekondi Takoradi (Western Region)	1.4	0	0.1	0
Gomoa East (Central Region)	0.6	0	0	0
Accra Metropolitan Area (Greater Accra Region)	6.2	16.2	1.6	4.4
Ho Municipal (Volta Region)	0.2	0	0.1	0
Kwaku North (Eastern Region)	0.5	0	0.9	0.9
Kumasi Metropolitan Area (Ashanti Region)	6.7	20.3	1	6.1
Techiman (Brong Ahafo Region)	0.6	0	1.2	1.3

(continued)

Table 5.6 (continued)

Variable	*Proportion of households by trip mode*			
	All modes	Car	Bicycle	Motorcycle
Tamale Metro (Northern Region)	2.1	0.7	5.4	10.9
Bolgatanga Municipal (Upper East Region)	1.2	0.7	2.6	1.3
Wa Municipal (Upper West Region)	0.2	1.4	0.1	1.7
Vehicle Ownership				
Bicycle	26.78	17.57	100	56.64
Motorcycle	8.04	7.43	17.3	100
Car	5.19	100	3.41	4.8
Bus	0.56	2.7	0.5	3.93
Truck	0.49	2.03	0.7	2.62

5.3.1 Methodology for Household Vehicle Fleet Ownership

The multinomial logit is used to model the household vehicle fleet ownership. Each household (represented by index n) is assumed to have six alternatives (represented by index i). The alternatives are zero vehicles, only car, only motorcycle, single bicycle, multiple bicycles, and mixed fleet. Each alterative provides a certain level of utility to a household ($U_{i,n}$). This utility depends on various factors such as household size structure, income, and residential location. The net utility from all these observable factors is represented by $V_{i,n}$. . Additionally, the model also incorporates the effects of unobserved factors not considered in the utility function ($\varepsilon_{i,n}$). Thus,

$$U_{i,n} = V_{i,n} + \varepsilon_{i,n}$$

where,

$U_{i,n}$ = total utility of household n for choosing vehicle fleet type i,

$V_{i,n}$ = deterministic or observed component of the utility (dependent on household structure, income, etc.)

$\varepsilon_{i,n}$ = random or unobserved component of utility for household n for choosing vehicle fleet type i.

The MNL model assumes that households choose the alternative that maximizes their utility. Consistently, the probability of a household n choosing vehicle fleet type i can be calculated by the logit formula:

$$P_n(i) = \frac{\exp(V_{i,n})}{\sum_{j \in C_n} \exp(V_{j,n})}$$

where *Cn* represents the choice set of the household (the six alternatives).

5.3.2 Methodology for Trip Frequencies by Mode

Negative binomial regression models were estimated to model the trip frequencies by mode (count data). If $Y_{i,n}$ is the number of trips by individual i by mode n, the expected value of the trips is related to explanatory factors $X_{i,n}$ such as age, gender, income, residential location, etc. as shown below:

$$E[Y_{i,n}] = \exp(\beta X_{i,n})$$

In the above equation, β is the vector of the estimated coefficients on the explanatory variable. In general, if the coefficient is positive, it indicates that the corresponding variable is associated with more trips and if the coefficient is negative, the corresponding variable is associated with fewer trips. The negative binomial regression also recognizes the over dispersion in the data (variance > mean).

For each of the workers and non-workers, six models were estimated, one each for car, bicycle, motorcycle, foot, taxi, and bus trips. The models for trips by foot, taxi, and bus were estimated for all adults as these modes are available for all. The models for trips by car, motorcycle, and bicycle were estimated for those adults whose household had at least one of the corresponding modes of travel. As already discussed, the survey provided data on daily demand for walk trips and weekly demand for all other modes. It is useful to reiterate that this is neither a "classical trip generation model" (as it does not model trip frequencies by purpose) nor a "classical mode choice model" (as it does not model the choice of mode for each trip).

5.4 Empirical Results

5.4.1 Results for the Household Vehicle Fleet Ownership

The model for household vehicle fleet ownership was estimated using the SPSS statistical software with the following six choice alternatives: no vehicles, only car, only motorcycle, only single bicycle, only multiple bicycles, and a mixed fleet. The choice of no vehicles was taken as the reference alternative (i.e., the utility of this alternative is fixed to 0 and the utilities for other alternatives are interpreted relative to the utility of having no vehicles). The model results are shown in Table 5.7. All results presented are statistically significant at 95% confidence. A "–" in the table indicates that the corresponding coefficient was statistically insignificant, and "ref" indicates that the level of the corresponding categorical variable was chosen as the reference category. For example, households with 6+ persons were taken as the reference level for household size, and therefore, the coefficients on other household size variables capture the effects of a certain level of household relative to the largest size households. The model results are discussed next. We have also sought to provide plausible explanations (authors opinions) to justify the directionality of impacts (i.e., the sign of the estimated model coefficients).

The household vehicle ownership patterns are impacted by various socio-economic factors. Households with fewer members have a lower utility for owning multiple bicycles. Married couple households have a greater utility to own a motorcycle than unmarried couple households. Additionally, households with predominantly female adults have a lower utility to own two-wheelers (bicycles or motorcycles) compared to households with an equal proportion of male to female adults. Also, predominantly male households have a greater utility to own a motorcycle. These gender impacts are generally as expected as females are usually more concerned about safety and appearance in traffic thus less likely to ride bicycles. In addition, there is also a social stigma attached to women riding bicycles. It is, however, useful to note that these social myths and other cultural norms have evolved over the years and there is a possibility of increase in bicycle ownership in predominantly female adult households.

The effect of the total number of children and workers in a household appear reasonable. Households with more adult workers are more likely to own multiple bicycles to serve many workers. Also, the larger the number of children in a household, the more likely it is for that household to own mixed fleet of vehicles. As would be expected, lower income households are less likely to own cars. Household with lower education levels in general have a lower utility for car and motorcycle ownership.

The location of the household was found to be a very strong predictor of the household fleet composition. The strongest predictors were dummy variables representing the different regions of the country. For example, households living in wealthier parts of the country such as Greater Accra, Ashanti, and Eastern regions have a higher utility for car-ownership relative to households in other parts of the country. Acheampong (2020) also reported a strong influence of income and spatial structure on car ownership of workers in urbanized Ghana. At the same time households living in the wealthier parts of the country have a lower utility for motorcycles and bicycles compared to the poorer parts of the county (Northern, Upper East, and Upper West regions). On the other hand, households in the northern part of the country (Northern, Upper East, and Upper West regions) are more likely to own a bicycle, motorcycle, or a mixed fleet. In addition to increased affordability of cars in wealthier regions like Greater Accra, ownership of two wheelers is lower in these regions potentially because of safety reasons as well. The road networks were not designed with cyclists in mind and the two-wheelers must compete with other (aggressive) road users like trucks, buses, cars, and even road-side hawkers (Eriksson et al. 2009; Asafo-Adjei et al. 2017). These factors could generally deter households in the greater Accra region from owning two-wheelers. However, in recent times, there has been a significant increase of two-wheelers in the capital region and across all regions as these vehicles are often used for door-to-door delivery of food, and other personal packages. A dummy variable indicating whether the household lived in an urban or rural area was also included in the model, but this was found to be statistically insignificant.

Table 5.7 Model for household vehicle fleet ownership

Variables	Car		Motorcycle		Single Bicycle		Multiple Bicycles		Mixed Fleet	
	Coeff	Sig	Coeff	Sig	Coeff	Sig	Coeff	Sig	Coeff	Sig
Constant	−24.059	0	−0.073	0.942	0.531	0.433	−1.645	0.127	1.9	0.052
Household (HH) size										
HH size = 1	–	–	–	–	–	–	−2.555	0.003	–	–
HH size = 2	–	–	–	–	–	–	−1.109	0.05	–	–
HH size = 5	–	–	–	–	−0.51	0.017	−0.638	0.049	–	–
HH size = 6+	Ref	Ref	Ref	Ref	Ref	Ref	Ref	Ref	Ref	Ref
Married couple household	–	–	0.927	0.001	–	–	–	–	–	–
Predominantly female households	–	–	−0.891	0.002	−0.717	0	−0.683	0.009	−0.903	0.002
Predominantly male households	–	–	–	–	–	–	0.635	0.009	–	–
Equal proportion of male to female adult	Ref	Ref	Ref	Ref	Ref	Ref	Ref	Ref	Ref	Ref
Total number of children			–	–	–	–	–	–	0.168	0.033
Total number of workers	–	–	–	–	–	–	0.258	0.004	–	–
Highest Education										
Less than middle school graduate	−2.111	0	−2.093	0	–	–	–	–	–	–
Middle school graduate/BECE	−1.583	0	−1.193	0	–	–			–	–
Some High School	–	–	−1.174	0.035	–	–	–	–	–	–
SSCE certificate/A level certificate	−1.258	0.008	−1.865	0.005	–	–	–	–	–	–
Training College/Polytechnic	−1.4	0	−0.976	0.004	–	–	–	–	0.894	0.026
Bachelor's Degree	−0.989	0.005	–	–	–	–	–	–	–	–
Others	Ref	Ref	Ref	Ref	Ref	Ref	Ref	Ref	Ref	Ref

(continued)

Table 5.7 (continued)

Variables	Car		Motorcycle		Single Bicycle		Multiple Bicycles		Mixed Fleet	
	Coeff	Sig	Coeff	Sig	Coeff	Sig	Coeff	Sig	Coeff	Sig
Household Monthly Income										
Income less than 200.01 ghc	−3.406	0	–	–	−0.552	0.039	−0.854	0.037	−1.651	0
Income between 200.01 ghc and 600 ghc	−2.106	0	–	–	–	–	–	–	–	–
Income between 600.01 ghc and 1000 ghc	−1.367	0	–	–	–	–	–	–	–	–
Income greater than 1200 ghc	Ref	Ref	Ref	Ref	Ref	Ref	Ref	Ref	Ref	Ref
Household Location										
HH lives along a major road	–	–	1.025	0.05	–	–	1.064	0.011	–	–
Others	Ref	Ref	Ref	Ref	Ref	Ref	Ref	Ref	Ref	Ref
Regional Location										
Western Region	22.483	0	−3.702	0	−2.548	0	−3.038	0	−5.417	0
Central Region	23.62	0	−3.885	0	−3.361	0	–	–	−6.341	–
Greater Accra Region	24.543	0	−3.082	0	−2.597	0	−2.192	0.002	−4.558	0
Volta Region	22.767	0	−1.986	0	−1.593	0	−1.376	0.043	−3.71	0
Eastern Region	23.983	0	−4.025	0	−2.884	0	−3.534	0	−4.83	0
Ashanti Region	24.039	0	−3.229	0	−2.702	0	−2.973	0	−5.186	0
Brong Ahafo Region	23.15	0	−1.87	0.005	−1.006	0.1	–	–	−3.406	0
Northern Region	–	–	–	–	–	–	–	–	−1.258	0.007
Upper East Region	–	–	–	–	–	–	–	–	−1.365	0.007
Upper West Region	Ref	Ref	Ref	Ref	Ref	Ref	Ref	Ref	Ref	Ref

(continued)

Table 5.7 (continued)

Variables	Car		Motorcycle		Single Bicycle		Multiple Bicycles		Mixed Fleet	
	Coeff	Sig	Coeff	Sig	Coeff	Sig	Coeff	Sig	Coeff	Sig
Number of observations	5921									
Log likelihood, constant only	7417.898									
Log likelihood, final model	5706.772									

Given that significant differences in urbanization across the regions (see Fig. 5.1), it is likely that the regional dummy variables are also reflective of urbanization.

Overall, the model indicates significant differences in household vehicle ownership patterns across the different regions of the country. The choices are also influenced by socio-economic factors and potentially the cultural norms associated with the various groups.

5.4.2 *Results for Trip Frequencies by Mode*

The negative binomial regression models developed trip frequencies by the mode using the STATA statistical software and the models were subject to several iterations to obtain our best models. The best model incudes explanatory variables significant at 95% confidence or higher.

The models for **working adults** are presented in Table 5.8. The models for the frequency of walk, taxi, and transit trips were estimated using data for all working adults (9648 persons) while the models for car, motorcycle and bicycle trip frequencies were estimated using the subsample of persons whose household had at least one of the corresponding vehicles (467 persons for car, 2748 persons for bicycle, and 840 for motorcycle).

The negative coefficients on household size for trips made by car, foot, and taxi show that the number of trips by these modes decreases with increasing household size (the corresponding coefficients in the model are −0.14, −0.018, and −0.054). Workers in households with more adults make fewer trips by bicycle and motorcycle. Also, persons in households with more workers make fewer bus trips. The gender composition of a household also plays a significant role in the trip generation by car, bicycle, motorcycle, and taxi. Female adult workers make fewer trips by these modes than male adult workers. The strongly significant negative coefficient on the gender variable for bicycle and motorcycle modes potentially reflects the imports of cultural norms and safety perceptions limiting women's use of these modes for travel. The age of an adult worker has a significant impact on trip generation by car, bicycle, motorcycle, and foot older workers make more trips by car and fewer trips by foot, bicycle, and motorcycle. Agyemang (2017) also reported that older persons travel more by car.

Education level plays a significant role in trip generation. Workers with a bachelor's degrees make more trips by car, taxi, and bus than workers with lower educational levels. Findings by Amoh-Gyimah and Aidoo (2013) also identified that households with a diploma degree or higher make more car trips compared to households of lower educational status. Workers who are currently in school make more trips by bicycle and foot and fewer trips by taxi. Furthermore, the results suggest that all workers with a middle school certificate or higher who do not own any vehicle (car, bicycle, motorcycle) are more likely to make more trips by taxi and bus although working adults with bachelor's degree would make more taxi trips.

Table 5.8 Trip Generation models by mode for working adults

Variables	Car		Bicycle		Motorcycle		Foot		Taxi		Bus	
	Coeff	T-stat	Coeff	T-stat	Coeff	T-stat	Coeff	T-stat	Coeff	T-stat	Coeff	T-stat
Constant	1.213	2.3	2.621	4.68	5.1	12.72	1.886	21.8	1.792	7.4	0.918	12.03
Socio-Economic Variables												
Household size	−0.14	−3.26	–	–	0.101	2.28	−0.018	−5.06	−0.054	−6.43		
Total number of adults in household	–	–	−0.199	−5.99	−0.516	−5.68						
Total number of employees/workers in household	–	–	–	–	–	–					−0.08	−4.48
Female	−0.785	−4.15	−1.731	−16.84	−2.129	−12.93			−0.098	−2.26		
Age in years	0.044	4.06	−0.015	−3.45	−0.024	−3.23	−0.0005	−6.94				
Household member currently attending school	–	–	0.954	3.82	–	–	0.194	4.39	−0.377	−3.44		
Less than middle school graduate	–	–	−0.374	−2.64	0.46	2.11			0.381	6.08	0.276	4.27
Middle school graduate/BECE	–	–	−0.845	−5.94	–	–	−0.075	−3.4	0.59	10	0.498	8.35
Some High School			−0.886	−2.09	–	–			0.402	2.64	0.391	2.42
Vocational Training	–	–	–	–	–	–	−0.162	−2.55	0.939	6.46	0.516	3.3
SSCE certificate/A level certificate	0.674	2.61	−0.677	−3.28	0.582	2.4	−0.068	−2	0.666	7.94	0.663	7.73
Training College/Polytechnic	1.186	3.42	−1.241	−3.31	0.806	2.3			0.695	5.2	0.544	3.82

(continued)

Table 5.8 (continued)

Variables	Car		Bicycle		Motorcycle		Foot		Taxi		Bus	
	Coeff	T-stat	Coeff	T-stat	Coeff	T-stat	Coeff	T-stat	Coeff	T-stat	Coeff	T-stat
Bachelor's Degree	1.045	4.19	−2.362	−5.06	–	–	−0.131	−2.22	0.82	5.88	0.453	3.06
Employment Status												
Self Employed	–	–	–	–	–	–	0.072	4.01				
HH with monthly income less than 200.01 ghc	−2.226	−4.39	0.304	3.05			0.077	3.82	−0.542	−9.36	−0.294	−6.03
HH with monthly income between 200.01 ghc and 600 ghc	−1.053	−4.81	–	–	0.403	2.55			−0.165	−3.03		
HH with monthly income between 600.01 ghc and 1000 ghc	–	–	–	–	–	–	−0.071	−2.46				
HH with monthly income between 1000.01 ghc and 1200 ghc	−0.911	−2.94	–	–	–	–	−0.138	−2.8				
HH with monthly income greater than 1200 ghc	–	–	–	–	–	–						
Never married	–	–	0.759	2.86					−0.258	−4.54		
Married	0.552	2.2	0.461	2.22	0.658	2.94						
Divorced/Separated	1.206	2.05										

(continued)

Table 5.8 (continued)

Variables	Car		Bicycle		Motorcycle		Foot		Taxi		Bus	
	Coeff	T-stat	Coeff	T-stat	Coeff	T-stat	Coeff	T-stat	Coeff	T-stat	Coeff	T-stat
Vehicle Ownership												
Car owned by household	–	–	−0.622	−3.23	−0.697	−3.02			0.141	2.24		
Motorcycles owned by household	–	–	–	–	–	–	0.044	2.07	−0.224	−3.7	−0.121	−1.96
Bicycles owned by household	–	–	–	–	–	–	0.024	2.02	−0.174	−9.09	−0.095	−3.62
Bus owned by household	–	–	–	–	2.922	2.98	−0.424	−5.1	−0.523	−2.71		
Truck owned by household					−2.985	−2.18					−0.609	−2.08
Residential Location												
Household lives in city centre	−1.973	−4.9	–	–	–	–	0.516	11.92	0.36	2.76		
Household lives in town	–	–	–	–	–	–			0.67	7.36		
Household lives in a suburban area	–	–	–	–	–	–	0.258	12.23	0.777	8.32	0.186	3.49
Household lives along a major road	–	–	–	–			0.099	2.99	0.565	5.16	−0.189	−2.26
Household lives on farm	–	–	–	–	−0.842	−3.21	−0.256	−7.34	0.273	2.36	−0.348	−3.79

(continued)

Table 5.8 (continued)

Variables	Car		Bicycle		Motorcycle		Foot		Taxi		Bus	
	Coeff	T-stat	Coeff	T-stat	Coeff	T-stat	Coeff	T-stat	Coeff	T-stat	Coeff	T-stat
Sekondi Takoradi (Western Region)	–	–	–	–	–	–			0.435	2.53	0.576	3.14
Accra Metropolitan Area (Greater Accra Region)	1.533	5.13	−1.097	−2.6	–	–	0.346	7.56			0.863	7.74
Ho Municipal (Volta Region)									1.321	2.53		
Kwahu North (Eastern Region)	–	–	–	–	–	–	−0.897	−8.12			−0.955	−3.44
Gomoa East (Central Region)	–	–	–	–	–	–	−0.443	−3.42				
Kumasi Metropolitan Area (Ashanti Region)	–	–	–	–	−1.018	−2.43			0.253	2.95	0.701	7.52
Techiman (Brong Ahafo Region)	–	–	–	–	–	–					−0.722	−2.95
Tamale Metro (Northern Region)	–	–			1.044	4.09			1.128	7.03	−0.923	−4.75
Bolgatanga Municipal (Upper East Region)									−0.862	−2.94		
Wa Municipal (Upper West Region)	–	–	−1.8	−2.59	–	–			−1.453	−2.68	0.973	2.34

(continued)

Table 5.8 (continued)

Variables	Car		Bicycle		Motorcycle		Foot		Taxi		Bus	
	Coeff	T-stat	Coeff	T-stat	Coeff	T-stat	Coeff	T-stat	Coeff	T-stat	Coeff	T-stat
Time to walk to the nearest taxi rank (minutes)	–	–	0.051	2.61	–	–	0.026	6.96	−0.149	−16.15		
Time to walk to the nearest bus stop (minutes)	−0.206	−2.7	−0.042	−1.96	–	–					−0.09	−8.84
Time taken to walk to the nearest station/boarding point	–	–	–	–	–	–			−0.007	−3.57	–	–
Sample Size	467		2748		840		9648		9648		9648	
LL(B)	−1001.712		−4985.698		−1651.129		−28,331.415		−15,933.337		−15,737.146	
LL(c)	−1074.43		−5160.922		−1761.364		−28,720.217		−16,589.564		−16,146.289	
Pseudo R square	0.067		0.034		0.0626		0.0135		0.0396		0.0253	

Workers with a household income less than 1200ghc ($624 per 2012 cedi to dollar conversion rate) make fewer car trips than high-income workers. Workers with a household income between 600.01 ghc and 1200ghc make fewer foot trips than workers with a household income less than 200ghc. This result is also consistent with the finding by Acheampong (2020) that high-income households are more likely to make car work trips than middle-income and low-income households.

It was interesting to find out that married working adults generate more trips by car, bicycle, and motorcycle than unmarried workers. Also, single working adults generate fewer taxi trips and more bicycle trips. Divorced or separated working adults make more car trips than self-employed workers who are more likely to make bicycle and foot trip trips, whereas apprentices were more likely to make bicycle trips.

The vehicle ownership of a household also significantly affects the frequency of bus trips. As expected, adults in households who own a car are less likely to make bicycle and motorcycle trips. In terms of residential location, adult workers that reside in the city centre make fewer trips by car than similar adults who live outside the city centre. Such adults also generate fewer trips by taxi except for adults that live on the farm. In contrast, adults living in the city centre generate more foot trips although the number of foot trips shows a positive correlation across all the residential location variables except households that live on a farm. The findings from Haybatollahi et al. (2015) also indicate that people in highly dense areas of the city centre often walk or bike. Working adults in the Accra Metropolitan Area (AMA) make more car trips than workers in all the other major metropolitan areas in the country. Also, the number of motorcycle and taxi trips increases, whereas the number of bus trips decreases for working adults in the Tamale Metro area. This result is also consistent with the findings of Abane (2011). This is because the main mode for travel in the northern part of the country is captured by the Tamale Metro area is the motorcycle.

The models for trip frequencies by mode for non-working adults are presented in Table 5.9. The models for walk, taxi, and transit trip frequencies were estimated using data for all non-working adults (2849 persons) while the models for car, motorcycle, and bicycle trip frequencies were estimated using subsample of persons whose household had at least one of the corresponding vehicles (148 persons for car, 763 persons for bicycle, and 229 for motorcycle).

Non-working adults make fewer foot, taxi, and bus trips as the household size increases (the corresponding coefficients in the model are −0.024, −0.073, and −0.056). Also, non-workers in households with more adults make fewer bicycle trips. Females generate substantially fewer bicycle and motorcycle trips than males. Females also make fewer bus trips than men. Age is also a significant determinant in trip generation as older adults make fewer car, bicycle, and foot trips.

The effect of the adult's highest educational level and the household's income on the trip frequencies are as anticipated. Household members currently attending school generate more trips by all modes except car and bus. The marital status of an adult and the household residential location are rather strong predictors of the trip generation by car. Single adults make fewer car trips and more motorcycle trips depending on their location. Adults living in the towns and suburbs generate the most car trips although adults living in the suburbs also make more motorcycle, foot, taxi,

Table 5.9 Trip generation models by mode for non-working adults

Variables	Car		Bicycle		Motorcycle		Foot		Taxi		Bus	
	Coeff	T-stat	Coeff	T-stat	Coeff	T-stat	Coeff	T-stat	Coeff	T-stat	Coeff	T-stat
Constant	2.98	2.47	4.1	7.56	−6.063	−2.66	2.339	22.7	−1.185	−3.68	0.629	2.84
Socio-Economic Variables												
Household size	–	–			–	–	−0.024	−3.83	−0.073	−4.09	−0.056	−3.15
Total number of adults in household			−0.223	−3.67								
Female			−1.423	−7	−1.986	−3.65	–	–	–	–	−0.22	−2.05
Age in years	−0.043	−2.39	−0.045	−7.48			−0.012	−14.12	–	–	–	–
Household member currently attending school	–	–	0.879	3.36	2.067	3.43	0.15	3.26	0.498	3.51	–	–
Less than middle school graduate	–	–	−0.647	−2.22	–	–	–	–	0.525	3.54	0.497	3.27
Middle school graduate/BECE	–	–	−1.611	−4.94	–	–	–	–	0.519	3.75	0.674	5
Some High School											0.861	2.13
Vocational Training	–	–			–	–	–	–			1.003	2.69
SSCE certificate/A level certificate	–	–	−1.448	−3.22	–	–	–	–	0.638	3.62	0.556	3.03
Training College/Polytechnic	1.006	2.37	–	–	4.033	3.02	–	–	0.651	2.12	–	–
Bachelors Degree	1.121	2.86	–	–	–	–	–	–	0.671	2.06	–	–

(continued)

Table 5.9 (continued)

Variables	Car		Bicycle		Motorcycle		Foot		Taxi		Bus	
	Coeff	T-stat	Coeff	T-stat	Coeff	T-stat	Coeff	T-stat	Coeff	T-stat	Coeff	T-stat
HH with monthly income less than 200 ghc					2.39	3.2						
HH with monthly income between 200.01 ghc and 600 ghc					1.941	2.72			0.328	2.76		
HH with monthly income between 600.01 ghc and 1000 ghc	–	–	–	–	2.178	2.75	−0.113	−2.1	0.391	2.49		
HH with monthly income between 1000.01 ghc and 1200 ghc	–	–	–	–	–	–	–	–	0.875	3.64		
HH with monthly income greater than 1200 ghc	–	–	–	–			–	–	0.662	3.28		
Never married	−3.456	−4.31	–	–	2.868	2.26	–	–	–	–		
Married					3.159	2.6					0.271	2.47
Vehicle Ownership												
Car owned by household	–	–	–	–	–	–	–	–	–	–		
Motorcycle owned							0.106	2.19				

(continued)

Table 5.9 (continued)

Variables	Car		Bicycle		Motorcycle		Foot		Taxi		Bus	
	Coeff	T-stat	Coeff	T-stat	Coeff	T-stat	Coeff	T-stat	Coeff	T-stat	Coeff	T-stat
Bicycle owned					0.509	2.23			−0.129	−2.52		
Bus owned by household	0.574	2.4	–	–	–	–	–	–	−1.73	−2.9	−0.987	−2.26
Residential Location												
Household lives in an urban area												
Household lives in city centre	–	–	−1.286	−2.03	–	–	0.456	5.26	0.674	2.57		
Household lives in town	3.961	2.417	–	–	–	–	−0.121	−2.26	0.894	5.2		
Household lives in a suburban area	3.813	2.847	–	–	1.835	3.19	0.134	2.53	1.01	5.71	0.295	2.44
Household lives along a major road									0.704	3.39		
Household lives on farm	–	–	–	–	–	–	−0.37	−4.93	–	–		
Sekondi Takoradi (Western Region)	–	–	–	–	–	–	0.355	2.38	0.81	2.05	1.03	2.53
Accra Metropolitan Area (Greater Accra Region)	1.501	4.28	–	–	2.146	2.29	0.251	3.15	0.578	2.7	1.063	4.66
Kwahu North (Eastern Region)	–	–	–	–	–	–	−1.295	−4.64	–	–	–	–

(continued)

Table 5.9 (continued)

Variables	Car		Bicycle		Motorcycle		Foot		Taxi		Bus	
	Coeff	T-stat	Coeff	T-stat	Coeff	T-stat	Coeff	T-stat	Coeff	T-stat	Coeff	T-stat
Kumasi Metropolitan Area (Ashanti Region)	–	–	–	–	−2.816	−1.95	–	–	0.81	4.05	1.078	5.09
Tamale Metro (Northern Region)	–	–	–	–	–	–	–	–	1.1	3.43	−2.48	−4.63
Bolgatanga Municipal (Upper East Region)			–	–	–	–	–	–	–	–	–	–
Time to walk to the nearest taxi rank (minutes)	–	–	–	–	0.208	2.19	0.021	2.76	−0.155	−6.97	−0.095	−2.26
Time to walk to the nearest bus stop (minutes)	–	–			−0.372	−3.3	−0.02	−2.37	–	–	−0.052	−2.1
Sample Size	148		763		229		2849		2849		2849	
LL(B)	−328.612		−1370.465		−309.743		−8024.228		−3637.324		−3445.976	
LL(c)	−361.715		−1416.507		−339.765		−8198.831		−3806.916		−3554.867	

and bus trips. It was interesting to find out that non-working adults who live on the farms make fewer foot trips.

Household vehicle ownership impacts trip frequencies by all modes except by bicycle. Interestingly, non-working adults that own motorcycles make more foot trips. Out of the 21% of non-working adults that are seniors, (adults 65 years and above) majority of them make mostly foot trips; this could be a possible explanation of why non-working adults generate more foot trips. This result is also consistent with the findings from Hu et al. (2013) whose study on understanding the travel behaviour of older adults in a developing country shows that almost half of the older adults in China choose to walk. Non-working adult households that own at least one bicycle make more motorcycle trips and fewer taxi trips. Such adults with this trip pattern most likely own motorcycles as well. Adults from households that own a bus make fewer bus and taxi trips but more car trips. A plausible explanation to this travel trend could be that households which own busses are more likely able to afford to own cars and may use their buses for commercial purposes whiles they use their personal cars for most of their trips.

In terms of the metropolitan location of the household, the only significant variable for car trips is the adults that live in AMA. Such adults generate more car trips in addition to motorcycle, foot, taxi, and bus. However, adults in the Kumasi Metropolitan Area (KMA), a comparable large metropolitan area to AMA, generate fewer motorcycle trips but more taxi and bus trips.

Overall, the negative binomial regression models indicate very strong effects of various socioeconomic and location factors on trip generation by mode for both workers and non-workers. While the models for foot, bus, and taxi trips were estimated using larger samples (all workers and non-workers), the models for trips by the other modes were estimated using smaller samples (especially those for car and motorcycle trips). Thus, caution must be administered in interpreting these model results.

5.5 Summary and Conclusions

Understanding the travel behaviour of people is critical to forecasting travel demand. However, in many developing countries like Ghana, there have been few empirical efforts in understanding the travel patterns using statistical models applied to survey data. This study contributes to the literature by developing statistical models for household vehicle fleet ownership and person-level trip frequencies for Ghana. The data used for this study came from the 2012 Ghana Transport Indicator Database Survey. To the best of our knowledge, apart from the summary reports developed by Ghana Statistical Services, the disaggregate survey data have not been used for any developing models for travel demand. Therefore, results from this data serve as a useful contribution to the Ghana Statistical Services, the Ministry of Transport, and other governmental transportation agencies in developing national transportation policies.

A multinomial logit model was estimated to model household vehicle fleet ownership. Six choice alternatives were considered: (1) no vehicles, (2) only car, (3) only motorcycles, (4) single bicycle, (5) multiple bicycles, and (6) mixed fleet (one or more of cars, bicycles, motorcycles, and other vehicles). While a vast majority of past vehicle ownership models have focused only on car ownership, our study considers all types of vehicles together. Such a model can capture trade-offs made by households in owning different types of vehicles. Some of the currently zero-vehicle households may choose to own motorcycles while others may choose to own cars. The estimated model allows us to capture the probability of these choices as a function of several factors such as income, education, household structure, and residential location. The model also highlights significant heterogeneity in vehicle fleet choices across Ghana suggesting the need for local, context-sensitive policies to enhance mobility while ensuring sustainability. Finally, the vehicle fleet ownership patterns are also found to be an important predictor of trip frequencies (travel demand).

Negative binomial regression models were estimated to model person trip frequencies by mode (car, walk, bicycle, motorcycle, taxi, and bus). Models were estimated separately for workers and non-workers. While past research has focused more on workers and their commute, our study also provides travel demand estimates for non-working adults. The models demonstrate that factors such age, gender, household composition, income, and residential location all have statistically strong impacts on choices about trip frequencies by mode. To a large extent, these results appear intuitively reasonable based on common sense expectations and socio-cultural underpinnings of Ghana. The impacts of each of these factors are also different between workers and non-workers. Capturing this difference is important to accurately predict the impacts of employment trends on future travel demand.

Overall, the models from this study can be used to forecast the impacts of changes in socio-economic characteristics of the population on both vehicle ownership and travel demand (trip frequencies). It is strongly recommended that future surveys consider collecting data in the travel diary format in which each household is asked to report the trips made by all household members on one/two specific days identified by the surveyor. Such an approach would be necessary to model all dimensions of travel demand such as frequency of trips by purpose, destination choice, mode choice, and time of day choice. Second, a national plan for periodic (3–5 years) travel surveys would be important. Given the rapid rate of urbanization, technology changes, and shifts in cultural norms and attitudes being witnessed across the world, a program of panel data collection would be critical to capture the dynamics in behavioural changes. Finally, it would also be important to increase the size of data samples. The models presented in this paper showed strong differences in behaviours across the different parts of Ghana. Having adequate data from the different regions would be important to develop policies appropriate for the different parts of the country.

Finally, it important to acknowledge that this study was entirely conducted prior to the COVID pandemic and using data from 2012. It has been widely acknowledged that the pandemic has had transformative changes in travel behaviours across the entire world. The impacts in the context of Ghana are already being documented. Barbieri et al. (2021) compared the impacts of COVID-19 on the mobility of commuters before and during the COVID-19 restrictions across ten countries including Ghana. Their findings show that, although majority of Ghanaians shifted to remote working, Ghana had the least work-from-home rate amongst all the ten countries sampled. Based on data collected to assess the choice of public transport prior to and during the COVID-19 pandemic, Sogbe (2021) reports a decline in public transport ridership with an increase to the patronage of taxis. This decline was largely because commuters placed a larger value to the cleanliness of the public transport vehicles and to social distancing. In the broader perspective, it is not evident that the post-pandemic travel behaviour patterns will necessarily return to those during pre-pandemic times. Therefore, the need for panel data to monitor dynamic trends in behaviour is critical more than ever. The analysis methods such as those presented in this study when applied to the new data can shed robust insights to support transportation planning decisions for Ghana in the future.

References

Abane AM (2011) Travel behaviour in Ghana: empirical observations from four metropolitan areas. J Transp Geogr 19(2):313–322

Acheampong RA (2020) Spatial structure, intra-urban commuting patterns and travel mode choice: analyses of relationships in the Kumasi Metropolis. Ghana. Cities 96:102432

Adogla-Bessa D (2016) Target private car owners to reduce traffic congestion—consultant. Citi 97.3 FM—Relevant Radio. Always. Retrieved July 31, 2019, from http://www.citifmonline.com/2016/12/20/target-private-car-owners-to-reduce-traffic-congestion-consultanttarget-private-car-owners-to-reduce-traffic-congestion-consultant/

Agyemang E (2017) Mode choice for long distance trips: evidence from the greater Accra metropolitan area of Ghana. J Transp Geogr 64(C):150–157

Amoh-Gyimah R, Aidoo EN (2013) Mode of transport to work by government employees in the Kumasi metropolis, Ghana. J Transp Geogr 31:35–43

Ankomah M (2019) An analysis of transportation demand patterns in Ghana. Doctoral dissertation submitted to the University of Florida, Gainesville, FL, USA

Asafo-Adjei C, Iyer-Raniga U, Aranda-Mena G (2017) Mobility and access to transportation for the rural poor: examining the role of intermediate means of transport (IMTS) in Ghana. In: 13th international postgraduate research conference 2017, p 881

Barbieri DM, Lou B, Passavanti M, Hui C, Hoff I, Lessa DA, ... Rashidi TH (2021) Impact of COVID-19 pandemic on mobility in ten countries and associated perceived risk for all transport modes. PloS One 16(2):e0245886

Bansal P, Kockelman KM, Schievelbein W, Schauer-West S (2018) Indian vehicle ownership and travel behavior: a case study of Bengaluru, Delhi and Kolkata. Res Transp Econ 71:2–8

Eriksson G, Davidsson T, Lundgren P (2009) Minor field study on traffic safety in Ghana: Pedestrian and cyclist facilities and access in central Accra

Fiagborlo JD (2017) The economic cost of traffic in Accra. Modern Ghana. Retrieved July 31, 2019, from http://www.modernghana.com/news/755613/the-economic-cost-of-traffic-in-accra.html

Gatersleben B (2011) The car as a material possession: exploring the link between materialism and car ownership and use. Emerald Group Publishing Limited, In Auto motives

Haybatollahi M, Czepkiewicz M, Laatikainen T, Kyttä M (2015) Neighbourhood preferences, active travel behaviour, and built environment: an exploratory study. Transp Res Part F: Traffic Psychol Behav 29:57–69

Hu X, Wang J, Wang L (2013) Understanding the travel behavior of elderly people in the developing country: a case study of Changchun, China. Procedia Soc Behav Sci 96:873–880

Li J, Lo K, Guo M (2018) Do socio-economic characteristics affect travel behavior? A comparative study of low-carbon and non-low-carbon shopping travel in Shenyang City, China. Int J Environ Res Public Health 15(7):1346

Martin EW, Shaheen SA, Zohdy IH, Hamilton BA (2016) Understanding travel behavior: Research scan. https://www.fhwa.dot.gov/policy/otps/travel_behavior_research_scan.pdf

Masoumi HE (2018) Editoriale Urban travel behavior in the Middle East and North Africa. TeMA J Land Use Mobility Environ (01/18):3–6

Moody JC (2019) Measuring car pride and its implications for car ownership and use across individuals, cities, and countries (Doctoral dissertation, Massachusetts Institute of Technology, Department of Civil and Environmental Engineering)

Potoglou D, Kanaroglou PS (2008) Modelling car ownership in urban areas: a case study of Hamilton. Canada. J Transp Geogr 16(1):42–54

Potoglou D, Susilo YO (2008) Comparison of vehicle-ownership models. Transp Res Rec 2076(1):97–105

Sogbe E (2021) The evolving impact of coronavirus (COVID-19) pandemic on public transportation in Ghana. Case Studies on Transport Policy

Srinivasan S, Srinivasan KK, Dhakar NS (2012) Modeling travel behavior: the developing country context, from travel behavior research in an evolving world: Selected papers from the 12th international conference on travel behavior research, Pendyala R, Bhat CR, pp 311–330

Train K, Lohrer M (1982) Vehicle ownership and usage: an integrated system of disaggregate demand models. Cambridge Systematics, Incorporated

Vehicles in use. OICA (n.d.) Retrieved July 17, 2019, from https://www.oica.net/category/vehicles-in-use/

Marian Ankomah is a civil engineer at Fekams Company Limited in Accra, Ghana. Dr. Ankomah's primary research interest is in traffic impact studies, traffic safety analysis, traffic operation analysis and travel demand forecasting in developing countries. Her doctoral research focused on the travel demand patterns in Ghana by examining how the vehicle fleet composition of a household together with the household and individual characteristics influence the travel behavior of road users. She has a Ph.D. and M.S. from the Department of Civil Engineering at University of Florida, and a B.S. (Civil/Environmental Engineering) from Youngstown State University.

Sivaramakrishnan Srinivasan is an Associate Professor in the Dept. of Civil and Coastal Engg. (CCE) with an affiliate appointment in the Dept. of Urban and Regional Planning (URP) at the University of Florida. His research interests include Transportation Safety, Time Use and Travel Behavior Analysis, and Transportation Demand Forecasting. Dr. Srinivasan chaired the Transportation Research Board (TRB) Committee on the Impacts of Information and Communication Technologies on Travel Choices (2012–2018). He was also a member of the TRB committee on Traveler Behavior and Values. Dr. Srinivasan is a member of the editorial advisory board of the journal "Transportation Research Part B".

Part II
Transport Poverty, Equity and Inclusion

Chapter 6
Transport Poverty and Social Divisions in African Cities: An Introduction

Karen Lucas

This section of the book explores the highly prevalent issue of transport poverty, which is closely associated with the developed and developing spatial of African cities as they have been discussed in the previous section. It comprises four case study chapters, each looking at various aspects of the transport poverty problems and how this affects different social groups and communities. In Chap. 7, Moller-Jensen and Agergaard present the mobility experiences of women in Accra, Ghana, examining an important nexus between their mobility regimes, gendered responsibilities and their employment and livelihood activities. Chapter 8 by Anciaes and Bradbury considers the problems associated with community severance for populations living in proximity to new and upgraded road networks, which is a spiralling problem as these massive infrastructure projects are being introduced in many African cities. In Chap. 9, Médard and Boniface Ngah offer a case study of cities in the Cameroon to model the relationship between the distance travelled by the poor and the mode of accessing educational and health facilities using National Household Consumer Survey data highlighting the need for more inclusive urban planning of these social infrastructures. In the final Chapter 10 for this section, Yankson offers a framework to assess how different modes of transport affect the state of the physical environment, public health and social equity through a comparative study of Ghana and Namibia. He concludes by finding that in future African urban transport policies and planning should move away from an overreliance on road transport and towards more significantly supporting the existing prevalence of walking and non-motorized travel in African cities.

Before turning to these chapters, we offer a brief introduction to the wider issues of transport poverty and how this is currently manifested within African cities in order to set a context for the more specific case studies that follow.

K. Lucas (✉)
Department of Geography, School of Environment, Education and Development, The University of Manchester, Oxford Road, Manchester M13 9PL, United Kingdom
e-mail: karen.lucas@manchester.ac.uk

R. A. Acheampong et al. (eds.), *Transport and Mobility Futures in Urban Africa*, The Urban Book Series, https://doi.org/10.1007/978-3-031-17327-1_6

6.1 Defining Transport Poverty

Transport poverty is a generic term used to discuss multiple problems with transport systems in terms of an unequal distribution of their benefits and burdens across people and places. Lucas et al. have identified that a person is in transport poverty when:

1. *There is no transport option available that is suited to the individual's physical condition and capabilities.*
2. *The existing transport options do not reach destinations where the individual can fulfil his/her daily activity needs, in order to maintain a reasonable quality of life.*
3. *The necessary weekly amount spent on transport leaves the household with a residual income below the official poverty line.*
4. *The individual needs to spend an excessive amount of time travelling, leading to time poverty or social isolation.*
5. *The prevailing travel conditions are dangerous, unsafe or unhealthy for the individual.*
6. *Inability to participate in transport decision-making and lack of recourse to legal justice*

(Lucas et al. 2016: 356)

However, this definition overlooks the collective and systemic issues that lead to these individualized outcomes such as hypermobility, car-dependency, spatial dissonance and community severance, with all exclusions having a role to play within the transport poverty debate. Oviedo et al. (2018) have grouped these different aspects of transport poverty into three broad dimensions: (i) the material assets required to secure an adequate standard of living and the skills, abilities and conditions to secure such assets; (ii) personal and social relational factors, such as social networks and responsibilities and obligation; and (iii) subjective dimensions associated with perceptions, values and experiences, such as the freedom to travel when and where you want to, safely, affordably and with a reasonable degree of comfort and dignity.

From an operational perspective, the 'problems' of transport poverty in African cities is also shaped by often competing disciplinary perspectives. For example, transport engineers frame it is a deficit of basic transport infrastructures; roads, rail, ports and airports, and the technologies that support them, while transport economists are mostly concerned with the inefficiencies of these systems in terms of the timely delivery of goods, services and labour and their economic consequences. Environmentalists tend to worry about the land-take levels of pollution and emissions leading to environmental destruction these systems are causing, and their impact on climate change, whereas social and health scientists focus more upon how transport resources and the benefits and disbenefits of transport systems are distributed across different population groups. Increasingly, they also ask questions about the social outcomes of these different distributions in terms of people's livelihoods and well-being, and whether or not these can be seen as morally fair in line with the social norms that prevail within the society (e.g. Lucas et al. 2020).

In the African context, the most prevalent policy discourses consist of finding ways to increase the speed and ease with which people can travel by building new transportation infrastructures, through more economically efficient, environmentally benign and humanly safe operations and improved governance of these transport system (Uteng and Lucas 2018). This approach is mainly advocated by politicians and global investors who perceive that over-coming the physical deficit in transportation infrastructures and resources presents the best way forward. However, such policies pay little or no attention to the everyday human needs and capabilities of existing populations (Levy 2019) or to the well-being of future generations.

While a deficit of transport infrastructures and transit services in African cities maybe the primary concern of transport professionals, it can often be the excessive presence of these that mostly concerns local citizens. Too many roads with too many vehicles on them and drivers paying too little due care and attention to pedestrians, cyclists and other non-motorized road users is the most vocal complaint of the vast majority of people who do not themselves own cars (Porter et al. 2019). Many people also say they have to travel too far to access their key activity destinations, such as employment, education and healthcare services, with most people walking for all or most of these distances. Otherwise, they must pay an excessive proportion of their household budget to access the informal motorized services that are available to them, whilst still others are overexposed to traffic-related risks such as pedestrian fatalities and injuries, noise and air pollution or incidents of crime and personal harassment whilst travelling (for some case studies of transport poverty in African cities see Lucas et al. 2019 and associated country-specific case studies at https://intalinc.leeds.ac.uk/media/africa/).

6.2 The Social Distributions of Transport Poverty

The case study chapters which follow reveal that transport poverty mirrors and intersects with existing social divisions within African societies, as Porter et al. identify:

> Class, income/socio-economic status, gender, age, household composition, bodily abilities (physical and mental health), religion/cultural attributes, employment status and livelihood type, residential and work locations: all are likely factors in the mix shaping everyday travel opportunities and experiences. Intersectionality - the interconnected nature of these social categorizations that can contribute to overlapping and interdependent systems of discrimination and disadvantage - adds further complexity: the older woman who needs a wheelchair to get around and is resident in a poor neighbourhood experiences mobility very differently from the wealthy young woman who owns and drives her own car. (Porter et al. 2019: 10)

Figure 6.1 demonstrates the underlying causal pathways in this intersectional relationship between transport and material poverty. In the first place, the majority of the poorest citizens of African cities live in informal and slum settlements, many of which are located long distances from the centralized economic and social activities of the city (Venter 2011). Even those, which are located more centrally, tend to

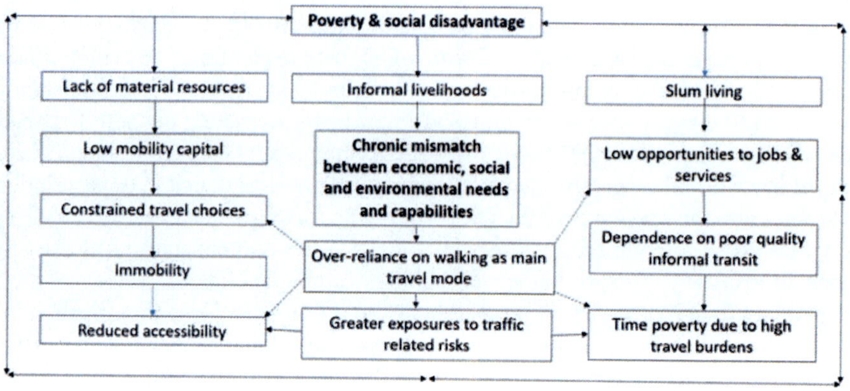

Fig. 6.1 The key dimensions and dynamics of transport poverty in African cities

lack formal pathways and transport infrastructures (Lucas et al. 2019). This causes a spatial dislocation between key destinations, such as employment centres, schools, colleges, hospitals and markets and the home locations, often leading to long journey distances and travel times (Lucas 2011). Furthermore, case studies have identified difficult physical travel conditions within informal settlements, especially during times of flooding, rendering many residents completely immobile, so that they cannot even leave their homes for days or even weeks to access basic goods and activities.

Second, these accessibility deficits combined with low personal capital to secure adequate mobility resources and network capital lead to immobility, on the one hand, and an over-reliance on walking and inadequate and expensive informal services, on the other (Salon and Gulyani 2010; Venter 2011). Travel demand surveys conducted in Nigeria, in Abuja, Kano and Lagos, indicate the high cost of transport: 49, 40 and 33 percent, respectively, of household budgets that were used to cover the cost of transport fares in these cities (Lall et al. 2017: 77). Many of the poorest citizens cannot afford these travel costs on a regular basis and so do not move far from their home locations, thus often restricting their ability to secure a reasonable livelihood (Lucas 2011; Lucas et al. 2020).

Third, the high reliance on walking leads to excessive exposures to traffic-related risks. The World Health Organization data on physical activity identifies that, as average across all African cities, including 61% of men and 51% of women, walk for over 55 min a day. This varies significantly between countries, ranging from an average high of three hours per day for women in Niger to as low as 25 min for men in Sierra Leone (see Fig. 6.2). But this comes with an extremely high exposure to traffic-related risks (Vanderschuren and Zuidgeest 2017). On average, 90,322 people died while walking or cycling across all Africa in 2019, 93% of whom were when walking, and a further 42,723 pedestrians and cyclist suffered serious injuries from traffic collisions. Road traffic injuries are highest amongst the poorest, vulnerable road users, especially children and young people. In a survey of nine countries (Walk

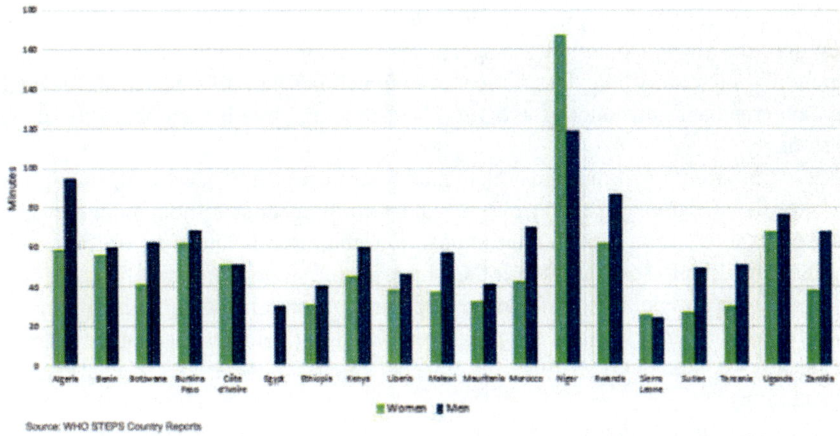

Fig. 6.2 Time spent walking per day in different African countries for men and women

21; 2021), policymakers estimated that 95% of walked urban streets are below the minimum level of standard for pedestrian safety and comfortable walkability.

Fourth, there is poor access to formal public transit services. UN Habitat estimates that on average only 35% of Africans are in reach of formal public transport within 500 m of their homes, compared with 49% as a global average. This means that they are dependent on walking for at least part of every journey, even if they record motorized transit as their main mode of transport. Many people can walk 5 miles or more before reaching even an informal transit stop, largely due to the impenetrable conditions of informal and slum settlements, which mean that even small vehicles cannot enter them (Lucas 2011).

6.3 Transport Poverty as a Failure of Transport Governance

Identifying the pervasive and ubiquitous problem of transport poverty in African cities is not to suggest that policymakers and planners are doing nothing to improve transport systems. For example, a recent survey of African transport authorities found that many African countries have developed transport policies to improve conditions for pedestrian and other non-motorized travellers. Nevertheless, the majority report that most of these haven't been converted into real changes on the ground because they either lack the capacity and/or knowledge of how to enact them (Walk 21; 2021).

Simultaneously, many African cities are in the throes of introducing new roads, rail, metro and bus rapid transit systems, and 'smart' technologies, or they soon plan to do so. Innovations such Uber, Ocada and 'boda boda' style motorcycle taxi services are also emerging from the ground up to compensate for the lack of traditional public transport services and people's increasing basic need to be mobile. Sometimes these new services are welcomed by transport and city planners in the absence of funding for more traditionally planned solutions, and sometimes they are rejected and banned as dangerous and unwanted in the already chaotic and overloaded urban transport system.

It is natural for individuals to seek to maximize their ease of access to destinations through the adoption of speedier, cheaper and more direct door-to-door mode choices, as well as to try to minimize their exposure to the hazards and risks of the traffic system. In African cities, as all over the world, people strive to improve their lives by moving from unreliable, unsafe and costly transport options, which remain largely outside their control, towards the private ownership of vehicles that can be at their disposal 24 h a day and will flexibly transport them wherever and whenever they wish to travel.

The allocation of transport resources is complex in any context, and even more so in the African context, where the multiple actors who are involved have hugely varying degrees of power and control over the decision process. It can been argued that the development of urban infrastructures continue to follow the trajectories of colonial planning (Van der Straaten and Hasenöhrl 2017). Critical commentator Chatterton (2018) suggests that investment in developing cities has been mainly in the interests of 'big capital' and global growth, rather than to meet the needs of the people who live within them. Transport systems are often seen as the catalyst for urban development and similar criticisms could be levelled at the investments that have been poured into them in recent years, both by the Development Banks and Chinese Government.

As such, investments are primarily technical and economically focused, whereby the city is perceived primarily as an enabler of mainstream economic development supported either by state-led infrastructure projects designed to accommodate motorized mobilities. These new projects primarily only ever serve a minority of the population living in African cities because they are mostly aimed at serving commuting trips within and to the central business districts, where mostly skilled, white-collar jobs are based and connected to the locations where the new middle-class elite residents live. On the whole, they will not be available to the urban poor either because they are unaffordable or because they do not operate in the informal and peripheral areas (Vanderschuren et al. 2017).

This project development focus may be partly because transport decision-makers, city planners and other influencers of the transport system in African cities are self-serving. Most are car drivers, and so have little understanding or respect for walkers who are poor and powerless, so deemed unworthy of policy attention. This had led to a set of held values and beliefs within city planning circles that African urban transport systems need to deliver high levels of car-based access and increased travel speeds for the elite few, over and above protecting the safety and social well-being of the majority of travellers (Walk 21; 2021). There is also a general lack of knowledge about how to plan for walking and other non-motorized modes and so policies and budgets avoid these perceptually 'difficult' domains, which is often not about pouring more money at the problem but using existing resources differently.

The external influence of Development Banks and other non-national funders can also mean that local decisions are over-ridden by the interventions of these more powerful actors and agencies, which might have quite different goals to addressing local needs. Although the Sustainable Development Goals (SDGs) have the rhetoric of local inclusivity in the policy and planning process, MacKintosh (1992) argues that there is an absence of authentic involvement in their implementation at state and city level. Consequently, policies and plans mostly fail to adequately account for local mobility and accessibility needs, especially as experienced by the most vulnerable and marginalized.

In terms of addressing transport poverty, this strongly suggests a need to redefine delivery success and a value-shift from performance measured of quantity, i.e. from road miles built and journey time savings, towards the articulation of desired social values of inclusion through improved accessibility, enhanced safety through significantly reduced pedestrian deaths and injuries safer and generally more comfortable and connected travel environments. A mobility future based upon unfettered car use and an individualized transport system, entirely overlooks the issue of who will have access to this system of provision (the minority elite) and who will be left out of it (the majority of women, children, young people, older disabled people, low-income populations and urban peripheral households.

Not everyone can, or will ever, be able to drive or to afford private mobility options and so there will always be a need and demand for walking, cycling and mass public transport alternatives. These alternative travel options cannot operate to their maximum efficiency if they must share the same road-space with competing private vehicles. Restricted financial resources, physical land space and the urban morphologies of many African cities cannot adequately function economically, environmentally or socially based on an individualized model of transport provision similar to that of American cities. This means that private vehicle ownership and use needs to be restricted in order to ensure a fairer distribution of mobility and accessibility across the whole population.

Just as in Global North cities, speeding up travel to reduce individual travel times will also encourage further urban sprawl. More affluent populations will locate in attractive new residential developments close to the CBD, whilst poorer residents living in informal settlements close to the city centre will be pushed out of these locations and into the urban periphery where they will have less access to livelihoods and services. This unequal physical access excludes them from deriving the economic benefits that accrue from these projects, causing knock-on problems of reduced opportunities for social development and SDG equity goals. The absence of adequate traffic laws and transport governance leads to further overall chaos in the transport system that are unsuited to even the limited private vehicle options that are available.

References

Chatterton P (ed) (2018) Unlocking sustainable cities. A manifesto for real change. Pluto Press, London

Lall SV, Henderson JV, Venables AJ (2017) African cities: opening doors to the world. World Bank, Washington DC. https://openknowledge.worldbank.org/handle/10986/25896

Levy C (2019) Travel choice reframed: "Deep distribution" and gender in urban transport. In: Scholten CL, Joelsson T (eds) Integrating gender into transport planning. Palgrave MacMillan, Switzerland

Lucas K (2011) Making the connections between transport disadvantage and the social exclusion of low-income populations in the Tshwane Region of South Africa. J Transp Geogr 19(6):1320–1334

Lucas K, Mattioli G, Verlinghieri E, Guzman A (2016) Transport poverty and its adverse social consequences. Proceed Inst Civil Eng Transp 169(6):353–365 13

Lucas K, Tsoneva E, Abane A, Amoako-Sakyi R, Owusu S, Porter G, Jennings G, Behrens R, Zuidgeest M, Venter C, Asenime C, Oviedo D, Davila J, Mitullah W, Opiyo R, Basil P, Lwasa S, Kasaija P, Evans J (2019) Transport and social exclusion in Five African Cities. Volvo Research and Education Foundation (VREF) http://www.vref.se/download/18.45182a5f16a84e95fac8b2b5/1561370471910/Transport%20and%20Social%20Exclusion%20in%20five%20African%20Cities%20-%20June%202019.pdf

Lucas K, Akyelken N, Stanley J (2020) Social assessment of transport project in Global South cities using community perception of needs. In: Hickman R, Mella Lire B, Givoni M, Geurs K (eds) Companion to transport, space and equity. Edward Elgar, Cheltenham, pp 180–196

Mackintosh M (1992) Introduction. In: Wutys M, Mackintosh M, Hewitt T (eds) Development policy and public action. Oxford, Oxford University Press

Oviedo D, Levy C, Davila J (2018) Constructing well-being, deconstructing urban (im)mobilities in Abuja, Nigeria. In: Uteng TP, Lucas K (eds) Urban mobilities in the global south, pp 173–194. Routledge, Abingdon

Porter RE, Abane A, Lucas K (2019) User needs and practices in African Cities. A Position Paper 2019 VREF Program Mobility and Access in African Cities: Future Urban Transport. http://www.vref.se/publications/researchsynthesisreports/synthesisreports1/achievingequitablesustainablemobilityandaccessinsubsaharanafricaanoverviewofthestateofknowledge.5.74c15c6d173d8c1d0b2849c9.html

Salon D, Gulyani S (2010) Mobility, poverty and gender: travel 'choices' of slum residents in Nairobi, Kenya. Transp Rev 30(5):641–657

Uteng TP, Lucas K (2018) The trajectories of urban mobilities in the Global South. An introduction. In: Uteng TP, Lucas K (eds) *Urban* mobilities in the Global South, pp 1–18. Routledge, Abingdon

Vanderschuren M, Zuidgeest M (2017) Road safety and non-motorised transport in African cities. In: Mitullah W, Vanderschuren M, Khayesi M (eds) Non-motorised transport integration into urban transport planning in Africa. Routledge, Abingdon, pp 57–72

Vanderschuren M, Jennings G, Khayesi M, Mitullah W (2017) Introduction: challenges and opportunities for non-motorised transport in urban Africa. In: Mitullah W, Vanderschuren M, Khayesi M (eds) Non-motorised transport integration into urban transport planning in Africa. Routledge, Abingdon, pp 57–72

Van der Straaten and Hasenöhrl (2017) Connecting the empire: new research perspectives on infrastructures and the environment in the (Post) Colonial World. NMT 24(4):355–391. https://pubmed.ncbi.nlm.nih.gov/28251245/

Venter C (2011) Transport expenditure and affordability: the cost of being mobile. Dev South Afr 28(1):121–140

Walk 21 (2021) Valuing walking in Africa: decision-makers policy brief for high volume transport project. Walk 21 March 2021

World Health Organisation (WHO) Road traffic injuries report. https://www.who.int/news-room/fact-sheets/detail/road-traffic-injuries

Chapter 7
Mobility Regimes and Equity in Urban Transport: Examining Women's Mobility Experiences in Accra

Maya Møller-Jensen and Jytte Agergaard

Abstract In this chapter, we engage with the theme of equity in urban transport systems in Sub-Saharan Africa and discuss the ways in which constrained everyday mobility bears upon the issue of women's empowerment. Drawing on three months of fieldwork by the first author and taking our point of departure in the conceptualization of a *mobility regime*, we examine how spatial, economic, and social dimensions combine to sustain certain structures related to working women's mobility in Accra, the capital city of Ghana. Within the context of increasing urbanization, underinvestment in urban planning and limited traffic regulation, the chapter examines a nexus between livelihood activities, gender, and employment, and foregrounds a specific importance placed on work by women in Accra—both in terms of fulfilling their responsibilities as mothers and of engaging in the development of the city—which to a large extent shape a mobility regime characterized by distinctly female attitudes towards and use of mobility. In conclusion, we point to the mental and physical barriers for women as agents of change if they are disadvantaged in transport systems and as such to the importance of understanding gendered mobility regimes when planning for equity in transport systems in fast growing cities in Sub-Saharan Africa. Doing so, we argue, will require an exploration of how people move, their reasons for doing so and their experience of this movement as well as, critically, how these differ for gender and family contexts.

Keywords Mobility regime · Urban transport systems · Gender · Ghana · Equity in transport

M. Møller-Jensen (✉) · J. Agergaard
Department of Geosciences and Natural Resource Management, University of Copenhagen, Copenhagen, Denmark
e-mail: mmj@ign.ku.dk

J. Agergaard
e-mail: ja@ign.ku.dk

R. A. Acheampong et al. (eds.), *Transport and Mobility Futures in Urban Africa*, The Urban Book Series, https://doi.org/10.1007/978-3-031-17327-1_7

7.1 Introduction

Within the last decade, scholars, practitioners and politicians have increasingly stressed the deep intertwinement between equity in transport systems and the well-being of urban residents (Sheller and Urry 2006; Cresswell 2010; Uteng and Lucas 2018; Croese et al. 2020). This is especially true in Sub-Saharan Africa, where travellers through the urban landscape are confronted with the immediate need to overcome obstacles in the transport system presented by unreliable institutions and limited transport infrastructures (Berdiel 2018). Indeed, as low-density sprawl and unplanned development has come to characterize most Sub-Saharan African cities, a substantial travel burden is put on residents trying to access work and services located in central areas. This is expressed in high fare costs, uncomfortable journeys, poor connections and safety and security issues that all mitigate travel experiences along the lines of income, race, age, working status, gender etc. (Behrens 2020). Developing equity in transport systems in support of sustainable development in Sub-Saharan African cities has thus become a key political and economic goal (Pirie 2009; Behrens 2020). In recent years, it has, however, become increasingly clear that doing so will require homegrown solutions that reflect the diversity and uniqueness of urban transport contexts in the region, including their entanglement with wider considerations surrounding inclusions and exclusions embedded in *mobility regimes* of the urban landscapes (Uteng and Lucas 2018; Porter et al. 2020).

Since the 1990s, scholars have increasingly employed a *mobility lens* to make sense of the historic and contemporary importance of movement for individuals and societies (Sheller and Urry 2006; Cresswell 2010). Underscoring mobilities as a holistic concept concerned with both patterns, representations and practices of movements on different scales (cf. Cresswell 2010), such studies have shown mobilities as a 'key difference and otherness producing machine of our age' (Salazar and Smart 2011, v): bringing dignity and comfort to some, while severely constraining potential economic and social development for others.

One of the first ways this lens was articulated was through the concept of a *mobility regime,* coined by Sørensen (1999) at the end of the twentieth century. Using the concept to highlight the historical and cultural basis of urban mobility, Sørensen argues that transportation needs and practices must be understood through several dimensions, including the physical shape of a city, the available transport and communication systems, the relationships between mobility and economic, social and cultural activities and the meanings attributed to mobility (ibid., 4). Collectively, these dimensions create a mobility regime, which in subtle ways shapes not only how people move, but also influence their reasons for doing so and their experience of this movement. As such, a mobility regime can neither be conflated with the transport system in a given city, nor is it a political level in itself. Rather, unpacking the mobility regimes in which people move can be used to work politically with equity in transport systems, in the sense that it highlights that governance within urban transport is not solely a question of planning, but rather entails the development of new ideologies and vision (ibid.).

Drawing on this concept of a mobility regime, this chapter will examine experiences of working women commuting through Accra on a daily basis, showing gendered vulnerabilities and inequalities that need to be considered when planning for equity in transport systems. In Accra, preceding analyses have highlighted the high degree of visibility of women in the street scene (Overå 2007; Darkwah 2007; Thiel and Stasik 2016). Furthermore, according to Grieco et al. (1996), the economic crisis of the 80 s and 90 s sharply increased urban households' dependency on women's work and prompted their need for mobility at a time when the transport infrastructure was deteriorating due to privatization and decentralization policies. These insights have later been supported by other studies, which stress the paramount importance of high levels of mobility for most livelihood activities in the city, including those of women, and emphasize how severe obstacles in the public transport system are challenging livelihoods for many social groups (Yankson et al. 2015; Esson et al. 2016; Cobbinah et al. 2018; see also Porter et al. 2011 for a perspective on rural Ghana).

Focusing on working women and their experiences and understanding of their own mobility in the city, this chapter contributes to these studies by bringing forth a gendered lens that adds new empirical and analytical insights to the role of mobility and livelihood in Accra. This analytical concept of 'a working woman', should not be understood to refer to a certain type of labour or level of employment. Rather, as will become clear later, being a working woman means being actively engaged in some form of income generating activity related to certain gendered economic and domestic responsibilities while at the same time denoting a sense of independence. Furthermore, most working women share a common mobility experience—namely that of having to move for work—which analytically brings to the fore the relationship between mobility and livelihoods and empirically puts into focus a major nexus of the mobility regime shaping women's experiences in the city. Indeed, in this chapter we show the importance placed on 'mobility for work' by women in Accra while highlighting their difficulties of movement. This schism, we argue, has consequences for the progressive development of Accra as a city, in that it leads women to mentally feel that they are not a part of the city, while practically it robs them of the time to be so.

The chapter begins with an introduction to the mobility regime concept and ends with a discussion of the difficulties of alleviating gender disparities within urban transportation in the African urban context. It builds on the mobility experiences of 47 women engaged with by the first author through three months of fieldwork in Accra during the autumn of 2018. Living in very different areas and moving through different trajectories in the city, the women interviewed spent an average of two hours on transportation every day, with many spending much more. Eight had access to a private car, five could barely afford public transport and thus moved primarily on foot, and the rest used either minibuses, taxis or a combination of the two. All names of interviewees are pseudonyms and recognizable details have been altered to ensure anonymity.

7.2 Grasping a Mobility Regime: Reading Mobility Experiences

Current mobility research that seeks to go beyond vehicles and infrastructure are indebted to Sørensen's conception of mobility regimes and its focus on the subjective dimensions of mobility (Uteng and Lucas 2018, 6). Touching on ideas later developed further by scholars like Cresswell and Uteng (2008) as well as by Porter and Oviedo within the specific context of livelihoods and well-being in Global South countries (e.g. Oviedo and Sabogal 2020; Porter et al. 2007), Sørensen (1999) notes as his point of departure that 'modern people are not mobile just because it is possible or because it is expected' (ibid., 3). Thus, he argues, researchers have to be cautious not to perceive mobility as being basically an instrumental issue and an object of rational economic decision-making. Rather people have diverse reasons for moving in the way that they do, and in order to support or challenge those reasons we need to raise questions directed at the dimensions shaping them. These dimensions are, as noted, not limited to infrastructural provisions or transport modes; they to an equal extent involve the meanings attributed to mobility in a given context and the subjective experience of being mobile (ibid., 3–4). Together, these dimensions form mobility regimes with broadly traceable histories and geographies, which sustain certain mobility practices and identifications in favour of others.

When studying mobility in Sub-Saharan Africa—that is in the context of rapidly transforming urban landscapes, and governance systems that lack the institutional capacity or fiscal resources to provide scheduled public transport networks at scale—one needs to be careful not to exaggerate neither the decoupling of mobility practices from economic decision-making nor the mode choices available to especially vulnerable population groups. However, what Sørensen argues is that the only way to imagine radically different future transport systems is to take a subjective approach to answering the question of why people move in the way that they do. Therefore, according to Uteng and Lucas (2018), using the concept of a mobility regime ushers the discussion of sustainability and equity in transport systems into the wider realm of identity formation, freedom and rights to the city (ibid., 6).

Moving the discussion in this direction is exactly what we aim to do in our subsequent analysis for this chapter, as we make use of the mobility regime approach to unpack the field of power, which forms the entanglement of patterns, representations and practices of movement for women in Accra. The empirical data underpinning this analysis is collected by the first author through interviews, written and photo diaries, ethnographic descriptions and participant observation while working in a shop in the central Makola Market. The data documents 47 women's use of vehicles (private cars, busses, taxis etc.) for primarily work journeys, their experiences of being on the move and their relationships with family members and colleagues/employers as mediated by mobility needs and desires. The data was analysed using matrix coding combining descriptive thematic and deductive analytical coding, which made it possible to discern a paramount schism between the women's desire for mobility and their ability to be mobile, as well as its economic and social constraints.

To bring this schism to the fore, in what follows, we first analyse the obstacles women experience when commuting on long journeys in the city. Then we explore why women desire mobility in spite of these obstacles, showing first the importance of commuting for the women and highlighting mobility as a resource, and considering secondly the identification and self-identification of mobility practices, through an analysis of how the women perceive themselves as mobile individuals. Together, these dimensions shape a mobility regime with a distinctly female attitude towards and use of mobility and discrete considerations around comfort and dignity. This mobility regime, we argue, entangles with the progressive development of the city and women's role in this development, and must be understood carefully when developing future transport systems.

7.3 Meeting Obstacles in the Urban Landscape During the Daily Commute

The geographical and infrastructural dimensions of the mobility regime women move in Accra, is characterized by increasing urbanization, underinvestment in urban planning and limited traffic regulations, leading to a transport situation, in which it is becoming increasingly difficult to commute safely and effectively through the urban landscape (Yankson and Bertrand 2012; Hart 2016; Agyemang 2017).

Located within the Greater Accra Region and situated along the coast of Guinea, Accra is the capital city and dominant urban centre of Ghana in political, economic and cultural terms. In 2020, an estimated 4.7 million people lived within the part of the region called the Greater Accra Metropolitan Area (GAMA) (GSS 2020), an area which has witnessed a rapid spatial growth during the last 20 years, especially at its fringes, driven largely by the acquisition and development of land for residential purposes (Yankson and Bertrand 2012). This spatial growth has led to constantly expanding peri-urban areas from where an estimated 2 million commuters travel to the central business district on a daily basis (Accra Metropolitan Assembly 2019). Still located within the old city core, the CDB sprawls out from just off the coast and reaches towards Accra's inner Ring Road. Key commercial activities and administrative functions are highly concentrated here, while the functional integration between the city centre and the rapidly expanding fringes are of vital importance (Esson et al. 2016). However, the new residential areas emerging in the periphery are often characterized by poor accessibility, reliance on a limited number of highly congested arterial roads and limited connectivity to other parts of the city (Cobbinah et al. 2018; Møller-Jensen et al. 2020).

For the women interviewed, travelling through this urban landscape presented difficulties of planning and reaching desired destinations on time. Indeed, commuting from one such peripheral community to her work in the central market, Julie, a 23-year-old single mother recounted how it could easily take her two hours to reach work, as she first had to walk from her house to the main road, queue for public transport, and then travel through heavy traffic. This was common for the women, as public transport nodes are customarily located along major arterial roads, such that the women first had to reach these nodes via walking or taxi, before they could join the descend towards the city. For Julie, this meant that, in order to be certain of getting to work by 7 o'clock, she had to leave the house at 5 and, to her great distress place her son in the care of the security guard at his school until the school opened. She comments:

> It is very bad because you don't get the kind of sleep you want. If I was staying at Accra here, I wouldn't wake up around 4. I would wake up at 5.45, take my bath and then come here. But my place, if you don't calculate your time, you won't come to work. Sometimes too, the money that you have for the road is not enough, but you still have to manage it so that every day you get to work. Sometimes they increase the fares without you even being aware of it. So, you are not aware of the price, and when you sit in the car and they tell you 'Oh, we are going to increase the prices'. It is bad. Then we only get to eat something small.

Embarking on this journey, Julie, like most residents of peri-urban Accra (Agyemang 2017), relied on the operations of minibuses in which seats for 15–21 people had been installed, and as evident from her comment, the unpredictability of these were a large stressor in her life. Locally known as *trotros* the quality of theses buses varies immensely, with some almost falling apart from rust and lack of maintenance, and others sporting TV screens showing commercials on the road. They operate on a fill-and-run basis, which causes long wait time outside of peak hours, and their operational system is managed in a complex collaboration between the Accra Metropolitan Assembly and the Ghana Private Road Transport Union (Esson et al. 2016), with prices regulated relative to the price of gasoline. Price changes are then communicated to the public via official canals, but according to many of the interviewed women, these price changes always came with a transition period in which trotro drivers could exploit uncertainties around the new prices to charge higher fares; and as Julie's comment suggests, a raise in transport prices necessarily reduces the amount of money left to buy food. Therefore, the interviewed women heavily criticized this system for its high prices and poor operational and safety standards. Specifically, they lamented the discomforts of riding in these minibuses as well as the indignity in having to argue in the trotro about the price of the fare, as well as speed and waiting time or abrupt route changes.

In 2007, partially in response to public complaints, the development of a Bus Rapid Transit (BRT) system was started in Accra. The project was partially funded by the World Bank, and its aim was to better connect such peri-urban areas as Sowutoum, with the central business district of the city (Agyemang 2015). In the beginning of the fieldwork for this study in late-August 2018, however, only a few BRT-buses serviced just one of the main corridors leading into the city, while the majority of the bus-fleet was parked in stations due, according to our information, to a problem with insurance and a lack of involvement with the project on the part of the new government. By the end of the fieldwork, the operations of the system were completely halted following severe strikes from drivers. This, however, came as no surprise to the interviewed women, who explained that transport operations provided by the government were not to be trusted. As one of them said, 'we Ghanaians are not used to government things and when the government runs anything, we don't value it'. Accordingly, the other partially citywide bus system, known as *Metromass*, was not frequented by any of the women, which was, however, also due to the limited routes covered by the service, as well as the stated level of distrust in government-run services.

Such experiences of mobility obstacles in the urban landscape made the women characterize the public transport system as unpredictable, and as a result, uncertainty and stress confounded their mobility experiences. Indeed, the large margins and flexibility needed to ensure timely arrival caused stress and discomfort in the sense that it forced the women to wait at the road side before sunrise for an empty seat, it required them to run quickly between cars, whether carrying goods or accompanying children and it made them constantly aware of the possibility of missing out on economic and social opportunities because of delays in movement. Furthermore, congestion caused uncertainty to permeate the women' mobility experiences both time-wise and cost-wise, frustrating the women's ability to manage and control their everyday lives by constantly making proposed plans unattainable, disrupting other activities and limiting their ability to plan ahead. Such issues were not always in themselves gendered or reserved for women. However, as we will explore in the following sections, in this specific cultural and social context, these infrastructural and geographic conditions intersected with certain experienced needs for mobility and the relation of these needs to both domestic and economic responsibilities, creating for the women a distinctly female mobility experience.

7.4 Tackling Economic and Domestic Responsibilities in Everyday Life

> Women work more than men because they work for money – but women, you're working whilst taking care of the home. You have to work to support the man, but at the same time, you have to take care of the home. So it makes it difficult. You have to see to it that everything is clean, you have to cook, you have to do this, you have to do that. So women, yes, women work a lot. (Regina, 43 years old)

As Regina's comment suggests, in Accra women are the primary caretakers of the home, but they are by no means expected to be stay-at-home-moms. Rather, all interviewed women partook in income generating activities while also being responsible for domestic work such as cleaning, grocery shopping and caring for children. More importantly, however, in many households, a strict division of financial responsibilities existed between men and women, with women usually holding responsibility for providing a large proportion of the family's living expenses while men take care of major cash expenditures such as rent, school fees, and the acquisition of larger 'equipment' including kitchen tools, furniture and cars (cf Darkwah 2007).

This gendered division of economic responsibilities is historically linked to the predominance in southern Ghana of the matrilineal Akan family and clan system, known for the system of spousal separation (Grieco et al. 1996, 5), as well as the colonial government's preferential training of men for administrative positions, paving the way for women taking the lead role as traders (Thiel and Stasik 2016, 8, 9). For our study, however, the women primarily explained it with reference to the dignity of economic freedom between husband and wife, as well as by pointing to the unfairness of putting the complete financial burden on men, illustrating a shifting of gender relations instigated by economic and development policies of the 80s and 90s (Overå 2007).

Therefore, to meet their financial responsibilities, all of the women worked in income generating activities, which required them to commute through the city on a regular basis, either to their offices and shops, or to bring goods home for selling. Because of their responsibilities as caretakers, however, they to a higher extent also had to travel to markets across the city to buy food supplies, clothing etc. and to fetch children from school. To fulfil these responsibilities, planning was an important skill for women in Accra, including planning how to manage the cost of expenditures, schedules for when to frequent different markets and organizing food for family members. However, as highlighted above, in this planning transportation emerged as a major source of uncertainty, which required large margins and flexibility to manage. Indeed, just as Julie left her home as early as possible to beat the morning traffic, other women postponed their return journey to after the evening rush. Charlotte, who worked in human relations and drove a private car, explains:

> We close at 4:30 but I have never closed at 4:30. Sometimes I leave the office around 8:00 p.m. Because of things I have to finish and submit, and because of the traffic as well. By 8:00, the traffic has gone down. But then that means I have to communicate with my children, what they have to eat in the house, communicate with my husband what he also has to eat. Make sure that everything is sorted out. And then I retire. And then it starts again the next day.

As mentioned above, few mothers stay at home with their children in Accra (cf. Darkwah 2007, 206), and for the interviewed women, the financial responsibilities of being both a breadwinner and mother took precedence over the hands-on physical care of their children. Therefore, all of the women stressed that their long hours at work were a benefit to their children, which consequently gave the time spent in traffic somewhat of a purpose, despite their frustrations about planning issues and uncertainty discussed in the previous section. Remaining immobile was neither an option nor a desire for the women, as the relationship between mobility and economic activities collapsed their needs and desires to be mobile to the extent that their responsibilities required it. Mary, a 24-year-old mother concludes:

> You need to move around. I don't know how you can be productive if you don't move. You don't have everything in your room. Everything is out there. Your school, your work. Everything is out there, so you need to move. It is very important. You cannot be productive, you cannot do anything.

As such, the women all considered commuting to be a crucial part of being a mother in Accra, and of being able to tackle simultaneously the economic and domestic responsibilities attached to being a woman. This need emerged in their mobility experiences, however, not only as an economic dimension. Rather, the need and desire for mobility also emerged as a moment of identification; an expectation from self and others of mobility and of being actively engaged in the city.

7.5 Identification and Self-identification: Being a Mobile Individual

Indeed, to the interviewed women, political gender equality was assumed[1] and several of the women made specific mention of the trope of the *industrious African woman* to underscore the great importance women played in public life. This trope has been gradually developed in both scholarship and policy development since the United Nations Decade for women, 1975–1985, and the Women in Development movement (WID), which focused on giving women a place within the existing development paradigm (Moser 1989). This was done by pointing to the fact that women were an untapped resource for development (ibid), and since then, the role of women in the economic and social development of their countries and communities as well as their position to invest and save money, have been increasingly underscored by development agencies (Cornwall 2003).

Highlighting themselves as industrious and as drivers of change, the role of women in developing Accra was something all of the informants stressed, and something that gave them pride. They all highlighted the ability of women to find creative ways to earn money and support their families, as well as the work ethics of women, for example, by highlighting the heavy loads carried by women head porters. According to Darkwah, work has intrinsic value in Accra in that it is seen as morally suspect to be idle (Darkwah 2007, 210), and the importance of this sentiment is evident in a comment from Love, who sold water bags in traffic:

> I want to work, so that I can set an example for my children. Because then they grow up, and they can come and meet me and say, my mommy wasn't lazy. My mommy was working (Love, 23 years old).

As highlighted above, however, being mobile was of paramount importance for achieving a productive economic life, and as such, the meaning of mobility to a high degree became conflated with the meanings attributed to work, stated plainly by Julie: "being able to move, is being able to work". The dignity the women felt about being able to create a living for themselves as well as the importance of being able to move to achieve this is highly visible in this comment by Josephine, a successful shop owner in Makola market:

> Here in Ghana, as you can see, the women move a lot. They are doing their best. They are self-reliant. Women are in charge of families. Most women do not sit at home doing nothing. Everyone is trying to do something. They contribute a lot to their families. To support their husbands. And to the economy as well. Because they are all doing something. If they don't do as well, they do petty trading. They buy stuff to sell on the market, they try to learn how to sow, they do so many things.

In the face of such a coupling of being able to commute effectively with the identification of an empowered and industrious women, the obstacles faced in the transportation system represented an extensive barrier for development to the women. Furthermore, however, this coupling was not the only identification attributed to *mobile women* permeating the mobility regime.

In her work on mobility constraints and their implications for rural women and girls in sub-Saharan Africa, Porter (2011) argues, that a general association between promiscuity and movement prevails in southern Ghana, and that this association can be triggered to cast women's mobility as synonymous with being an uncaring wife, if her trip does not adapt to the confines of male concerns and expectations (ibid). This moral suspicion of women's movement was also directly expressed by some of the women in this study, as they thoroughly dismissed having a desire for travelling for reasons other than work or going to the market by stating that 'they were not that kind of women'. For most of the women, this was, however, cast as a view among men that they did not agree with, as in this statement by Sophie, a student and mother of three:

> Sometimes they say, if you see a woman driving, she doesn't respect the man. When we are driving, when we see ourselves, we don't respect men anymore. That's why they prefer the men go for this thing. But it's not true that women don't respect [men when they drive]. It's the men's fault. It's their fault, because they see their wife as only a wife.

This judgement of women 'on the move' was, however, difficult to shake for the women, and even on public transport, several women recounted incidents in which they were faced with judgements on their ability to handle themselves in a traffic situation. Mary, who after the interview laughingly explained that she recently had to stop seeing a man because he told her she was a 'hen behaving like a cock', commented:

> On the trotro, when you are about to sit in the front, some of the men will say, 'Who is this woman who wants to sit in the front?' Once I asked why, and the response I got was, that as they drive and they speed, women don't have the energy to take in the speed, so they end up yelling when they are not supposed to yell or screaming at those meandering or even when they are overtaking a vehicle. So, if it is a man, they have the courage to take in everything that is happening in the road, so the women should sit in the back.

As such, while women in Accra have relatively unrestricted physical access to public spaces compared to the 'landscapes of power' (Bagheri 2019) that shape women's rights to the city in other parts of the world, being on the road did not exempt the women in this study from being identified as weaker than or dependent on men, nor from receiving instructions about how to behave properly as a woman. As the above comment suggests, however, such confrontation emerged less out of a moral governance of female mobility, and more as a consequence of the exclusively male dominance of the public transport system, which cast all practices related to transportation as too tough or too difficult for women to handle (cf. Thiel and Stasik 2016; Hart 2016). This was also the case with driving the new emerging transport mode of the city, namely commercial motorbikes known as *Okadas,* which both men and women considered as too dangerous for women to use. Consequently, of the women engaged with, Mary was the only one who frequented this transportation mode, and even she recounted being overcharged when using them, because, as she was told, she was a woman and therefore 'knew less about transportation' (for a thorough analysis of Okada use see Oteng-Ababio and Agyemang 2012). Justine, a student at the University of Ghana, recounts similar sentiments when considering acquiring a car:

> Having a car is the dream. But getting the car, you need a man to go. Because a woman going to buy a car, they are going to increase the price for you, because they think you don't know the price and the market. So they can just give you any price. But for a man, he is going to bargain, and he knows. So it is easy to get, but you have to get a man. It is bad. It is bad, but that is what they are doing. So you have to get a man to go with you.

For most of the women engaged with, a critical disjuncture thus existed between the identifications they themselves attributed to their mobility, which cantered on being productive and empowered, and the masculine connotations attributed to the transport system, which confounded their experiences with a broader sense of vulnerability related to the real and perceived weakness of women. This in turn affected how the women engaged with the city, in that it made the city feel hostile to them, contributing to a distrust in the city's ability to provide much needed strategies for development, and, by extension, good living conditions and opportunities for its residents.

7.6 Towards Equity in Urban Transport

In this chapter, we have shown how women in Accra experience both physical and social obstacles while commuting and how these disrupted mobilities are associated with discomforts and senses of indignities for women while on the move. Shaping these experiences are both material and geographical conditions of the transport system in Accra, as well as women's reliance on their ability to move for work as they straddle between economic and domestic responsibilities, associated with certain cultural and historic values attached to women's work. Furthermore, these experiences intersect with women's self-identification as being industrious drivers of change for their country, highlighting a male-centeredness of the transport system and shaping a practical barrier to involvement in city life, as well as a mental barrier of exclusion for the women.

As such, this analysis demonstrates how several dimensions intersect to create a mobility regime with clear disadvantages for women, having implications for lasting social and economic constraints and distorting progressive visions of the future—expressed by the women as a disbelief in any improvements of Accra's mobility system. These insights, however, also highlight the complexity of establishing equity in transport as well as the importance of considering not only technical solutions but also identity formations, rights to the city and the embeddedness of gender in practices of everyday life when designing transport systems.

In studies from developed cities, women's complex travel patterns as a result of caregiving and domestic responsibilities are well documented (Peters 2013), and in recent years, this focus has also emerged in research on the cities of Africa and Asia. This research has shown mobility to be a multi-faceted phenomenon that significantly impacts the overarching aim of women's empowerment (e.g. Phadke et al. 2009; Bagheri 2019; for Sub-Saharan Africa specifically see Salon and Gulyani 2010; and Vanderschuren et al. 2019), and it has pointed out that even in cities where women's possibilities for mobility have been uniquely expanded in recent years through cultural change and new modes of transport, women still circumscribe geographies of avoidance, with spatial and temporal constraints generated through extant social norms and concerns about safety (e.g. Butcher 2019; Turdalieva and Edling 2017). This relates to our analysis in showing that such constraints are not simple 'gender differences', if such a thing exists. Rather they represent lasting problems of gender inequality in urban transport systems, which continue to persist, despite at least a decade of researchers and planners calling for the transport sectors to address the needs of socially vulnerable and disadvantaged population groups, including the specific needs of women while in transport (e.g. Pirie 2009, 23).

As noted by Ureta (2008), the social exclusion produced through unequal mobility patterns does not necessarily require absolute immobility on the parts of some groups (i.e. women). Rather, some travellers may be obliged to devote most of both their social and financial resources for movement on mandatory trips, where after they may not be able to afford or access the transport system anymore. As we have seen, this was to a great extent the case for the women in Accra for whom travelling for caregiving responsibilities, including income generating activities, was a given. However, this rarely left them neither time nor money to travel on their own desires and caused their routes of movement to become 'contracting to the known' (Butcher 2019, 8). As such, beyond moving through the usual routes, most of Accra's city landscape seemed impenetrable for the women, which further ads to their experiences of vulnerability and exclusion through mobility. Therefore, even though our analysis has mainly focused on mobility for work, we agree with Porter et al. (2020) when they argue that understanding such 'other' forms of mobility will be key for understanding the depths of inequality and vulnerability associated with mobility regimes.

For our analysis, the focus on mobility regimes has provided an important analytical lens through which we have been able to read women's commuter experiences and practices as embedded in a system of ordered ways of moving in the city. As Sørensen (1999) emphasizes, the mobility regime approach alerts us to the powerful mechanisms converging to cast some mobility forms as better than others (i.e., for work rather than leisure) and some individuals as more mobile than other (i.e., men being strong enough to take in traffic, and women not). These are insights about the social discipline and conformity scripting the terms of usages for the transport system, which needs to be considered in future transport planning. Furthermore, the mobility regime approach has contributed to a comprehensive understanding of how several dimensions interplay to create a mobility system that in the case of Accra disfavours women's needs for mobilities. This has implications for policies and practices concerned with the establishment of sustainability and equity in transport systems, in that it highlights the multiple dimensions shaping exclusions through mobility practices.

For the past decades, the attempts to establish an efficient, cheap and modally complementary and integrated transport network in Accra has faced considerable difficulties (Esson et al. 2016): the recent halting of the BRT-system being a point in case. However, even if these new modes of transport will be successfully implemented, the present analysis warns that there is no guarantee that this will bring about more equitable mobility experiences. Rather, achieving true gender equitability requires both a head on confrontation with suppressive gender norms as well as a wider consideration of the role and responsibilities of women in family life and society; that is with the cultural and social reproduction and contestation of regimes of gendered entitlements and exclusions more widely. Thus, we assert that planning for equitable and sustainable transport systems in fast growing cities in Sub-Saharan Africa should be based on critical explorations of the how gender and family contexts interact with needs for mobility and transport systems to create advantages and disadvantages along lines of gender. Likewise, we argue for a need of raising awareness of the ways in which mobilities maintain exclusionary practices (for women), through, in particular, the multitude of social and historic relations of inequalities embedded in how and why people move.

Notes

i. Since Ghana's independence in 1957, women's issues and issues of gender equality have received increasing attention in national discourses and policy agendas. For a critical discussion of political gender equality in Ghana see Oduro and Ackah (2017).

References

Accra Metropolitan Assembly (2019) Accra resilience strategy. 100 Resilient Cities

Agyemang E (2015) The bus rapid transit system in the greater Accra metropolitan area, Ghana: looking back to look forward. Norsk Geografisk Tidsskrift-nor J Geograp 69(1):28–37. https://doi.org/10.1080/00291951.2014.992808

Agyemang E (2017) Mode choice for long distance trips: evidence from the greater Accra metropolitan area of Ghana. J Transp Geogr 64(October):150–157. https://doi.org/10.1016/j.jtrangeo.2017.09.003

Bagheri N (2019) Tehran's subway: gender, mobility, and the adaptation of the "Proper" Muslim woman. Soc Cult Geogr 20(3):304–322. https://doi.org/10.1080/14649365.2017.1356364

Behrens R (2020) Achieving equitable, sustainable mobility and access in Sub-Saharan Africa. An overview of the state of knowledge. Volvo Research and Educational Foundations, Gothenburg, Sweden

Berdiel F (2018) 'Undertheorized mobilities'. In: Uteng TP, Lucas K (eds) Urban mobilities in the global south. Transport and Mobility. Routledge, Abingdon, Oxon; New York, NY, pp 195–215.

Butcher M (2019) "Sir, it was my right of way!" examining cultural change and the contested entitlements of automobility. Mobilities 14(6):795–808. https://doi.org/10.1080/17450101.2019.1635337

Cobbinah PB, Poku-Boansi M, Adarkwa KK (2018) Develop first, make amends later: accessibility within residential neighbourhoods in Ghana. J Housing Built Environ 33(1):69–89. https://doi.org/10.1007/s10901-017-9544-7

Cornwall A (2003) Whose voices? Whose choices? Reflections on gender and participatory development. World Dev 31(8):1325–1342. https://doi.org/10.1016/S0305-750X(03)00086-X

Cresswell T (2010) Towards a politics of mobility. Environ Plann D Soc Space 28(1):17–31. https://doi.org/10.1068/d11407

Cresswell T, Uteng TP (2008) Gendered mobilities: towards an holistic understanding. In: Cresswell T, Uteng TP (eds) Gendered mobilities. Ashgate Publishing Limited, Hampshire, England, pp 1–15

Croese S, Green C, Morgan G (2020) Localizing the sustainable development goals through the lens of urban resilience: lessons and learnings from 100 resilient cities and cape town. Sustainability 12(2):550. https://doi.org/10.3390/su12020550

Darkwah A (2007) 'Work as a duty and as a joy: understanding the role of work in the lives of Ghanaian female traders of global consumer items. In: Harley S, Cleary E (eds) Women's labor in the global economy: speaking in multiple voices. Rutgers University Press, Piscataway, United States, pp 206–220

Esson J, Gough KV, Simon D, Amankwaa EF, Ninot O, Yankson PWK (2016) Livelihoods in motion: linking transport, mobility and income-generating activities. J Transp Geogr 55(July):182–188. https://doi.org/10.1016/j.jtrangeo.2016.06.020

Grieco M, Apt NA, Turner J (1996) At Christmas and on rainy days: transport, travel and the female traders of Accra. Avebury, Aldershot, Brookfield, USA

GSS (2020) Projected population by age and sex, 260 Districts, 2020. statsghana.gov.gh/nationalaccount_macros.php?Stats=MTA1NTY1NjgxLjUwNg==/webstats/s679n2sn87

Hart J (2016) Ghana on the go: African mobility in the age of motor transportation. Indiana University Press, Bloomington

Møller-Jensen, M, Agergaard J, Andreasen MH, Oteng-Ababio M, Yankson P (2020) Urban expansion and consolidation in Accra's peripheries. IGN Report. University of Copenhagen, Copenhagen. https://static-curis.ku.dk/portal/files/253241102/IGN_Working_Paper_CLIMACCESS_December_2020.pdf.

Moser CON (1989) Gender planning in the third world: meeting practical and strategic gender needs. World Dev 17(11):1799–1825. https://doi.org/10.1016/0305-750X(89)90201-5

Oduro AD, Ackah CG (2017) Closing the gender gaps in Ghana. In: Aryeetey E, Kanbur R (eds) The economy of Ghana sixty years after independence. Oxford University Press, New York, pp 317–32. https://doi.org/10.1093/acprof:oso/9780198753438.003.0019.

Oteng-Ababio M, Agyemang E (2012) Virtue out of necessity? Urbanisation, urban growth and Okada services in Accra, Ghana. J Geograp Geol 4 (February). https://doi.org/10.5539/jgg.v4n1p148

Overå R (2007) When men do women's work: structural adjustment, unemployment and changing gender relations in the informal economy of Accra, Ghana. J Mod Afr Stud 45(4):539–563. https://doi.org/10.1017/S0022278X0700287X

Oviedo D, Sabogal O (2020) Unpacking the connections between transport and well-being in socially disadvantaged communities: structural equations approach to low-income neighbourhoods in Nigeria. J Transp Health 19(December):100966. https://doi.org/10.1016/j.jth.2020.100966

Peters D (2013) Gender and sustainable urban mobility. Official thematic study for the 2013 UN habitat global report on human settlements. https://doi.org/10.13140/RG.2.1.4746.9287

Phadke S, Ranade S, Khan S (2009) Why loiter? Radical possibilities for gendered dissent. In: Butcher M, Velayutham S (eds) Dissent and cultural resistance in Asia's cities. Routledge, London

Pirie GH (2009) Virtuous mobility: moralising vs measuring geographical mobility in Africa. Afrika Focus 22(1):21–35. https://doi.org/10.21825/af.v22i1.5042

Porter G (2011) I think a woman who travels a lot is befriending other men and that's why she travels: mobility constraints and their implications for rural women and girls in Sub-Saharan Africa. Gend Place Cult 18(1):65–81. https://doi.org/10.1080/0966369X.2011.535304

Porter G, Abane A, Lucas K (2020) User diversity and mobility practices in Sub-Saharan African Cities: understanding the needs of vulnerable populations. The State of Knowledge and Research. Volvo Research and Educational Foundations, Gothenburg, Sweden

Porter G, Blaufuss K, Acheampong FO (2007) Youth, mobility and rural livelihoods in Sub-Saharan Africa: perspectives from Ghana and Nigeria. Africa Insight 37(3):420–431

Porter G, Hampshire K, Abane A, Tanle A, Esia-Donkoh K, Amoako-Sakyi RO, Agblorti S, Owusu SA (2011) Mobility, education and livelihood trajectories for young people in Rural Ghana: a gender perspective. Child Geograph 9(3–4):395–410. https://doi.org/10.1080/14733285.2011.590705

Salazar NB, Smart A (2011) Anthropological takes on (Im) mobility. Identities 18(6):i–ix. https://doi.org/10.1080/1070289X.2012.683674

Salon D, Gulyani S (2010) Mobility, poverty, and gender: travel "Choices" of slum residents in Nairobi, Kenya. Transp Rev 30(5):641–657. https://doi.org/10.1080/01441640903298998

Sheller M, Urry J (2006) The new mobilities paradigm. Environ Plan A Econ Space 38(2):207–226. https://doi.org/10.1068/a37268

Sørensen KH (1999) Rush-hour blues or the whistle of freedom? Understanding modern mobility. STS-working paper 3/99. Norwegian University of Science and Technology, Trondheim, Norway

Thiel A, Stasik M (2016) Market men and station women: changing significations of gendered space in Accra, Ghana. J Contemp Afr Stud 34(4):459–478. https://doi.org/10.1080/02589001.2017.1281385

Turdalieva C, Edling C (2017) Women's mobility and "Transport-related social exclusion" in Bishkek. Mobilities 0(0)1–16. https://doi.org/10.1080/17450101.2017.1388348

Ureta S (2008) To move or not to move? Social exclusion, accessibility and daily mobility among the low-income population in Santiago, Chile. Mobilities 3(2):269–289. https://doi.org/10.1080/17450100802095338

Uteng TP, Lucas K (eds) (2018) Urban mobilities in the global South. Transport and mobility. Routledge, Abingdon, Oxon, New York, NY

Vanderschuren MJWA, Phayane SR, Gwynne-Evans AJ (2019) Perceptions of gender, mobility, and personal safety: South Africa moving forward. Transp Res Rec 2673(11):616–627. https://doi.org/10.1177/0361198119854087

Yankson P, Namangaya A, Mainet H, Esson J, Hoppe Andreasen M, Gough K (2015) City dynamics-mobility and livelihoods of urban residents. Københavns Universitet; Faculty of Science; Department of Geosciences and Natural Resource Management, Copenhagen. oai:pure.atira.dk:publications/d0382fd8–1475–452f-b163–86a40364d762

Yankson PWK, Bertrand M (2012) Challenges of urbanization in Ghana. In: Ardayfio-Schandorf E, Yankson PWK, Bertrand M (eds) The mobile city of Accra, vol 25. CODESRIA, Dakar, Senegal, pp 25–46

Maya Møller-Jensen held a position as Research Assistant at the Department of Geosciences and Natural Resource Management (Geography Section), University of Copenhagen, at the time of writing and is currently employed as a Research Assistant at the Centre for Social Data Science, University of Copenhagen. Maya has a background in anthropology with a research focus on cities and infrastructure, and has conducted both anthropological and geographical fieldwork in Ghana on issues of mobility, transportation and urban development, the former with a specific focus on women's mobility experiences.

Jytte Agergaard holds a position as Associate Professor at the Department of Geosciences and Natural Resource Management (Geography Section), University of Copenhagen. Trained as a human geographer, Jytte's research focuses on human migration, mobility and urban transformation and how these processes intersect in producing social transformation, e.g. in relation to gender and family relations. Jytte has conducted fieldwork in Nepal, Vietnam, South Africa, Tanzania and Ghana and is currently involved in two research projects in Ghana; Climate Change Resilience in Urban Mobility (CLIMACCESS–https://ign.ku.dk/english/climaccess/) and a newly started project: Climate Resilience Across the Rural-Urban Continuum (RurbanClimate).

Chapter 8
Community Severance in Urban Africa

Paulo Anciaes and Annabel Bradbury

Abstract Community severance arises when transport infrastructure and motorised traffic are physical or psychological barriers to the movement of pedestrians. This is increasingly the case of many African cities, due to the expansion and upgrade of the road network and the increase in the number of motorised vehicles. These are becoming major barriers to walking, the main mode of transport for large sections of the population. This chapter examines how the problem of community severance is felt in African cities and its consequences on mobility, accessibility, safety, livelihoods, health, wellbeing, and inequalities. We first review existing evidence on community severance across urban Africa, focusing on how the problem reproduces and reinforces social inequalities. We then analyse the extent of severance in a medium-sized city (Praia, Cabo Verde), which is facing fast growth of population, built-up area, road infrastructure, and motorised traffic. Finally, we discuss the potential consequences of inaction in the face of current pressures and the potential for transformation to improve the mobility of pedestrians, proposing pathways and policy solutions to the severance problem.

Keywords Community severance · Barrier effect · Pedestrians · Walking · Social equity

8.1 Introduction: What is Community Severance?

Urban roads and other transport infrastructure are crucial for economic development. They connect people with distant places in the city, where they can access employment, education, shopping, leisure, and other opportunities. However, transport infrastructure can also disconnect people and places. As linear structures, they

---

P. Anciaes (✉)
Centre for Transport Studies, University College London, London, UK
e-mail: p.anciaes@ucl.ac.uk

A. Bradbury
Ricardo Energy and Environment, Didcot, UK
e-mail: Annabel.Bradbury@ricardo.com

R. A. Acheampong et al. (eds.), *Transport and Mobility Futures in Urban Africa*,
The Urban Book Series, https://doi.org/10.1007/978-3-031-17327-1_8

Fig. 8.1 The road as a barrier—Thika Highway, Nairobi. © Cardo (Reproduced with permission)

often cut across neighbourhoods and block the movement of pedestrians. Sometimes, this is because of the infrastructure design. For example, motorways and railways usually have walls or fences preventing pedestrians from crossing except at a small number of points. Other times, it is because of vehicles using the infrastructure: motorised traffic can be so dense or so fast that road crossing becomes dangerous and unpleasant, even when crossing facilities are present. Transport infrastructure and traffic can thus become physical and psychological barriers separating communities (Fig. 8.1). This disconnection effect of transport on pedestrians is known as community severance (Anciaes et al. 2015; Bradbury 2014).

This chapter examines how the problem of community severance is felt in African cities and its consequences on mobility, accessibility, safety, livelihoods, health, well-being, and inequalities. The hypothesis is that recent trends in most African cities are characterised by four factors leading to community severance: (1) an expanded and upgraded road network and (2) increased motorised traffic volumes, but still (3) a dominance of walking in the modal share, and (4) poor pedestrian infrastructure. This is compounded by aspects specific to the African continent (or to low-income countries), including the high relevance of walking to urban livelihoods, land use patterns that originated from colonial systems, hot climate, poor public transport systems, and funding of urban mobility projects closely linked to international agencies.

We acknowledge the diversity of urban environments in such a large continent—the problem is not felt in the same way in all countries or in cities of all sizes. Our aim was to draw on aspects that apply across many African cities, while mentioning differences across cities. We have also attempted to give a balanced view of existing knowledge, drawing not only from English language but also from French and

Portuguese language sources, and to produce new knowledge about a context that is often neglected: the capital of a small island country (Praia, Cabo Verde).

The rest of the chapter is split into three sections. Section 8.2 looks back to what is already known about community severance in African cities, focusing on how the problem reproduces and reinforces inequalities. Section 8.3 looks more closely at community severance in Praia, a city experiencing a fast growth of population, built-up area, roads, and motorised traffic. Section 8.4 looks ahead, placing severance in the context of mobility futures in urban Africa. We discuss possible consequences of inaction and the potential for transformation and then propose pathways and policy solutions to the problem.

8.2 Looking Back—What Do We Know About Community Severance in Africa?

8.2.1 Unequal Infrastructure Provision

African cities are growing fast in population, built-up area, income, and motorisation rates. This growth has been reflected in larger volumes of cars, motorcycles, and informal paratransit vehicles on the road. In most cases, this has led to traffic congestion, because there is not enough road capacity to accommodate the increase in the number of vehicles. However, where and when roads are not congested, the problem becomes high traffic speeds, as speed limit enforcement tends to be weak. As an example, Damsere-Derry et al. (2008) observed 60–98% of vehicles above the speed limit on urban roads in Ghana. In both cases (high traffic volume and speed), pedestrians find it difficult to cross the road—and severance arises. In some cities, this is aggravated by road design. For example, a report by the World Bank (2020) on Maseru, Lusaka, and Harare identified squares as a problem for pedestrians. Some were built in the colonial period and were large (to accommodate monuments in the centre island) and closely spaced, becoming barriers to pedestrian movement.

To accommodate the growth in road traffic demand, new roads have been built in many cities, and other roads have been upgraded with more traffic lanes and better surfaces. However, more lanes also mean longer crossing distances for pedestrians, and better surfaces increase both traffic volume and speed. Nevertheless, the severance problem still exists even when surfaces are not improved, as the combination of mud and traffic on unpaved roads during the rainy season makes crossing the road difficult and unpleasant. There is also severance during road construction (which often lasts for long periods). In some cases, the upgrade of roads is started but is never completed because of a lack of funding or unexpected technical complications. Mains (2019, Chap. 2) describes an example of an incomplete road in Jimma (Ethiopia) which was left with wide ditches on either side, creating a physical barrier, especially for elderly people, whose houses became obstructed.

New controlled-access roads also become physical barriers separating neighbourhoods. For example, Ndiaye (2018) describes severance caused by a new motorway crossing through a densely populated area in Dakar. The proportion of nearby residents who stated they only make 'rare' visits to meet others increased from 26 to 81% after the road was built. This suggests that the presence of barriers weakens people's ability to maintain social networks in their local area.

Some larger cities in the continent have also introduced bus rapid transit and light rail systems, to improve accessibility and relieve congestion on roads. The effectiveness of these modes, compared to conventional buses and paratransit, relies on the continuity of the infrastructure, which is achieved with the physical segregation of the road lanes used and with a reduced number of intersections with other roads (ITDP 2016). Inevitably, this also acts as a barrier to pedestrians. For example, after the launch of the Addis Ababa Light Rail Transit system in 2015, 23% of local residents reported a decrease in social interactions (Deyas and Woldeamanuel 2020).

The contradiction is that new and improved roads have poor or no pedestrian infrastructure, despite the fact that many people in African cities still rely on walking as their main mode of transport. As an example, the Thika Superhighway, an 8-lane road in Nairobi, was originally built with few crossing facilities (Maina and Wachira-Towey 2020). In New Juaben (Ghana), Obeng-Atuah et al. (2017) estimated that the average spacing of crossing facilities on main roads was 157-1067 m and the red-phase time for pedestrians at signalised facilities was 76–110 s—insufficient for seamless walking trips. However, the mere existence of crossing facilities would not be enough to remove the barrier effect of busy roads. Marked or signalised crossings are unhelpful for pedestrians if drivers do not comply with them. For example, Masaoe (2017) found that only 19% of drivers yielded to pedestrians at a marked crossing in Dar-es-Salaam. The alternatives—grade-separated facilities like footbridges and underpasses—are even less helpful. In African cities, as elsewhere, these facilities are inconvenient and intimidating: Maina and Wachira-Towey (2020) report footbridges in Nairobi being used by motorcycle taxis during the day and as a hideout for gangs at night. Steps and steep ramps are unsurpassable barriers for many pedestrians with mobility restrictions. For these reasons, pedestrians avoid grade-separated facilities if possible. In the study of Mfinanga (2014) in Dar-es-Salaam, footbridges and underpasses were the preferred crossing type for only 26% and 15% of pedestrians, respectively.

Non-crossing infrastructure also has its problems. Footways and street lighting are the exception, not the norm on most roads in African cities. As an example, an inventory made by Obeng-Atuah et al. (2017) in New Juaben (Ghana) showed that out of 14 roads, only 6 had footways and only 3 had lighting. Elsewhere, even when infrastructure is provided, it can fail to meet pedestrian demand or protect pedestrians from motorised traffic. The study of Mitullah and Opiyo (2017) in Nairobi revealed a series of problems. Pedestrian infrastructure followed the routes of busy arterial roads, rather than quiet roads. Even where paved footways existed, pedestrians often preferred to walk on nearby unpaved paths, because the footways were too close to the road carriageway. In other cases, they were forced to walk on the carriageway itself, because footways were too narrow or were being used by street vendors or by vehicles (parked or moving).

8.2.2 *Unequal Risk*

That roads are a barrier to pedestrians in African cities seems obvious from most statistics, which invariably show a large number of pedestrian fatalities (although almost no studies compare the relative risk of a walking trip with the risk of trips by other modes). As in other parts of the world, pedestrian collision risk is partly explained by high traffic speeds on roads crossing populated areas. But there are specific factors, common to most low-income countries. One factor is walking patterns. As noted earlier, pedestrians often walk on the road carriageway, as it is the only option. Some walk long distances (if public transport is absent or unaffordable). Others walk at night, along roads with poor lighting, and face headlight glare from vehicles. And some spend the whole day near or on major roads, hawking goods to drivers, due to the absence of designated areas or other forms of livelihood. A second factor increasing the risk for pedestrians is driving patterns. Urban transport is dominated by disorganised but highly competitive paratransit systems. Drivers of paratransit vehicles tend to drive fast, to increase the number of runs, and often encroach on pedestrian space when cruising for passengers.

Poor pedestrian infrastructure also leads to risky crossing behaviour, including jumping barriers and crossing several lanes of fast-moving traffic on motorways. The study of Sinclair and Zuidgeest (2016) in Cape Town found that 50% of pedestrians always crossed motorways at grade (i.e. not using footbridges) and another 24% crossed sometimes. Behrens and Makajuma (2017) found that 82% of pedestrians in Nairobi and 85% in Cape Town crossed (non-motorway) roads away from signalised crossings. The lack of pedestrian compliance may be because the crossings are not aligned with pedestrians' desired lines and, in the case of at-grade crossings, because some drivers do not stop for pedestrians. There are almost no studies quantifying how pedestrians balance risk and convenience when crossing roads, or estimating the monetary value of increased perceived safety on African urban roads. The exception is the stated preference study of Mofadal et al. (2015) in Khartoum and Nyala (Sudan), which found that 92–95% of pedestrians attached a (hypothetical) monetary value to being able to cross the road using safe crossing facilities.

8.2.3 *Unequal Mobility*

The considerations above assume that crossing the road is possible, although risky. But crossing may not be possible in the locations where pedestrians need to cross, leading to detours and delays in walking trips. For example, Ndiaye (2018) reports a case where walking distances increased sevenfold after the construction of a new motorway in Dakar. The risk may also lead some people to stop walking. In Yaoundé, Zogo et al. (2017) estimated that propensity to walk was 31% lower when people felt unsafe to walk because of traffic. Whichever the effect of traffic on walking (delays to existing trips or suppression of trips), the result is a loss of mobility (ability to move around), which in turn results in a loss of accessibility (ability to go to places in the city). Both contribute to a higher risk of social exclusion. While there are studies relating mobility and accessibility and social exclusion in African cities, the role of community severance is an unexplored topic.

There is more evidence on the indirect effects of severance on health. In a study in Maiduguri (Nigeria), Oyeyemi et al. (2012) found that perceptions about less safety from traffic for pedestrians were associated with lower levels of physical activity. In South Africa, Malambo et al. (2017) found that perceptions of high traffic levels were related to less walking for leisure (and related physical activity) among urban residents.

The losses in mobility and accessibility, and their indirect effects on social exclusion and health, are not equally distributed, reinforcing existing inequalities. The poor, children, the elderly, people with disabilities, and women are the most affected.

The poor are disproportionately affected by severance due to historical, economic, and geographic reasons. First, there is spatial segregation of income and ethnic groups in many African cities, a legacy of colonialism and apartheid. The poor tend to live on the fringes of cities. This means not only longer trips, with longer walking sections, but also walking nearer to arterial roads. Second, it is more likely for the areas abutting busy roads to be poor areas, as richer households can afford to pay a premium for housing in areas away from traffic. Furthermore, the alignment of new roads often crosses poor areas (Ndiaye 2018). Third, the poor are more reliant on walking due to lack of access to private cars, costs of public transport, and, in some areas, inexistence (or distance to) public transport routes. For example, in Nouakchott, Diagana (2010) found that 36–49% of residents in two communities walked for all trips (not only local trips or short trips to public transport stops). Fourth, losses of mobility have a high negative impact on the livelihoods of the poor, as many poor households rely on street vending and on informal job opportunities (and information about those opportunities) provided by the community. Restricted access to the parts of the city beyond busy roads means fewer opportunities.

For children, the restrictions brought about by severance are twofold. Children who are allowed to walk without the supervision of an adult are vulnerable to collision risk when crossing roads. But those who are not allowed to walk lose independent mobility and opportunities for physical activity. Most children in African cities are in the first group and walk on their own to school, because parents have no access to a car and have time constraints due to conflicting responsibilities. Some studies have looked at inequalities in exposure to collision risk. In a survey in Cape Town, among low-income households, 88% of children walked to school and 67% of seven-year-olds travelled unsupervised. In sharp contrast, among high-income households, 87% of children went to school by car and 0% of seven-year-olds travelled unsupervised (Behrens and Muchaka 2011). In Ouagadougou, children who play on the streets were more likely to have been involved in a traffic collision than others (Ouedraogo and Bonnet 2019). Other studies looked at the negative effects of risk avoidance. In Dar-es-Salaam, the study of Bwire (2011) suggested high risk but also perceived loss of independent mobility among children: 40% of those who were not allowed to cross main roads stated they would like to do so. In Nairobi, Muthuri et al. (2016) found that a lack of crossing facilities was associated with less walking and a lower probability of meeting physical activity guidelines among children.

Older people and those with disabilities are also vulnerable to severance caused by road traffic. Amosun et al. (2007) showed that 30% of elderly pedestrians could not cross the road within the allocated time in signalised crossing facilities in Cape Town. 45% of participants also reported apprehension. This may lead to suppressed walking trips, both for transport and for leisure, decreasing independent mobility. There is evidence of an association between lack of traffic safety and lower physical activity among older people in several cities—see for example Oyeyemi et al. (2019) in Maiduguri (Nigeria). Less walking and independent mobility then affect health and wellbeing.

Women are also disproportionately affected by severance. The livelihoods of many women involve informal selling on street markets, which means higher exposure to road traffic. The need to walk to several street markets, and lower access to motorised transport, also mean longer walking trips, often along major roads (because they are the only roads linking peripheries with central areas). Severance is compounded by fear of crime—a major deterrent to the mobility of women in all countries. Women may choose to walk on roads and use crossing points that are less safe from traffic simply because they are safer from crime. For example, in a study in Cape Town, safety from crime was the main reason influencing the choice of crossing point along a road for 65% of women (compared with 53% of men) (Behrens and Makajuma 2017).

8.3 Looking Closely—Community Severance in Praia, Cabo Verde

As it is obvious from the group of cities mentioned in the previous section, existing evidence covers mostly large cities (e.g. Dakar, Addis Ababa, Nairobi, Dar-es-Salaam, Cape Town, Khartoum, Yaoundé, and Ouagadougou). This section complements that evidence by looking in some detail at community severance in a smaller city (Praia, capital of Cabo Verde). The methods used are descriptive and illustrate the extent of the problem, even in a small city, rather than attempting an explanation of the causes and consequences of the problem, which the literature reviewed above already provides extensively. The statistics mentioned in the next three paragraphs come from the Praia municipal master plan (CMP 2016), except where noted.

Cabo Verde is a small, lower-middle-income island country off the coast of Senegal. Praia is the largest city (Fig. 8.2). In 2010, it had about 132,000 inhabitants, 27% of the country's population. According to forecasts, by 2023, it will have

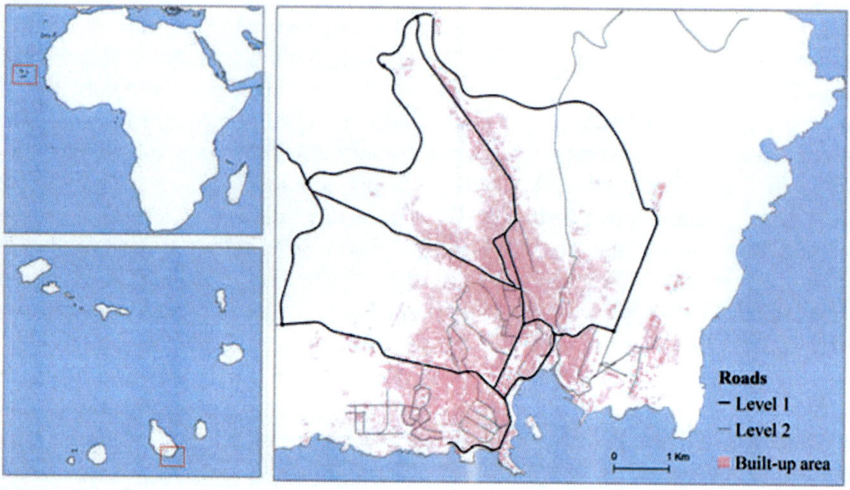

Fig. 8.2 Praia: context, built-up area, and major roads

188,000 inhabitants. Informal settlements represent 57% of urban space and are located either in inhospitable areas (e.g. hills, floodplains) and/or in the periphery. They have little or no employment or urban facilities, which are mainly concentrated in the central areas. As an example, the North part of Praia (which includes mostly poor neighbourhoods) has 36% of the population but only 24% of shops, 15% of public services, and 9% of health facilities in the city. Travel outside the local area is therefore a necessity for most people.

According to the latest population census, only 19% of households own a car, but this value varies from 2 to 89% (in the poorest and richest neighbourhood, respectively). Unlike most cities in Africa, the formal bus network, provided by private companies, is more important than paratransit (which is illegal for intra-urban transport). However, some areas are more than 2 km away from the nearest bus stop (Anciaes et al. 2014). Walking varies from 6% to 11–15% of the trips (for the highest- and lowest-income individuals, respectively) (Anciaes et al. 2014). In informal settlements, streets are the main places for relaxing and socialising, given the lack of suitable squares or green areas (Furtado 2008). However, these settlements have poorer walking conditions than more affluent neighbourhoods (Anciaes et al. 2017).

Walking is becoming less safe and less pleasant due to the increase in traffic levels—the number of vehicles has been increasing at a rate of 8.8% per year. Commuting trips per person may be double that in other capital cities: typically, car owners return home for lunch. The growth in travel demand has led to the expansion and improvement of the road network. Radial national roads connect Praia with the rest of the island and have high volumes of inter-urban paratransit vehicles and freight vehicles. There are also busy arterial roads inside the city. Several roads have been constructed or reconstructed recently and are now paved, allowing for high speeds. In some sections, they are more than 30 m wide and have four traffic lanes, with a median strip. These roads have a large imprint on the urban fabric, separating neighbourhoods (Fig. 8.3), and are a barrier to pedestrians because of high traffic volumes (or speeds) (Fig. 8.4).

Due to the topography of the city (hills, plateaus, and valleys), some roads have few access points to local streets and the existing ones are often along steep slopes. Unpaved (cobblestoned) arterial roads are also problematic because some sections have poor drainage and open ditches for utilities. Bus routes run along major roads, which means that, away from the centre, they do not provide inter-neighbourhood connections, which are made by walking—across those major roads.

Even on good-condition asphalt roads, footways are either non-existent or are in bad condition. In the centre, local streets are also a barrier to pedestrians due to illegal car parking on footways—the only car parking area that existed in the city was removed in the 1990s when car ownership was still low. Crossing facilities, where they exist, are either inconvenient or dangerous (Fig. 8.5).

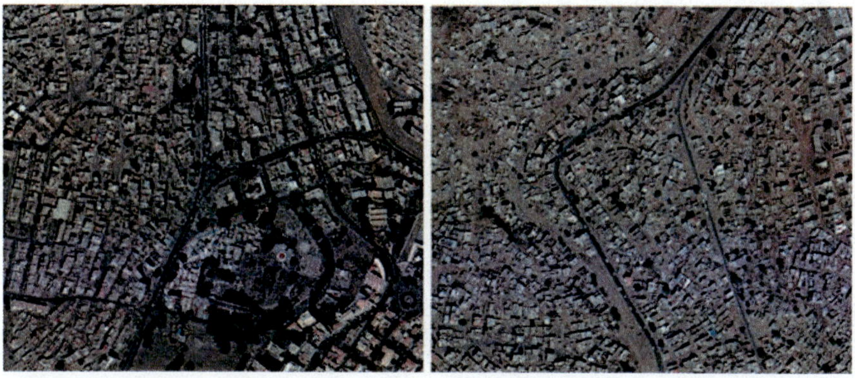

Fig. 8.3 Roads separating neighbourhoods. Left: central areas. Right: informal settlements. *Note* Extracts from Ortophotomap, provided by Praia Municipal Government

Fig. 8.4 Aspects of severance in Praia

Fig. 8.5 Crossing facilities in Praia: footbridge and signalised crossing

Problems related to walking caused by busy roads are particularly impactful for three groups: older people (only 5.6% of the population are aged above 65, but many live alone and have mobility restrictions); people with disabilities (3% of the population, only 24% of them in employment); and women (heads of 40% of households, and facing cost and time restrictions to travel, thus being more likely to walk).

Severance caused by roads is likely to reduce the walking accessibility of local residents, but there is no available data quantifying this reduction. We have estimated the magnitude of severance by calculating an indicator of severance for each building, covering the whole city. The indicator assumes that severance restricts walking access to buildings on the other side of roads. This captures both restricted access to social networks (in residential buildings) and to job opportunities, shops, or services (in non-residential buildings). We used spatial data provided by the Municipal Government on the location and hierarchical classification of roads and on location, shape, and number of stories of all 66,795 buildings in the city.

We then calculated, for each building, the total area of other buildings within 600 m straight-line distance. This was split by area on the same side and on the other side of major roads (Levels 1 and 2 of the road hierarchy). The area on the other side was discounted by a factor representing the barrier effect of the road, assumed to be 75% for Level 1 roads and 50% for Level 2 roads. The indicator of severance is finally the ratio of the discounted area on the other side of the road and the total area.

Figure 8.6 maps the indicator of severance across the city. There is some severance in most areas of the city, except for isolated places in the periphery. The maximum value of the indicator is 75%, i.e. roads are estimated to curtail more than three-quarters of the potential for pedestrian movement. The indicator is above 50% in four zones, as follows.

Zone A—the main shopping and services centre (see also the left side photos in Fig. 8.3 and Fig. 8.4). This is a densely developed area, with buildings on both sides of a busy road. The population has income above average. The area is also close

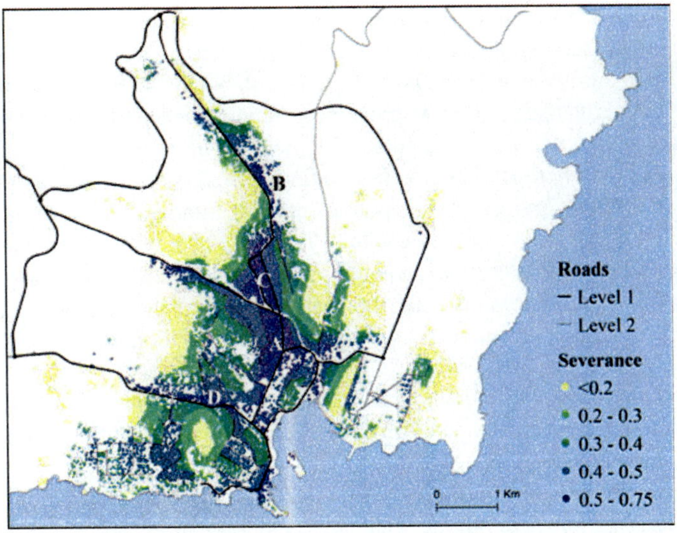

Fig. 8.6 Indicator of severance

to the main market in the city and so it is used as a link for many walking trips. The market can only be reached by crossing the road in a section with large traffic volumes generated by the inter-urban paratransit terminal, located nearby. Women are the majority of sellers in the market. The area also has day care facilities for older people, used by residents in other neighbourhoods, who have to cross busy roads to reach those facilities.

Zone B—northern suburbs (see also the left side of Fig. 8.5).

This area is crossed by the main road linking Praia with the rest of the island. The area was originally settled precisely because of the location near the road, so the population and (the few) facilities are distributed along a narrow strip around the road. The road is also a border between informal settlements on the east and formally planned areas on the west. The neighbourhood has a secondary school (on the west side), used by students from surrounding neighbourhoods, and a mobility rehabilitation centre, used by people from across the city and beyond. The indicators of severance are higher on the east side of the road.

Zone C—between centre and northern suburbs, just south of Zone B (see also the right side photo in Fig. 8.3). This area is surrounded by two branches of the same major road. It is a densely populated area but it has few facilities and the internal street network is irregular—so the main destinations for pedestrians are the facilities on the main road (including bus stops). Income is below average and 76% of household heads are women (Gonçalves 2007).

Zone D—west of the centre (see also the right side of Figs. 8.4 and 8.5).

This area is crossed by a busy road and has high indicators of severance on the north side of the road (as most buildings are on the south side). The road is the main link between Praia and the southwest part of the island, but also a major link within the city, connecting the main centre with the most important secondary centre, the main industrial area, and new car-oriented developments in the western suburbs. The part most affected by severance (an informal settlement) is poor and densely populated, with few shops and services, and no public spaces (Furtado 2008).

Overall, the analysis above showed that 'hotspots' of severance are mainly areas with high levels of need to cross major roads, as these roads concentrate local trip attractors (the neighbourhoods that the roads separate have few attractors themselves). Furthermore, the areas are mainly informal settlements or are used by vulnerable groups (e.g. poorer women and older people).

The calculation of the indicator of severance has some caveats. It is based on potential, rather than actual, need to cross the road. The latter would vary from individual to individual, depending on age, gender, disabilities, employment status, and personal preferences. Furthermore, the indicator is not split by different types of destinations (e.g. homes of other people, jobs, shops, or services). The barrier effect caused by each road was based on its functional classification (not width, traffic volume, the existence of crossing facilities, or pedestrian perceptions). Finally, we assumed a maximum distance of 600 m walking distance—adequate for most pedestrians but not some older ones. Previous research has not been conclusive regarding how acceptable walking distances depend on the characteristics of pedestrians and trips, especially in the urban African context.

Despite these caveats, the map in Fig. 8.6 clearly illustrates the extent of the severance problem across the city and the locations of the most-impacted areas. The implication for transport and urban policy is that projects for new or improved roads should account for the effects on communities severed by those roads. Severance is likely to increase in the future, as new national-level roads and an inner ring road will cut through dense neighbourhoods, with some areas becoming surrounded by busy roads on all sides. New and improved arterial roads are also planned, to decongest existing ones. Some will have viaducts and sunken sections, becoming physical and visual barriers to pedestrians. Some roundabouts will be enlarged and others redesigned, with at-grade pedestrian crossings replaced with underpasses. These projects will generally improve mobility in the city but may restrict the mobility of some groups. The needs of older people should be considered because forecasts indicate that they will be a greater proportion of the city's population than now. As income inequality grows, there is also a risk that new roads (and continued lack of public transport) contribute to the increased isolation of some neighbourhoods, leading to the social exclusion of their populations.

At the same time, it is reassuring that one of the proposed actions of the municipal master plan is to eliminate barriers posed by the built environment in order to improve pedestrian accessibility—a rare case of explicit reference to severance in master plans of African cities (CMP 2016, Vol. II, p. 27). The pedestrianisation of a street in the city centre and the requalification of the waterfront road have proved to be successful, increasing pedestrian footfall and use of streets for exercise. There is now a plan for a pedestrianised street in each neighbourhood. These types of interventions provide safe and pleasant places, away from road traffic, reducing barriers to pedestrian mobility and to the use of streets as social spaces.

8.4 Looking Ahead—Community Severance and the Future of Urban Mobility in Africa

This chapter showed that community severance poses distinctive challenges in African cities. Across the continent, pedestrians are being literally and symbolically pushed to the margins, as the number of motorised vehicles on the road increases. Roads become barriers to the movement of pedestrians and impede access to the opportunities that cities have to offer. This problem reinforces existing inequalities as it mainly affects vulnerable groups, including the poor, children, the elderly, people with disabilities, and women. In different ways, severance is a growing issue in all African cities. This chapter showed a glimpse of the diverse ways in which severance is felt by the population (by drawing from studies about various cities) and studied by researchers (by drawing from material in languages other than English). We found little evidence from North African cities, but that is because searches were conducted in English, French, and Portuguese, not Arabic. However, cities like Cairo, Algiers, and Casablanca have also seen massive construction of large roads cutting through residential neighbourhoods. Other cities not mentioned in our review but with severe problems of road-based severance include Lagos, Kinshasa, Johannesburg, and Abidjan. In this chapter, we have also looked in some detail at a rarely studied case: the capital city of a small island country.

It is clear from this chapter that community severance caused by roads and other major transport infrastructures creates challenges for the future of mobility in African cities. As car ownership and use grow, walking will become an even more marginal activity, causing problems such as insufficient physical activity, isolation, and lack of independent mobility for children, the elderly, and people with disabilities, especially those who do not have access to private cars. As African cities become larger, more roads and passenger transit infrastructures need to be built and more urban land will be encroached upon to accommodate increases in populations and their associated activities, causing urban sprawl, and even higher dependence on motorised transport to overcome the need for increased travel distances to navigate the city.

The consequences of inaction are clear when we look at cities in Asia and Latin America that are already at a more advanced stage of motorisation than most African cities. In those cities, walking became so risky and unpleasant that whenever possible, pedestrians shifted mode and become users of motorised vehicles themselves. In fact, the tendency in many low- and middle-income countries is for households to acquire motorcycles, and then cars, as their income grows. This increases their mobility and accessibility, reducing poverty, and contributing to economic growth. It also gives status and shields car drivers from crime and the risks and ignominies of being a pedestrian—see the detailed description of the case of Luanda in Pitcher and Graham (2006). However, car ownership also increases traffic volumes, which not only causes enormous problems of congestion and pollution, but also reinforces the barrier effect of roads and the separation of communities, affecting the same vulnerable groups who are already disadvantaged and not reaping the benefits of economic growth.

Despite the potential threats ahead, at the moment most motorised trips in African cities are still shared, in paratransit vehicles, and private car ownership and use are still relatively low. Most African cities, particularly small and medium-sized ones, are at a crossroads and can follow two possible pathways. The first pathway is the one that was followed by North American and Middle Eastern cities, and some Asian and Latin American cities, where population and income growth were accommodated by ever larger provision for motorised modes of transport. This meant more and larger roads, and roads over other roads, when these became too congested. Walking and street activities almost disappeared.

The alternative (and more hopeful) possible pathway is the one followed by some Northern European cities, Hong Kong, and Singapore, where car ownership and use never reached the high levels followed elsewhere, and have increased less and less until eventually they started declining. This was due to investment in public transport and, later, in safe walking infrastructure and attractive public spaces. Many African cities can short-circuit the evolution of their mobility systems and move from the current still low motorisation stage directly to a walking-centred stage, without passing through a high motorisation stage, preserving walking as an important mode of transport—but in better conditions.

A challenge to following a pedestrian-friendly urban mobility pathway is that community severance is still an underused concept in urban transport planning in Africa. Pedestrian problems are usually framed as a traffic safety issue. Consequently, a typical solution has been to segregate motorised and non-motorised road users, in a bid to reduce traffic collisions. In practice, this meant footbridges over motorways and busy roads. However, as noted in this chapter, these solutions rarely remove the physical and psychological barriers caused by roads.

Radical solutions such as removing or burying roads, now more palatable than before in other parts of the world, seem to be out of the question in contemporary urban Africa, for political and financial reasons. Reducing road capacity, by removing traffic lanes and reallocating space to pedestrians, also tends to be politically infeasible, given the current congestion problems on most roads. However, there is an extensive menu of other possible interventions to choose from to reduce severance. The obvious solutions are to add new at-grade crossing facilities (not footbridges or underpasses) and improve existing ones. Speed reduction is essential and can be achieved by lowering (and enforcing) speed limits. Traffic calming measures (e.g. raised speed humps, chicanes) increase safety but are more effective on minor roads in residential areas, not on arterial roads. Inevitably, any attempt to reduce speeds on arterial roads can lead to increased congestion where it is already high. This means that long-term solutions to reduce severance need to involve modes of transport that are less demanding on road space, i.e. mass public transport, complemented with measures to increase urban densities and reduce sprawl.

Reducing severance also calls for some changes in governance, with a higher emphasis on the coordination of different actors (e.g. national and local governments, road authorities, non-government organisations, and international agencies) and engagement with communities, to understand mobility needs, especially of vulnerable groups. Crowdsourced data can be useful, for example, to provide information on problems that pedestrians encounter and identify where to install or improve pedestrian infrastructure or apply traffic regulations. Lack of funding is a challenge. In several countries, the improvement of pedestrian infrastructure has been supported by international programmes. However, it is difficult to secure funding for retrofitting existing roads to make them more amenable to pedestrians, as this usually involves reallocating road space away from motorised vehicles.

The diversity of conditions across African cities means that there is no single solution to solve severance. Attention (and funding) is usually focused on megacities, but smaller cities—such as Praia, our case study—are also facing problems caused by rising traffic volumes and barriers to mobility caused by large roads. Another source of diversity is the intensity of the problem. In some cities, the severance problem is only emerging. In others, it is already difficult to revert, given the dependence on motorised traffic.

While most of the chapter paints a gloomy picture of the current situation for pedestrians in African cities, there are also reasons to be hopeful. Walking is increasingly being accepted as part of mobility plans and active travel strategies in some cities. For example, Cape Town and Nairobi have developed strategies aimed at non-motorised modes. In Nairobi, this included a commitment to allocate 20% of the road construction budget to those modes. This type of policy commitment is essential to ensure that the future of mobility in African cities includes walking—which is the same as saying that it includes everyone, regardless of income, gender, age, or ability to walk.

References

Amosun SL, Burgess T, Groeneveldt L, Hodgson T (2007) Are elderly pedestrians allowed enough time at pedestrian crossings in Cape Town, South Africa? Physiother Theory Pract 23:325–332

Anciaes PR, Jones P, Mindell JS (2015) Community severance: where is it found and at what cost? Transp Rev 36:293–317

Anciaes PR, Nascimento J, Silva S (2017) The distribution of walkability in an African city: Praia, Cabo Verde. Cities 67:9–20

Anciaes PR, Nascimento JM, Pinto ISM (2014) Public transport and livelihoods in a fast-growing lower-middle income city: Praia, Cape Verde. In: Royal geographic society-institute of british geographers 2014 conference, London, 27–29 Aug 2014

Behrens R, Makajuma G (2017) Pedestrian crossing behaviour in Cape Town and Nairobi: observations and implications. In: Mitullah WV, Vanderschuren M, Khayesi M (eds) Non-motorized transport integration into urban transport planning in Africa, Chapter 7, pp 22–56

Behrens R, Muchaka P (2011) Child independent mobility in South Africa: the case of Cape Town and its hinterland. Global Stud Child 1:167–184

Bradbury A (2014) Understanding community severance and its impacts on women's access and mobility in African countries. CPR 1895. Crown Agents, London
Bwire H (2011) Children's independent mobility and perceptions of outdoor environments in Dar Es Salaam City, Tanzania. Global Stud Child 1:185–206
CMP (Câmara Municipal da Praia) (2016) Plano director municipal 4a Versão (Municipal Master Plan). CMP, Praia. (in Portuguese)
Damsere-Derry J, Afukaar FK, Donkor P, Mock C (2008) Assessment of vehicle speeds on different categories of roadways in Ghana. Int J Inj Contr Saf Promot 15:83–91
Deyas GT, Woldeamanuel MG (2020) Social and economic impacts of public transportation on adjacent communities: the case of the Addis Ababa light rail transit. Res Transp Econ 84:100970
Diagana Y (2010) Mobilité quotidienne et intégration urbaine à Nouakchott: des difficultés d'accès aux transports urbains à l'expérimentation des stratégies d'adaptation (Daily mobility and urban integration in Nouakchott: from the difficulties of access to urban transport to the experimentation of adaptation strategies). PhD dissertation, Université Rennes 2, Université Européenne de Bretagne. (in French)
Furtado EAR (2008) O papel do espaço público na estruturação do tecido urbano–o caso de Tira Chapéu–Cidade da Praia (The role of public space to structure the urban fabric–the case of Tira Chapéu–Praia). Bsc. Dissertation, Universidade Jean Piaget de Cabo Verde. (in Portuguese)
Gonçalves AJF (2007) Construção espontânea na cidade da Praia–o caso do bairro Safende (Informal settlements in Praia–the case of Safende neighbourhood). Bsc. Dissertation, Universidade Jean Piaget de Cabo Verde. (in Portuguese)
ITDP (Institute for Transportation and Development Policy) (2016) The BRT standard. ITDP. https://www.itdp.org/2016/06/21/the-brt-standard
Maina GN, Wachira-Towey IN (2020) Challenges faced in the implementation of planned safety measures in mega infrastructure projects: a case study of road safety on Thika superhighway. Africa Habitat Rev J 14:2071–2080
Mains D (2019) Under construction-technologies of development in Urban Ethiopia. Duke University Press, Durham
Malambo P, Kengne AP, Lambert EV, De Villers A, Puoane T (2017) Association between perceived built environmental attributes and physical activity among adults in South Africa. BMC Public Health 17:213
Masaoe E (2017) Safety of vulnerable users on a road in Kinondoni municipality Dar es Salaam, Tanzania. In: Mitullah WV, Vanderschuren M, Khayesi M (eds) Non-motorized transport integration into urban transport planning in Africa., Chapter 6, pp 80–89
Mfinanga DA (2014) Implication of pedestrians' stated preference of certain attributes of crosswalks. Transp Policy 32:156–164
Mitullah WV, Opiyo R (2017) Non-motorized transport infrastructure provision on selected roads in Nairobi. In: Mitullah WV, Vanderschuren M, Khayesi M (eds) Non-motorized transport integration into urban transport planning in Africa., Chapter 7, pp 22–56
Mofadal AIA, Kanitpong K, Jiwattanakulpaisarn P (2015) Analysis of pedestrian accident costs in Sudan using the willingness-to-pay method. Accid Anal Prev 78:201–211
Muthuri SK, Wachira L-J, Onywera VO, Tremblay MS (2016) Associations between parental perceptions of the neighborhood environment and childhood physical activity: results from ISCOLE-Kenya. J Phys Act Health 13:333–343
Ndiaye I (2018) Coupure urbaine et autoroute à péage à Dakar: entre séparation des communautés et contraintes de mobilité? (Urban separation and toll highway division in Dakar: between community separations and mobility constraints?) Études Caribéennes 39–40. https://journals.openedition.org/etudescaribeennes/11487. (in French)
Obeng-Atuah D, Poku-Boansi M, Cobbinah PB (2017) Pedestrian crossing in urban Ghana: safety implications. J Transp Health 5:55–69
Ouedraogo M, Bonnet E (2019) Les enfants de moins de 15 ans face au risque d'accident de la route à Ouagadougou (Burkina Faso) (Children under 15 in face of road accident risk in Ouagadougou).

Revue Francophone sur la Santé et les Territoires-Dossiers Thématique: Mobilités Transports et Santé. https://ouest-edel.univ-nantes.fr/rfst/index.php?id=120. (in French)
Oyeyemi AL, Adegoke BO, Sallis JF, Oyeyemi AY, De Bourdeaudhuij I (2012) Perceived crime and traffic safety is related to physical activity among adults in Nigeria. BMC Public Health 12:294
Oyeyemi AL, Kolo SM, Oyeyemi AY, Omotara BA (2019) Neighborhood environmental factors are related to health-enhancing physical activity and walking among community dwelling older adults in Nigeria. Physiother Theory Pract 35:288–297
Pitcher MA, Graham A (2006) Cars are killing Luanda: cronyism, consumerism, and other assaults on Angola's postwar, capital city. In: Murray MJ, Myers GA (eds) Cities in contemporary Africa, Palgrave Macmillan, New York
Sinclair M, Zuidgeest M (2016) Investigations into pedestrian crossing choices on Cape Town freeways. Transp Res F Traffic Psychol Behav 42:479–494
World Bank (2020) Challenges in urban mobility and the way forward: a study of Maseru, Lusaka, and Harare cities. World Bank, Washington
Zogo VO, Epo BN, Meli CN (2017) La marche à pied dans les métropoles africaines: Le cas de Yaoundé (Walking in African metropolises: the case of Yaoundé). Cahiers Scientifiques Du Transport 72:143–159. (in French)

Paulo Anciaes is a Senior Researcher at UCL (University College London). An economist and geographer by background, he has over 20 years' experience at the intersection of transport, environment, and health. He has worked in several research and consultancy projects about community severance since 2014, developing tools to measure and value the negative impacts of busy roads on pedestrians. Paulo is a member of the Editorial Board of the Journal of Transport and Health and of the Sustainable Land Use and Transport Committee of the Association for European Transport. In 2019, he was awarded the UCL Excellence in Health Research prize.

Annabel Bradbury has worked in a transport research capacity for 24 years, with expertise in rural transport services, social inclusion and stakeholder consultation, spanning domestic and international markets. She is a subject matter expert in transport and development, with a Ph.D. on the contribution of Sustainable Livelihood Approaches to the effectiveness of transport planning in Africa, and is a widely published and cited expert. Annabel is a Fellow of the Chartered Institution of Highways and Transportation (CIHT) and was included in the 3rd edition of 'Remarkable Women in Transport' in 2021.

Chapter 9
Transport Poverty, Distance Covered to Access to Basic Infrastructures and Modal Choice in Urban Cities in Cameroon

Jules Médard Nana Djomo and Boniface Ngah Epo

Abstract This study analyses the relationship between the distance travelled by the poor and the mode of accessing basic urban social infrastructure in Cameroon. We adopt a recursive triprobit model where in a stepwise manner, and we identify joint probabilities for a poor individual to cover a given distance to basic infrastructure (i.e., educational and health infrastructures) using one of the three modes of transport (i.e., walking, motorcycle and car). We use the 2014 Cameroon Household Consumption survey compiled by the National Institute of Statistics. The results show that individuals in poorer households overwhelmingly choose to walk to public educational facilities closer to their areas of residence as well as health facilities situated further from their place of residence. Whist the dataset date back to 2014, recent city-specific datasets for the two largest cities—Yaoundé and Douala—suggest that mobility and transport-poverty characteristics have not significantly evolved since 2014, principally caused by insufficient investments in transforming urban cities. The main takeaway within a context of future cities is the glaring need for urban cities to become more inclusive through urban planning that develops urban infrastructures that facilitate walkability for poor households in cities in Cameroon.

Keywords Poverty · Distance travelled · Modal choice · Cameroon

9.1 Introduction

Transport poverty does not yet have an agreed definition or single measure, but generally refers to households and individuals who are unable to make the journeys they need. In this vein, transport poverty can refer to mobility poverty (Ahrend et al. 2014; Martens 2013; Velaga et al. 2012), accessibility poverty (Martens and Bastiaanssen 2014; Scheiner 2008), social exclusion (Hine 2009) or transport disadvantage (Currie et al. 2009). Despite these different approaches, transport poverty has

J. M. Nana Djomo (✉) · B. N. Epo
University of Yaoundé II, Yaoundé, Cameroon
e-mail: jdjomo81@yahoo.fr

R. A. Acheampong et al. (eds.), *Transport and Mobility Futures in Urban Africa*, The Urban Book Series, https://doi.org/10.1007/978-3-031-17327-1_9

not fully captured the attention of governments and researchers in developing countries, yet it is a problem that negatively affects the daily lives of millions of people and should significantly impact future cities. Moreover, given the persistent patterns of urbanization resulting in the unavailability of basic infrastructure, transport poverty is a much larger and attention-based problem. Thus, unlike other studies, this study has the distinction of examining the relationship between transport poverty, distance travelled and accessibility to basic infrastructure (education and health facilities). This study aligns with the observation that conventional approaches to dealing with urban mobility in the Global South need context-specific analysis and solution to resolve complexities and inequities in urban cities (Uteng and Lucas 2017).

With the phenomenon of rampant urban population growth, transport poverty has become a global concern. More than half of the world's population lives in urban areas, and this proportion is expected to increase. The growth of cities and the reduction of population density increase the distances travelled. This leads to more complex journeys and makes it more difficult to provide public transport away from city centres. Cameroon is no exception to this movement. To this end, the rate of urbanization in Cameroon is increasing steadily. It increased from 28.5% in 1976 to 37.8% in 1987 and 48.8% in 2005. In 2010, the urbanization rate was 52%, with 17 cities having at least 100,000 inhabitants. The cities of Douala (economic capital) and Yaoundé (political capital) which in 2005 alone housed 21.3% of the total population and 43.7% of the urban population, totalize in 2010 more than 5 million inhabitants, or about 23.5% of the total population of Cameroon (UN-HABITAT III 2015). One of the direct consequences of this trend is the overcrowding of the suburbs and an increase in precarious neighbourhoods landlocked in the middle of urban cities or rejected on the outskirts of the agglomeration.[1]

This rapid expansion of urban surface areas has contributed profoundly to the unplanned enlargement of cities. Strongly oriented by the preference for horizontal housing, uncontrolled urbanization has led to the densification of the city centres and the sprouting of peripheries of spontaneous habitat, lacking infrastructure and basic services. The development of cities always seems to be a step forward in the completion of land-use planning and the construction of public infrastructure.

In terms of access to transport infrastructure, the poorest populations, especially those living in outlying neighbourhoods, face significant difficulties in getting around, due to their residential location, deficient public transport, or the high cost of public transport. The situation of urban transport illustrates the absence of financial resources or the absence of a well-crafted strategy that can lead to the development of the transport sector. The corollary is the presence of a poorly developed road network, paved lanes limited to the centre and main centre-to-periphery links, individual motorization reserved for a minority, and the disappearance of public transport enterprises in most major cities in the 1980 and 1990s (Diaz Olvera et al. 2003).

Thus, functional urban transport is, in order of importance, taxis, motorbike taxis, buses and minibuses in most urban cities in sub-Saharan African countries

[1] View university of Sherbrooke databases (http://perspective.usherbrooke.ca/bilan/tend/CMR/fr/SP.URB.TOTL.IN.ZS.html).

like Cameroon. A 2002 study by Cameroon's Ministry of Transport estimated that there were 6,000 and 8,000 taxis in the cities of Yaoundé and Douala, respectively. However, these means of transportation though resourceful, do not always guarantee the quality of service or the safety of users. Urban transport is therefore marked by the predominance of the artisanal sector, which has supplanted companies structured to cover mobility needs since the years of crisis and structural adjustment programs in the 1990s.

At the same time, Cameroon is experiencing a comparative upsurge in the number of the poor, despite the decline in the rate of monetary poverty, which fell from 39.9% in 2007 to 37.5% in 2014 (National Institute of Statistics [INS] 2015a). Indeed, the number of poor, which was 8.1 million in 2014, had thus increased by 1.9 million compared to 2001, an average relative annual increase of 1.8% over this period (0.2% in urban areas). The INS estimates that one of the main drivers of the relative increase in the number of poor people is a strong population growth estimated at 2.6% per year (INS 2015a).

In view of the above, this paper has a dual purpose. Firstly, to show that poverty has an effect on distance travelled in urban cities in Cameroon. Secondly, to verify that poverty and distance travelled influence transport modal choices in urban cities in Cameroon. This dual effect marks the particularity of this study, which stands out from previous studies on urban mobility. The rest of the paper includes: the literature review, the methodology, the database, the results, discussions and a conclusion.

9.2 A Review of the Literature on Transport Poverty and Mobility Habits

Historically, the transport sector has been dominated by political, economic and technical considerations, with a poor understanding of the criteria of mobility, accessibility, social exclusion and transport disadvantage. Indeed, Gannon and Liu (1997) indicate that there were no guiding principles or systematic approaches to poverty problems in the World Bank transport operations, as transport was considered to have only an indirect relationship to poverty. However, due to research in the field, a large body of knowledge has been developed on transport poverty, particularly in urban cities. Studies highlighting the problems associated with transporting poor urban dwellers include Deaton (1987) and Kranton (1991) who both argued for the need for poor-friendly transport policies.

It should be noted that the concept of transport poverty has several interdependent approaches. These include: (i) mobility-related poverty issues perceived as a systemic lack of motorized transportation resulting in mobility difficulties, often associated with a lack of services or infrastructure (Moore et al. 2013); (ii) accessibility poverty captioned as the difficulty of reaching certain basic infrastructure (employment, education, health, etc.) at a given time, easily and at a reasonable cost (DfT—UK Department for Transport 2014; Abley 2010; SEU 2003; Harris 2001);

(iii) transport affordability often considered as the inability to cover the cost of transport, usually in reference to the car and/or public transport (Litman 2015; Serebrisky et al. 2009; Carruthers et al. 2005) and (iv) social exclusion (Hine 2009) and transport disadvantage (Currie et al. 2009) which are used often differently depending on the context, though overlapping.

To this end, the literature review has largely agreed that the relationship between poverty and access to urban facilities depends on several factors. Howe and Bryceson (2000) showed that walking accounted for the majority of journeys up to five km, in Temeke (47%) in Dar es Salaam and up to 8 km in Morogoro (67%) in Tanzania. In the same vein, poor workers walk more frequently (31%) compared to non-poor (13%) to go to their areas of employment, especially informal employment in Douala. These poor and non-poor workers use public transport in an unfavourable context: rush-hour traffic jams, vehicle scarcity, resulting in a lot of wasted time and accumulated fatigue (Plat et al. 2004). In terms of access to infrastructure, Diaz Olvera et al. (2003) demonstrates that in Dar es Salam the poor are at a disadvantage in terms of access to paved roads, water, schools, clinics and hospitals compared to the wealthiest sections of the population.

According to Starkey and Hine (2014), non-motorized transport accounts for 70% of trips in Dakar and Douala, 50% in Beijing and about 40% in Nairobi, Addis, Lagos and Berlin and about 35% in Barcelona, Tokyo and Madrid. The authors reveal that with the exception of Beijing and Tokyo, walking is the dominant form of non-motorized transportation in these cities. Cook et al. (2005) report that walking is used by at least half of the urban population and accounts for 80–90% of all the trips of the poor.

In addition to walking or cycling, low-income populations tend to have more limited transportation options, lower quality transportation services and worse conditions in terms of safety, reliability and comfort. The International Transport Forum (2017) reports that limited transportation options contribute to creating the "poverty trap" by hindering people's access to jobs, education establishments and social networks. In order to limit social exclusion from transport, Titheridge et al. (2014) highlighted the necessity to establish an affordable transport system for young people from disadvantaged backgrounds.

On social exclusion, Crisp et al. (2018) report that this form of exclusion can also result from the disadvantages associated to transport-related inaccessible opportunities such as the acquisition of goods and the use of services. Transport poverty is then a source of discrimination by the gender-specific nature of accessibility and by the affordability of transport. An ensuing consequence is that the poorest women tend to find work closer to home, often in the informed area, such as street vendors, caring for children, working as domestic servants or doing work such as cleaning offices or working in a factory (Zhong et al. 2003). In addition, Zhong et al. (2003) listed a wide range of problems faced by people with disabilities in terms of accessibility and affordability of transport.

A glaring observation is the very limited or no number of studies investigating transport poverty, distance travelled to access urban infrastructure and transport modal choices in urban African cities in general and Cameroon in particular. In addition, most large cities in central Africa mimic urban cities in Cameroon. Thus, by carrying out this study for urban cities in Cameroon, we may provide useful insight on how poverty-distanced travelled to access infrastructure and modal choices operate and this could inform policies which may be adapted to other similar urban cities.

9.3 Methodology

9.3.1 Regression Model Choice

With a dual objective of measuring the effect of poverty on distance travelled as well as assessing the effect of poverty and distance travelled on modal choice, it is necessary to develop an appropriate methodological framework.

In order to study the demand for urban mobility, several studies have used discrete choice models that result from the maximization of the commuter's random utility (Miranda-Moreno et al. 2011; Brownstone and Fang 2014; Milioti et al. 2015). These models calculate, based on the observations given, the probability that an individual will select a particular mode from a set of possible and mutually exclusive choices. In this manuscript, we simultaneously study three alternatives which included the poverty status of the individual (1 = yes and 0 = otherwise), the distance travelled (1 = equal to or greater than 1 km and 0 = otherwise) and the transport modal choice (1 = walking, 2 = using a motorcycle and 3 = using a car). In order to circumvent issues of estimation bias, the appropriate model is the triprobit recursive model that jointly estimates the probabilities of these three alternatives.

9.3.2 The Recursive Triprobit Model

The triprobit model allows a combined estimate of the different dimensions of the three alternatives (poverty, distance travelled to access urban infrastructures and modal choice) by taking into account the correlation between the error terms of equations used to estimate the different alternatives. However, this model does not assess the direct effects that these dimensions can have on each other. To do this, it is possible to estimate a recursive triprobit model of probit equations with endogenous binary regressors (Wilde 2000). Nonetheless, for the system to be identified, it cannot contain reciprocal relationships. In other words, if poverty is included as an endogenous explanatory variable in the modal choice equation, modal choice cannot be included in return as an endogenous explanatory variable in the poverty equation. Choices must therefore be made as to which links will be modelled.

The triprobit structure is similar to a *seemingly unrelated regressions* (SUR) model in that equations do not necessarily have to contain the same set of explanatory variables. To this end, the sequence of the model below is motivated by literature where we hypothesize in a stepwise manner that (a) the poverty status of individuals explains the distance travelled to access urban infrastructure and (b) both the poverty status and the distance travelled by the poor individual explain the modal choice adopted by the individual. The recursive triprobit can then be written as:

$$\begin{cases} y_{i1}^{*} = x_{i1}\alpha_1 + u_{i1} \\ y_{i2}^{*} = x_{i2}\alpha_2 + \beta_1 y_{i1} + u_{i2} \\ y_{i3}^{*} = x_{i3}\alpha_3 + \beta_2 y_{i1} + \beta_3 y_{i2} + u_{i3} \end{cases} \tag{9.1}$$

where the variables y_{ik} (i = 1, 2, …, n individuals and k = 1, 2, 3 equations) are continuous latent variables that represent poverty, distance travelled and modal choice which are binary variables revealing the occurrence of the event or not. X_{ik} are vectors of explanatory variables, and α_k and β_k are parameter vectors. We assume that the error terms are random normally distributed variables.

The right-hand side variables for the three outcomes are outlined as follows. We include household size, gender, age, level of education and the formal sector employment of the household head as determinants of poverty. Concerning the equation for distance travelled, we identify as covariates the level of education of the household head and household size with the poverty status considered as the instrumental variable. Regarding the modal choice equation, the variables formal sector employment of the household head, poverty and distance travelled are considered with the last two variables considered as instrumental variables. Tables A.1 and A.2 in the annex summarizes the different covariates of the three equations.

The parameters of the recursive triprobit model are estimated by computing the maximum probability associated with simultaneously choosing a given modality of the different alternatives (Cappellari and Jenkins 2003; Terracol 2002). This method has the merit of calculating linear predictions associated with the different equations and estimating the related marginal probabilities of the occurrence of the events. See Cappellari and Jenkins (2003) for detailed explanations.

9.4 Database and Descriptive Statistics

The data used in this paper is gotten from the 2014 Cameroonian Household Survey (CHCS IV) conducted by the National Institute of Statistics (INS). This survey covered ten regions, and its main objective was to provide information on household welfare status and other aspects on their living conditions and settings, and households' assessment of their situation (subjective poverty), among others. Till date there is no current household survey and overly, summary projections demonstrate that there have not been significant changes on issues of mobility and transport-poverty characteristics in the two largest metropoles like Yaoundé and Douala (see Yaoundé Urban Council, 2020) when compared to 2014 caused by insufficient investments in transforming urban cities. Tables 9.1 and 9.2 present descriptive statistics of the variables used in this study.

Statistics indicate that overall about 95% of travels to the nearest public primary schools are on foot with average distance travelled being about 1 km and the duration of travel around 17 min. The distance covered by individuals from poor household varies between 1 and 10 km with a duration of between 20 min to about two hours. For public secondary school, 77% walk the nearest public secondary school covering 1.44 km in about 22 min. For poor household, this distance ranges from 1.20 to 4 km with individuals using between 29 min to two hours.

Considering travels to hospitals, about 70% of households access the nearest public health centre on foot, covering about 1 km and spending on average 22 min. Individuals from poor households travel between 1.4 and 5 km using between 27 min and 6 h. In terms of access to a district hospital, 44% commute on foot covering 2 km and spending on average 16 min. Poor individual cover between 600 m and 5 km using between 45 min to 16 h.

Table 9.1 Descriptive statistics of continuous variables used in this study

Variable	Obs	Mean	Std. dev	Minimum	Maximum
Access to a public primary school					
Distance commuted (in km)	1,093	0.856	1.463	0.001	30
Distance commuted by the poor (in km)	1,093	0.191	0.629	0.000	10
Distance commuted by the poor on foot (in km)	1,093	0.177	0.610	0.000	10
Time spent by the poor on foot (in minutes)	1,093	3.643	10.47	0.000	111
Access to a public secondary school					
Distance commuted (in km)	998	1.435	1.532	0.001	20
Distance commuted by the poor (in km)	998	0.168	0.614	0.000	6
Distance commuted by the poor on foot (in km)	998	0.135	0.516	0.000	4
Time spent by the poor on foot (in minutes)	998	3.205	11.53	0.000	111
Access to an integrated health centre					
Distance commuted (in km)	2,207	1.040	1.402	0.001	20
Distance commuted by the poor (in km)	2,207	0.136	0.656	0.000	15
Distance commuted by the poor on foot (in km)	2,207	0.066	0.356	0.000	5
Time spent by the poor on foot (in minutes)	2,207	2.667	13.18	0.000	360
Access to a district hospital/subdivision medical centre					
Distance commuted (in km)	3,022	1.960	2.384	0.001	25
Distance commuted by the poor (in km)	3,022	0.202	1.132	0.000	25
Distance commuted by the poor on foot (in km)	3,022	0.051	0.316	0.000	5

Source Computed by authors

Table 9.2 Descriptive statistics of categorical variables used in this study

Variable	Access to a public primary school		Access to a public secondary school		Access to an integrated health centre		Access to a district hospital	
Poor	Freq	Percent	Freq	Percent	Freq	Percent	Freq	Percent
No	886	81.06	886	88.78	1,988	90.08	2,769	91.63
Yes	207	18.94	112	11.22	219	9.92	253	8.37
Total	1,093	100	998	100	2,207	100	3,022	100
On foot	Freq	Percent	Freq	Percent	Freq	Percent	Freq	Percent
No	64	5.86	228	22.85	652	29.54	1,696	56.12
Yes	1,029	94.14	770	77.15	1,555	70.46	1,326	43.88
Total	1,093	100	998	100	2,207	100	3,022	100
By motorcycle	Freq	Percent	Freq	Percent	Freq	Percent	Freq	Percent
No	1,045	95.61	842	84.37	1,732	78.48	1,904	63.00
Yes	48	4.39	156	15.63	475	21.52	1,118	37.00
Total	1,093	100	998	100	2,207	100	3,022	100
By car	Freq	Percent	Freq	Percent	Freq	Percent	Freq	Percent
No	1,083	99.09	937	93.89	2,072	93.88	2,552	84.45
Yes	10	0.91	61	6.11	135	6.12	470	15.55
Total	1,093	100	998	100	2,207	100	3,022	100
By bike	Freq	Percent	Freq	Percent	Freq	Percent	Freq	Percent
No	1,087	99.45	988	99.00	2,165	98.10	2,917	96.53
Yes	6	0.55	10	1.00	42	1.90	105	3.47
Total	1,093	100	998	100	2,207	100	3,022	100

Source Computed by authors

9.5 Results

9.5.1 Cross-Effect Analyse Between Poverty, Distance Travelled and Transport Modal Choice

Our results significantly suggest that variables not included in model and reflected by the error term reveal there is a strong association between poverty status, distance travelled and modal choice (see Tables 9.3 and 9.4). This indicates that other factors not included in our model-possibly associated with individual, community and policy tend to concur that there is a strong connection between these three issues. We also find that poverty, distance travelled, and modal choice are jointly related and it is more appropriate to study their relationship simultaneously for adequate possible policy suggestions. Indeed, the several forms of transport poverty (mobility-related poverty; transport accessibility poverty, transport affordability and social exclusion) affect distance travelled and modal choices in cities in Cameroun. Overly, poor households faced with transport poverty adopt strategies that maximise the household budgets to commute, principally opting to walk to access basic infrastructures.

We also observe a positive association between being poor and opting to walk whilst as the distance increases, poor households are less likely to commute. This translates the presence of transport affordability and mobility-related poverty in cities in Cameroon. This has led to the explosion in the usage of motorcycles by both poor and non-poor households as observed by its relatively high modal shares. Ongolo and Epo (2013) find that motorcycles accounted for about 30% of total inequality in transport mobility in the city of Yaoundé. Nonetheless, the reason for opting to use a motorcycle may be context specific with the main underlying motivation being accessibility to households. Poor households opt to use motorcycles as next best alternative after walking because they are relatively cheaper than taxis and can cover a longer distance. Both poor and moderately well-to-do households use motorcycles because they can reach their households located in peripheries that are poorly serviced in terms of road infrastructure or networks. This phenomenon is accentuated by uncontrolled urbanization where slums cohabitate with residential areas (Ongolo Zogo and Epo 2015).

In addition, education levels of poor households' heads play a significant role in their residential choices as well as where they rent. Individuals that have at least primary levels of education are closer to public educational facilities as well as integrated health centres compared to households having uneducated heads. On the contrary, we do not find such a scenario for district hospitals. Additionally, being poor presumably reduces the distance travelled to public schools because they tend to rent around these schools factoring in the decision that their kids go to these public schools on foot. On the other hand, the poor household is situated relatively far from public hospitals. This specificity for urban cities in Cameroun is driven by the fact that primary public schools are free of charge and therefore encourages poor households that cannot afford for expensive private primary schools to strongly choose to locate closer to these infrastructures.

Considering modal choices, poor household overwhelmingly prefers to commute on foot to access educational or health infrastructures. This result is consistent with findings by Howe and Bryceson (2000), Diaz Olvera et al. (2003) and Ahmed et al. (2008). Furthermore, transport accessibility poverty is further driven in cities in Cameroon because both central and local authorities have inadequately mitigated negative externalities of a dysfunctional transport sector, leading to an atomized market of small private operators that principally operate in the informal transport sector (Ongolo Zogo et al. 2017). Nonetheless, the growing decentralization framework process in Cameroon is apportioning significant responsibility to local and urban communities regarding ameliorating urban mobility. This framework should adequately benefit from financial support and reinforcement of the capacities of these councils to find solutions to mobility and urban displacement issues that incorporate growing urban poverty rates in their policy framework.

Table 9.3 Estimates associated to access to educational facilities

Variables	Access to a public primary school			Access to a public secondary school		
	Walking (1a)	Motorcycle travel (2a)	Car travel (3a)	Walking (1b)	Motorcycle travel (2b)	Car travel (3b)
Poor (1 = yes and 0 = rich)						
Male	−0.117 (0.094)	−0.130 (0.093)	−0.115 (0.094)	0.029 (0.110)	0.040 (0.117)	0.082 (0.110)
Primary education	−0.189* (0.100)	−0.186* (0.098)	−0.201** (0.102)	−0.260** (0.126)	−0.231* (0.128)	−0.362*** (0.122)
Secondary education	−0.825*** (0.141)	−0.783*** (0.143)	−-0.837*** (0.141)	−0.819*** (0.159)	−0.817*** (0.162)	−0.790*** (0.160)
Tertiary education	−0.941*** (0.314)	−0.906*** (0.310)	−0.934*** (0.319)	−1.206*** (0.324)	−1.194*** (0.331)	−1.333*** (0.301)
Working formal sector	0.198* (0.104)	0.192* (0.104)	0.201* (0.104)	0.147 (0.110)	0.146 (0.113)	0.120 (0.113)
Household size	0.051*** (0.013)	0.051*** (0.013)	0.051*** (0.013)	0.065*** (0.014)	0.064*** (0.015)	0.061*** (0.014)
Age	0.0046 (0.004)	0.005 (0.004)	0.004 (0.004)	−6.54e-05 (0.004)	−0.005 (0.004)	6.57e-05 (0.004)
Constant	−1.258*** (0.219)	−1.289*** (0.215)	−1.234*** (0.217)	−1.400*** (0.270)	−1.397*** (0.275)	−1.364*** (0.275)
Distance						
Primary education	−0.048 (0.085)	−0.062 (0.088)	−0.077 (0.088)	−0.093 (0.095)	−0.026 (0.105)	−0.202** (0.079)
Secondary education	−0.202* (0.110)	−0.175 (0.118)	−0.228** (0.111)	−0.205** (0.102)	−0.201* (0.114)	−0.106 (0.084)
Tertiary education	−0.226 (0.188)	−0.240 (0.194)	−0.287 (0.192)	−0.425*** (0.135)	−0.347** (0.148)	−0.298*** (0.107)
Household size	0.021* (0.012)	0.018 (0.012)	0.019 (0.012)	0.039*** (0.011)	0.038*** (0.012)	0.035*** (0.010)
Poor	−0.766** (0.308)	−0.749** (0.335)	−0.803*** (0.288)	−1.494*** (0.259)	−1.360*** (0.429)	−1.121*** (0.167)
Constant	0.081 (0.104)	0.099 (0.107)	0.117 (0.105)	0.268** (0.110)	0.238** (0.119)	0.268*** (0.096)
Modal choice						
Formal sector	−0.100 (0.118)	0.088 (0.133)	0.094 (0.230)	0.029 (0.083)	−0.086 (0.098)	0.053 (0.052)
Poor	0.730* (0.390)	−1.035*** (0.259)	−0.494 (0.719)	0.880*** (0.318)	−0.373 (0.417)	−1.874*** (0.208)
Distance	0.805** (0.396)	-0.292 (0.530)	−1.236** (0.515)	0.509* (0.287)	−0.285 (0.412)	−1.666*** (0.085)
Constant	0.791** (0.355)	−1.177** (0.465)	−1.310** (0.625)	0.244 (0.243)	−0.709** (0.334)	0.242*** (0.080)

(continued)

Table 9.3 (continued)

Variables	Access to a public primary school			Access to a public secondary school		
	Walking (1a)	Motorcycle travel (2a)	Car travel (3a)	Walking (1b)	Motorcycle travel (2b)	Car travel (3b)
/atrho21	0.618** (0.243)	0.604** (0.260)	0.644*** (0.232)	1.086*** (0.355)	0.914* (0.474)	0.718*** (0.109)
/atrho31	−0.465* (0.287)	0.786*** (0.262)	0.226 (0.427)	−0.277 (0.192)	0.0493 (0.226)	0.810*** (0.086)
/atrho32	−0.947*** (0.338)	0.655* (0.386)	0.868* (0.482)	−0.767*** (0.255)	0.528* (0.284)	2.841*** (0.485)
rho21	0.549*** (0.169)	0.539*** (0.184)	0.568*** (0.157)	0.795*** (0.130)	0.723*** (0.226)	0.616*** (0.068)
rho31	−0.435* (0.233)	0.656*** (0.149)	0.222 (0.406)	−0.270 (0.178)	0.049 (0.225)	0.669*** (0.048)
rho32	−0.738*** (0.154)	0.575** (0.258)	0.700*** (0.246)	−0.645*** (0.149)	0.484** (0.218)	0.993*** (0.007)
Observations	1,093	1,093	1,093	998	998	998
Log likelihood	−1477.01	−1430.20	−1298.59	−1485.95	−1392.76	−1194.32
Wald chi2(15)	99.39***	113.9***	90.99***	149.9***	93.91***	871.3***
Likelihood ratio test of rho21 = rho31 = rho32 = 0	chi2(3) = 11.61 Prob > chi2 = 0.01	chi2(3) = 10.70 Prob > chi2 = 0.01	chi2(3) = 6.31 Prob > chi2 = 0.09	chi2(3) = 12.36 Prob > chi2 = 0.01	chi2(3) = 7.65 Prob > chi2 = 0.05	chi2(3) = 18.24 Prob > chi2 = 0.004

Source Computed by authors. No education = refence for level of education. Institutional sector = reference for formal sector. Standard errors in parentheses. ***$p < 0.01$, **$p < 0.05$, *$p < 0.1$

9.5.2 *Predicted Probabilities of Combined Effects of Poverty, Distance Travelled and Modal Choices in Cameroon*

In this section, we substantiate on the probabilities associated with the different alternatives in accessing educational (see Table 9.5) and health facilities (see Table 9.6). We find that the likelihood for poor individuals to access public primary schools is 19 and 11% for public secondary schools. In terms of distance, the probability of a poor individual to reside at least 1 km away from a public primary school and public secondary school are 50% and 59%, respectively. To cover this distance by poor individuals, the prospect of walking to a public primary school is 87%, compared to 8.84% using motorcycle and 5.03% by car. Similarly, the probability of walking to a public secondary school is 73.23%, against 17% by motorcycle and 26.15% by car for poor individuals. The overwhelming takeaway is the preponderant role of commuting by foot for poor households and therefore the urgency to arrange urban settings to facilitate walking to public school facilities. This is possible in a context

Table 9.4 Estimates associated with access to health facilities

Variables	Access to a district hospital			Access to an integrated health centre		
	Walking (1a)	Motorcycle travel (2a)	Car travel (3a)	Walking (1b)	Motorcycle travel (2b)	Car travel (3b)
Poor (1 = yes and 0 = rich)						
Male	0.032 (0.078)	0.063 (0.078)	0.057 (0.076)	0.049 (0.0856)	0.060 (0.084)	0.025 (0.084)
Primary education	−0.201** (0.081)	−0.196** (0.081)	−0.236*** (0.081)	−0.181** (0.088)	−0.183** (0.088)	−0.213** (0.086)
Secondary education	−0.706*** (0.105)	−0.700*** (0.106)	−0.734*** (0.104)	−0.724*** (0.119)	−0.714*** (0.118)	−0.781*** (0.116)
Tertiary education	−1.323*** (0.283)	−1.293*** (0.274)	−1.374*** (0.267)	−1.237*** (0.304)	−1.254*** (0.306)	−1.306*** (0.294)
Working formal sector	0.310*** (0.087)	0.327*** (0.086)	0.332*** (0.085)	0.215** (0.092)	0.225** (0.090)	0.174* (0.093)
Household size	0.088*** (0.010)	0.088*** (0.010)	0.088*** (0.010)	0.092*** (0.012)	0.090*** (0.012)	0.090*** (0.012)
Age	0.007*** (0.003)	0.007*** (0.003)	0.007** (0.003)	0.003 (0.003)	0.003 (0.003)	0.002 (0.003)
Constant	−2.103*** (0.178)	−2.136*** (0.177)	−2.090*** (0.176)	−1.796*** (0.185)	−1.814*** (0.183)	−1.670*** (0.187)
Distance						
Primary education	0.036 (0.064)	0.118** (0.059)	0.158*** (0.054)	−0.129** (0.065)	−0.077 (0.067)	−0.136* (0.071)
Secondary education	0.034 (0.073)	0.114* (0.067)	0.193*** (0.063)	−0.132* (0.075)	−0.072 (0.078)	−0.185** (0.083)
Tertiary education	0.017 (0.095)	0.160* (0.088)	0.298*** (0.084)	−0.182 (0.111)	−0.069 (0.110)	−0.178 (0.122)
Household size	0.019** (0.009)	0.0145* (0.008)	0.012 (0.008)	0.004 (0.010)	0.001 (0.010)	0.0126 (0.010)
Poor	0.278 (0.307)	0.569** (0.252)	0.778*** (0.241)	0.798*** (0.296)	1.013*** (0.264)	0.439 (0.328)
Constant	0.224*** (0.065)	0.152** (0.059)	0.093* (0.055)	0.065 (0.066)	0.019 (0.067)	0.080 (0.072)
Modal choice						
Formal sector	−0.006 (0.053)	0.060 (0.045)	−0.064 (0.043)	−0.043 (0.062)	0.128* (0.067)	−0.087 (0.096)
Poor	0.918*** (0.223)	0.139 (0.235)	−1.454*** (0.202)	0.063 (0.387)	0.075 (0.383)	−1.257*** (0.231)
Distance	−1.278*** (0.462)	−0.108 (0.391)	1.775*** (0.114)	−0.538 (0.542)	0.438 (0.533)	0.802** (0.344)
Constant	0.589** (0.293)	−0.297 (0.303)	−1.775*** (0.064)	0.950*** (0.350)	−1.228*** (0.385)	−1.827*** (0.196)

(continued)

Table 9.4 (continued)

Variables	Access to a district hospital			Access to an integrated health centre		
	Walking (1a)	Motorcycle travel (2a)	Car travel (3a)	Walking (1b)	Motorcycle travel (2b)	Car travel (3b)
/atrho21	−0.081 (0.158)	−0.245* (0.139)	−0.351** (0.143)	−0.357** (0.180)	−0.489*** (0.181)	−0.142 (0.173)
/atrho31	−0.336*** (0.126)	−0.148 (0.117)	0.534*** (0.133)	−0.002 (0.190)	−0.037 (0.188)	0.746*** (0.229)
/atrho32	−0.092 (0.274)	0.774** (0.312)	−1.286*** (0.351)	−0.617* (0.350)	0.624* (0.345)	−0.008 (0.229)
rho21	−0.081 (0.157)	−0.240* (0.131)	−0.337*** (0.126)	−0.342** (0.159)	−0.454*** (0.144)	−0.141 (0.169)
rho31	−0.324*** (0.113)	−0.147 (0.115)	0.489*** (0.101)	−0.002 (0.189)	−0.0366 (0.188)	0.633*** (0.137)
rho32	−0.092 (0.272)	0.649*** (0.181)	−0.858*** (0.093)	−0.549** (0.245)	0.554** (0.239)	−0.008 (0.229)
Observations	3,022	3,022	3,022	2,207	2,207	2,207
Log likelihood	−4366.6	−4486.1	−3930.2	−3175.1	−3063.8	−2601.9
Wald chi2(15)	237.25***	220.24***	1089.03***	198.46***	214.31***	231.18***
Likelihood ratio test of rho21 = rho31 = rho32 = 0	chi2(3) = 10.41 Prob > chi2 = 0.01	chi2(3) = 6.38 Prob > chi2 = 0.09	chi2(3) = 30.32 Prob > chi2 = 0.0	chi2(3) = 9.89 Prob > chi2 = 0.01	chi2(3) = 8.98 Prob > chi2 = 0.02	chi2(3) = 9.89 Prob > chi2 = 0.01

Source Computed by authors. No education = refence for level of education. Institutional sector = reference for formal sector. Standard errors in parentheses. ***$p < 0.01$, **$p < 0.05$, *$p < 0.1$

where urban master plans are constructed to facilitate and encourage walking. In Cameroon like most cities in sub-Saharan Africa, cities are planned and constructed with little attention paid towards encouraging walkability. Urban commuters are discouraged from walking due to insecure walking paths that are poorly lighted and generally don't have leisure facilities.

Regarding access to health facilities, the possibility of poor households to a district hospital and integrated health center is around 8.35% and 10%, respectively. In terms of distance covered, 64 and 52% of poor households are situated at more than 1 km from district hospitals and integrated health centers, respectively. The likelihood of accessing these health facilities on foot is 44 and 73.3% for district hospitals and integrated health centers. Commuting to these services by motorcycle was 36.6% for district hospitals and 19% to integrated health centers. This suggests an urban layout where poor slumps may be closer to integrated districts which are fairly lower in grading than district hospitals. However, there are very low investments to facilitate walkability for poor individuals to access these health facilities thereby

Table 9.5 Predicted public school facility access probabilities

Variables	Access to a public primary school			Access to a public secondary school		
	Walking (1a)	Motorcycle travel (2a)	Car travel (3a)	Walking (1b)	Motorcycle travel (2b)	Car travel (3b)
Joint probability of being poor, covering at least 1 km and using one of the three modal choice (Cols. 1, 2 and 3)	0.102	0.046	0.015	0.061	0.019	0.053
Joint probability of being rich, covering <1 km and using one of the three modal choice (Cols. 1, 2 and 3)	0.008	0.452	0.458	0.031	0.379	0.393
Marginal success of being poor	0.190	0.189	0.189	0.112	0.112	0.111
Marginal success of covering at least 1 km	0.501	0.502	0.501	0.585	0.587	0.591
Marginal success of using one of the three modal choices	0.870	0.088	0.050	0.732	0.169	0.262
Standard error of the linear predictions of equation being poor	0.128	0.126	0.128	0.158	0.162	0.157

(continued)

Table 9.5 (continued)

Variables	Access to a public primary school			Access to a public secondary school		
	Walking (1a)	Motorcycle travel (2a)	Car travel (3a)	Walking (1b)	Motorcycle travel (2b)	Car travel (3b)
Standard error of the linear predictions of equation covering at least 1 km	0.125	0.133	0.122	0.101	0.123	0.080
Standard error of the linear predictions of equation using one of the three modal choices	0.284	0.325	0.493	0.171	0.229	0.085
Linear prediction for equation being poor	−0.968	−0.967	−0.970	−1.368	−1.368	−1.365
Linear prediction for equation at least 1 km	−0.003	0.0001	−0.003	0.209	0.218	0.230
Linear prediction for equation using one of the three modal choices	1.265	−1.458	−1.961	0.666	−0.972	−0.942

Source Computed by authors

accentuating mobility access difficulties. To contribute towards the noble goal of increasing access to health facilities, African cities should invest in encouraging and facilitating walking to health facilities. This should nonetheless be designed with the view of mitigating negative externalities like an increase in pedestrian accidents.

Table 9.6 Predicted public health facility access probabilities

Variables	Access to a district hospital			Access to an integrated health centre		
	Walking (1a)	Motorcycle travel (2a)	Car travel (3a)	Walking (1b)	Motorcycle travel (2b)	Car travel (3b)
Joint probability of being poor, covering at least 1 km and using one of the three modal choice (Cols. 1, 2 and 3)	0.015	0.021	0.009	0.019	0.011	0.013
Joint probability of being rich, covering <1 km and using one of the three modal choice (Cols. 1, 2 and 3)	0.170	0.282	0.138	0.048	0.381	0.397
Marginal success of being poor	0.084	0.084	0.086	0.099	0.099	0.099
Marginal success of covering at least 1 km	0.641	0.638	0.634	0.525	0.523	0.527
Marginal success of using one of the three modal choices	0.441	0.375	0.298	0.733	0.189	0.079
Standard error of the linear predictions of equation being poor	0.117	0.116	0.114	0.122	0.121	0.120

(continued)

Table 9.6 (continued)

Variables	Access to a district hospital			Access to an integrated health centre		
	Walking (1a)	Motorcycle travel (2a)	Car travel (3a)	Walking (1b)	Motorcycle travel (2b)	Car travel (3b)
Standard error of the linear predictions of equation covering at least 1 km	0.077	0.069	0.066	0.084	0.080	0.091
Standard error of the linear predictions of equation using one of the three modal choices	0.224	0.182	0.093	0.287	0.284	0.204
Linear prediction for equation being poor	−1.611	−1.606	−1.603	−1.462	−1.461	−1.458
Linear prediction for equation at least 1 km	0.363	0.360	0.355	0.070	0.071	0.071
Linear prediction for equation using one of the three modal choices	−0.158	−0.319	−0.798	0.645	−0.905	−1.586

Source Computed by authors

9.6 Discussion

Overly our findings indicate that the association between poverty status, distanced travelled and modal choices are endogenous and significantly related. Being endogenous entails for appropriate policy enactment, finding solutions to facilitate and optimize distances travelled and appropriate transport modal choices should ameliorate the welfare status of poor households. Consequently, rethinking future cities that facilitate walkability and access to infrastructure through spatial planning should be considered by policy actors as an important tool in reducing poverty and social exclusion for vulnerable groups of urban inhabitants (Titheridge et al. 2014).

We also highlight the observance that facets of transport poverty are showcased through questions of distanced travelled and modal choices as made evident through joint probabilities between poverty and distance travelled, poverty and modal choices and distance travel and modal choices. The implication is that cities in Cameroon and sub-Saharan Africa, uncontrolled urbanization largely accentuates difficulties related to distances travelled for poor households and choosing appropriate transport modes to commute. A possible consequence is the increase in three types of transport poverty. These include (a) mobility-related poverty associated with inadequate motorized transportation resulting in mobility difficulties and linked with poor urban city infrastructures; (b) accessibility poverty which indicates increases the cost and time of poor individuals to access basic infrastructure and (c) transport affordability issues due to high-cost transportation borne by poor individuals because the cover long distances.

Starkey and Hine (2014) indicate that in developing countries, a large number of poor individuals cover on average between 5 and 8 km either walking or using non-motorized transport. The long distances covered by walking increase the risk of possible accidents due inappropriate walking lanes and a poor cohabitation between agents occupying the sidewalks. In areas of the city where poor individuals reside, under investments in terms of accessing key residential infrastructures like education and health these areas increase the risk of disincentivizing poor individuals to travel and access these infrastructures and therefore further increasing urban poverty and causing spatial segregation between residences by welfare status. For instance, Diaz Olvera et al. (2003) demonstrated that in Dar es salaam, individuals with the least revenues tend to habitat in the unplanned segments of the city that are poorly endowed in terms of social infrastructures.

Differences in distance covered to access basis infrastructures reveal the disparity in travel habits between rich households that reside close to these facilities and poor individuals that walk to for long distances. This is because they have low revenues to pay for motorized transport modes. A possible corollary is a hyping up of the location of residences of poor individuals in slumps that develop close to these infrastructures to reduce mobility poverty. This phenomenon gained prominence in cities in Cameroon and sub-Saharan Africa with the collapse of formal transport systems, during the crisis years, that were put in place to help households with weak revenues. To attenuate distance covered, poor dilapidated residences sprout around basic infrastructures thereon crowding out land that was initially reserved for the expansion of these infrastructures and causing possible tension points between poor inhabitant and urban planning endeavors in these cities.

Indeed, there is a strong perception that urban transport planning and modelling in cities in Cameroon and sub-Saharan Africa gave a lot of weight to enabling commuters gain time and reduce cost for car owners to the detriment of other modes of transport like walking. Unfortunately, this form of urban planning neglected finding solutions for poor individuals that commute on foot to access infrastructures and services. Planning urban cities that contribute in attaining sustainable development goals of poverty reduction, ameliorating education and health as well as rendering cities more resilient cities should rethink conceptualizations towards a scenario where mobility-related poverty, accessibility poverty and transport affordability needs of less advantaged residents in bidonville are factored-in into urban planning to enable our cities to be more inclusive.

9.7 Conclusion

In a context of uncontrolled urbanization of urban cities in Cameroon due to rapid population growth, this study aimed to assess the joint effect of poverty, distance and modal choices in accessing education and health infrastructures. To achieve this objective, we adopt the recursive trivariate probit model which enables us to run simultaneously three equations. The first equation identifies determinants of poverty status. The second equation estimates covariates that affect distance travelled with poverty chosen as the instrumental variable. The third appraises the determinants of modal choices (walking, motorcycle or car) with poverty and distance travelled used as instruments. We use access to public primary or secondary educations to gauge educational infrastructure as well as access to public health units and district hospital to capture health infrastructure. The database used is constructed from the fourth Cameroon Consumption Household survey developed in 2014 by the National Institute of Statistics. Recent trends from urban development plans for metropoles like Yaoundé and Douala indicate that the profile of transport poverty and mobility habits have seldomly evolved due to under investments sustainable urban planning to transforming urban cities in Cameroon.

Poor individuals suggest overwhelmingly walking to access basic infrastructures and cover relatively long distances that don't encourage walkability. This highlights mobility-related poverty, accessibility poverty and transport affordability hurdles faced by poor individuals. The implication for these groups of people is a possible perpetuation of the poverty situation and social exclusion. In terms of urbanization, these dynamics produce non-inclusive cities with negative externalities like low levels of wellness of the city residents due to difficult access to basic infrastructures.

Policy orientations towards developing future cities that are resilient, sustainable and inclusive should comprise of finding adequate policy-mix or dosages between developing appropriate public transport system to correct for accessibility in covering long distances but also undertaking urban planning reforms that facilitate and encourage walkability in urban cities to ameliorate short distance travels for poor households to access basic facilities. The enactment of these policy mix could constitute in defining a framework where costly public transports are borne by the national budget whilst facilitating walkability is devolved to local cities and councils within the current decentralization framework with orientations guided towards ameliorating security, easiness to walk lanes around public basic education and health facilities.

References

Abley S (2010) Measuring accessibility and providing transport choice. Australian Institute of Traffic Planning and Management (AITPM), Brisbane, Australia

Ahmed QI, Lu H, Ye S (2008) Urban transportation and equity: a case study of Beijing and Karachi. Transp Res Part A: Policy Pract 42(1):125–139

Ahrend C, Schwedes O, Daubitz S, Böhme U, Herget M (2014) Kleiner Begriffskanon der Mobilitätsforschung. Fachgebiet Integrierte Verkehrsplanung, Technische Universität Berlin, Berlin, Germany, discussion paper (in German)

Brownstone D, Fang HA (2014) A vehicle ownership and utilization choice model with endogenous residential density. J Transp Land Use 7(2):135–151

Cappellari L, Jenkins SP (2003) Multivariate probit regression using simulated maximum likelihood. Stand Genomic Sci 3(3):278–294

Carruthers R, Dick M, Saurkar A (2005) Affordability of public transport in developing countries, vol TP-3. The World Bank Group, Washington, DC, USA, Transport Papers, pp 1–27

Cook C, Duncan T, Jitsuchon S, Sharma A, Guobao (2005) Assessing the impact of transport and energy infrastructure on poverty reduction. Asian Development Bank (ADB), Manila, Philippines. 290 p. http://www.adb.org/sites/default/files/pub/2005/assessing-transport-energy.pdf

Crisp R, Ferrari E, Gore T, Green S, McCarthy L, Rae A, Reeve K, Stevens M (2018) Tackling transport-related barriers to employment in low-income neighbourhoods. Joseph Rowntree Foundation. https://www.jrf.org.uk/report/tackling-transport-related-barriers-employment-low-income-neighbourhoods

Currie G, Richardson T, Smyth P et al (2009) Investigating links between transport disadvantage, social exclusion and well-being in Melbourne–preliminary results. Transp Policy 16(3):97–105

Deaton A (1987) The demand for personal travel in developing countries. Discussion paper. Infrastructure and Urban Development Department, World Bank, Washington DC, USA

Department for Transport (DfT) (2014) Accessibility Statistics 2013. Department for Transport, London, UK

Diaz Olvera L, Plat D, Pochet P (2003) Transportation conditions and access to services in a context of urban sprawl and deregulation: the case of Dar es Salaam. Transp Policy 10(4):287–298. Abstract: https://doi.org/10.1016/j.jtrangeo.2007.04.001

Gannon C, Liu Z (1997) Poverty and transport. Discussion Paper TWU-30, World Bank, Washington DC, USA. 65 p. http://siteresources.worldbank.org/INTURBANTRANSPORT/Resources/twu-30.pdf

Harris B (2001) Accessibility: concepts and applications. J Transp Stat 4(2/3):15–30

Hine J (2009) Transport and social exclusion. Elsevier, Newtownabbey, UK

Howe J, Bryceson DF (2000) Poverty and urban transport in East Africa: review of research and Dutch donor experience, International Institute for Infrastructural, Hydraulic and Environmental Engineering (IHE), the Netherlands for World Bank, Washington DC, USA. 147 p

International Transport Forum (2017) Income inequality, social inclusion, and mobility, Roundtable Report 164. https://www.itf-oecd.org/sites/default/files/docs/income-inequality-social-inclusion-mobility.pdf

Kranton RE (1991) Transport and the mobility needs of the urban poor: an exploratory study. World Bank, Infrastructure and Urban Development Department, Report INU 86

Litman T (2015) Transportation affordability. Evaluation and improvements strategies. Victoria Transport Policy Institute, Victoria, Canada

Martens K (2013) Role of the bicycle in the limitation of transport poverty in the Netherlands. Transp Res Rec 2387:20–25

Martens K, Bastiaanssen J (2014) An index to measure accessibility poverty risk. In: Proceedings of Colloquium Vervoersplanologisch Speurwerk, Eindhoven, the Netherlands

Milioti CP, Karlaftis MG, Akkogiounoglou E (2015) Traveler perceptions and airline choice: a multivariate probit approach. J Air Transp Manag 49:46–52

Miranda-Moreno L, Bettex L, Zahabi S, Kreider T, Barla P (2011) Simultaneous modeling of endogenous influence of urban form and public transit accessibility on distance traveled. Transp Res Rec J Transp Res Board 2255:100–109

Moore J, Lucas K, Bates J (2013) Social disadvantage and transport in the UK: a trip-based approach, school of geography and the environment. University of Oxford, Oxford, UK, TSU Working Paper Series, Ref. 1063

National Institute of Statistics (2015) Pauvreté et conditions de vie au Cameroun en 2014, L'Institut National de la Statistique (INS), Yaoundé, Cameroun

Ongolo Zogo V, Epo BN (2013) Suburbanization and inequality in transport mobility in Yaoundé, Cameroon: Drawing Public Policy for African Cities. Global Development Network (GDN) Working Paper Series No. 78, September 2013

Ongolo Zogo V, Epo BN (2015) Sources of inequality in the cost of transport mobility in the city of Yaoundé, Cameroon. Dev Southern Africa 32(2):229–239

Ongolo ZV, Epo BN, Nodem CM (2017) La Marche à Pied dans les Métropoles Africaines: cas de la Ville de Yaoundé. Cahiers Scientifiques Du Transport, Cahier Numéros 72:143–159

Plat D, Adolehoume A, Barry B, Boupda E, Diaz Olvera L et al (2004) Pauvreté et mobilité urbaine à Douala, Rapport de recherche

Scheiner J (2008) Accessibility, spatial context and location preferences: is there evidence for accessibility poverty? In: Becker U, Böhmer J, Gerike R (eds) How to define and measure access and need satisfaction. Transport Series of Dresden Institute for Transportation and Environment (DIVU), Dresden, Germany, pp 193–225

Serebrisky T, Gómez-Lobo A, Estupiñán N, Muñoz-Raskin R (2009) Affordability and subsidies in public urban transport: what do we mean, what can be done? Transp Rev: Trans Transdisc J 29(6):715–739

SEU (Social Exclusion Unit) (2003) Making the connections: final report on transport and social exclusion. Office of the Deputy Prime Minister, London, UK

Starkey P, Hine J (2014) Poverty and sustainable transport: how transport affects poor people with policy implications for poverty reduction. A literature review. ODI, UN-Habitat, DFID, UKaid, SLoCaT, London, UK

Terracol A (2002) Triprobit and the GHK simulator: a short note. http://fmwww.bc.edu/repec/bocode/g/GHK_note.pdf

Titheridge H, Christie N, Mackett R, Ovideo HD, Ye R (2014) Transport and poverty: a review of the evidence. UCL Transport Institute, London, UK. https://www.ucl.ac.uk/transport-institute/pdfs/transport-poverty

UN-Habitat III (Troisième Conférence des Nations Unies sur le Logement et le Développement Durable en milieu Urbain) (2015) Rapport National du Cameroun pour HABITAT-III. Conférence tenue du 13 au 15 octobre 2014 au Palais des Congrès de Yaoundé-Cameroun

Uteng TP, Lucas K (2017) Urban mobilities in the global south. Routledge. https://doi.org/10.4324/9781315265094

Velaga NR, Beecroft M, Nelson JD (2012) Transport poverty meets the digital divide: accessibility and connectivity in rural communities. J Transp Geogr 21:102–112

Wilde J (2000) Identification of multiple equation probit models with endogenous dummy regressors. Econ Lett 69(3):309–312

Zhong S, Wei H, Hou W, Cheng D (2003) A lifetime of walking: poverty and transportation in Wuhan. Economic Research Institute, Wuhan University

Jules Médard Nana Djomo is a Lecturer at the Faculty of Economics and Management of the University of Yaoundé II and the Research Officer at the Ministry of Transport in Cameroon. His areas of interest include welfare analysis, transport economics and governance.

Boniface Ngah Epo is a Professor and researcher in the Faculty of Economics of the University of Yaoundé. His areas of research include transport economics, development economics and welfare.

Chapter 10
Conceptualising an Urban Transport Framework for Enhanced Quality of Life in Sub-Saharan Africa: Case Studies of Ghana and Namibia

Eric Yankson

Abstract Unpacking the ramifications of urban transport for quality of life is pivotal for evaluating the pros and cons of national or local policy options. In the sub-Saharan African context, such an enquiry serves to comprehend how different modes of transport affect the state of the physical environment, as well as public health and social equity. This chapter therefore assesses the implications of urban transport for quality of life through a comparative analysis of Ghana and Namibia. The Ghanaian case study instantiates how the dominance of mini-buses (trotros), as well as private vehicles in major cities, results in outcomes such as traffic congestion and pollution. Moreover, the case of Namibia demonstrates a socio-spatial tapestry of inequality characterised by the prevalence of taxis and private vehicles in sprawled cities. The chapter articulates a conceptual framework premised on the tripod of environment, health and equity. The model argues that deleterious urban transport impacts such as pollution must be addressed through alternative modes and policy reforms. Also, the implications of transport options for the health of residents should define the choices made by the central government or local authorities. Additionally, the needs of groups such as pedestrians, cyclists, the physically challenged, children and low-income residents ought to be considered when designing transport options for use by urban residents. The chapter contributes to existing scholarship by unpacking the dialectical ramifications of urban transport for quality of life. This serves to enrich the discourse on sustainable development through the lens of transport policy options and outcomes.

Keywords Urban transport · Quality of life · Environment · Health · Equity

10.1 Introduction: Urban Transport and Quality of Life

The implications of urban transport for quality of life have received significant attention in the academic literature over the years. This notwithstanding, cities in North

---

E. Yankson (✉)
Namibia University of Science and Technology, Private Bag 13388, Windhoek, Namibia
e-mail: eyankson@nust.na

R. A. Acheampong et al. (eds.), *Transport and Mobility Futures in Urban Africa*, The Urban Book Series, https://doi.org/10.1007/978-3-031-17327-1_10

America and Europe appear to have attracted more scholarly focus than those in the Global South. Moreover, the comparative contexts of urban transport for human well-being in sub-Saharan Africa could benefit from more investigation. This chapter thus contributes to existing discourse by elucidating the urban transport system on the continent with the view of assessing implications for quality of life. Based on a comparative analysis of Ghana and Namibia, the discussion unpacks the historical and socio-spatial factors which shape the differential manifestations of sustainable transportation in the two countries. The goal is to assess options for enhanced well-being by users of the transport systems.

The term quality of life as used in this chapter refers to the level of human well-being experienced by residents in an urban setting. This may be evident in terms of the state of one's physical environment, as well as the quality of health. Moreover, the level of social inclusion plays an important role in shaping one's level of comfort in the urban environment. This chapter adopts a qualitative approach to evaluate the quality of life. The analysis relies on document and discourse analyses, as well as interviews to make its observations or arguments.

The environmental, health and equity aspects associated with urban transport portend significant ramifications for the quality of life. Specifically, the attainment of sustainable transport options may occur through more environmentally adept modes of travel such as walking and bicycling. There should also be the integration of transportation and land use plans to optimise planning effectiveness. Moreover, the utilisation of ecologically friendly energy sources to reduce reliance on fossil fuels is pivotal (Aftabuzzaman and Mazloumi 2011). The health impacts of urban transport policy options are also evident in terms of pollution, road accidents, absence of green spaces, negative climate impacts and poor physical exercise. These situations require a policy reorientation to upend the status quo (Khreis et al. 2017). Additionally, the subject of urban transport may be conceptualised as a social justice issue in terms of exposure to pollution, the allocation of space to different modes and time savings. Thus, even though pedestrians and cyclists tend to minimise negative ecological impacts, they are disadvantaged by the policy emphasis on motorised transport, as well as costs with respect to space utility and travel time (Gössling 2016).

Broadly speaking, quality of life may be assessed based on objective (quantitative) or subjective (qualitative) indices. Objective evaluation utilises data such as income and level of access to social services. Subjective analysis may however entail a review of existing documents or resident perceptions in terms of parameters such as amenity value and liveability (Khalil 2012; Biagi et al. 2018). Quality of life may also be determined based on *physical*, *environmental*, *mobility*, *social*, *psychological*, *economic* and *political* factors (El Din et al. 2013). The physical parameters comprise land use and infrastructure characteristics, while the environmental component entails neighbourhood attributes. Mobility embodies transport accessibility, while social elements encompass interpersonal relations and citizen engagement. The psychological aspects involve resident perceptions and place identity. Economic quality of life entails jobs and other opportunities available, while the political element involves policy choices as well as their implementation which affect resident liveability (El Din et al. 2013).

10.2 Conceptualising Urban Transport and Quality of Life in Africa

Within the sub-Saharan African context, a number of trends can be observed in terms of the implications of urban transportation for quality of life. To begin with, rapid urbanisation partly attributed to rural–urban migration has resulted in an ever-growing demand for transport services (Haq et al. 2012). Moreover, there is prioritisation of motorised as opposed to non-motorised forms of transport resulting in growing levels of pollution. The transport system across the continent is also characterised by a relatively poor road safety record as evident in the high number of accidents in several countries (Haq et al. 2012). In parts of West and Central Africa, low levels of vehicle ownership and high transport costs pose challenges to access (Olvera et al. 2013). Moreover, for places such as South Africa, urban transport is an equity issue which must address subjects of access and quality of service delivery, particularly in low-income neighbourhoods (Lucas 2011).

Urban transport has been variously conceptualised in the existing literature, especially within the lens of sustainability. For instance, Kane and Whitehead (2017) observed that a model for sustainable urban mobility may be premised on clean energy, ride-sharing, self-driving and other autonomous forms of transport, as well as increased urban density. Doi and Kii (2012) proposed a cross-assessment sustainable urban transport framework based on both visioning and consensus building. Within specific reference to sub-Saharan Africa, Jones et al. (2013) proposed a framework for sustainable urban transport premised on indigenous and scientific concepts, as well as a participatory planning mechanism. In order to reduce greenhouse gas emissions on the continent, transport policy options which emphasise multi-modality, non-motorised transit and reduced demand on the existing transport system may be envisaged (Creutzig et al. 2012). Moreover, innovative interventions in transport planning such as mixed land use and bus rapid transit can reduce the negative heath impacts of urban transport (Khreis et al. 2017). Also, green transport strategies which seek to advance the goals of mobility, equity and hence quality of life can be adopted (Carvalho et al. 2012).

As noted earlier, quality of life as used in this chapter refers to the level of human well-being experienced by residents in cities and towns. This chapter thus articulates that an urban transport framework for enhanced quality of life in an African context may be broadly distilled in terms of environment, health and equity (Table 10.1). Analysis of the environment helps to observe the ecological implications associated with transport modes and policy choices. Also, health is a sine qua non for evaluating the implications of urban transport for the life expectancy of residents. Additionally, equity helps to evaluate the differential impacts of transport on residents. Moreover, the concept entails differences in levels of accessibility to transport facilities by various demographic groups. Applications of these parameters are particularly important in the sub-Saharan African context given the tremendous negative ecological, health and socio-spatial impacts wrought on residents by urban transport systems.

Table 10.1 Urban transport framework for residential quality of life

Urban transport	Theoretical or Conceptual foundations	Quality of life		
		Environment	Health	Equity
Historical origins and contexts	Colonialism Apartheid Postcolonialism	Compact city; urban sprawl	Spatial expansion; low physical access based on urban form	Relatively less prioritisation of the needs of transport users
Dominant modes	Automobile dependency	Road and motorised transport	Congestion and pollution; respiratory and other diseases; low access to health facilities	Poor non-motorised and public transport facilities
Actors and interests	Public and private interests	Residents, community-based and neighbourhood groups, transport sector operators	Low- and high-income residents; marginalised groups	Pedestrians, cyclists, women, physically challenged persons, low-income residents, marginalised groups
Contemporary policy choices	Policy learning: multi-modality; non-motorised transport; public transport	Low carbon transport policies	Policies on minimising negative health impacts of transport	Policies on non-motorised and public transport
Sustainable transport and mobility futures	Ecological and social sustainability	Reduced pollution and less reliance on fossil fuels	High life expectancy from improved air quality and alternative transport forms	Enhanced accessibility for previously marginalised users of transport services

Source Author's construct based on document and discourse analyses, as well as interviews

The three broad parameters will be evaluated against various urban transport elements such as historical contexts and origins, dominant modes, actors and interests, contemporary policy choices, as well as sustainable transport and mobility futures. The historical contexts and origins provide situational analyses to unpack the evolution of urban transport. Moreover, dominant modes essentialise how the preferred or available transport forms translate into quality of life. Actors and interests embody the roles of stakeholders in shaping the transport sector as well as the degree to which they may be willing to cooperate in improving the system (Cleophas et al. 2019). Also, contemporary policy choices entail how government focus in modern times serves to deal with the pervasive problems of urban transport.

Moreover, sustainable transport and mobility futures determine the implications of government policy choices for urban transport going forward. Overall, the proposed framework argues that in the sub-Saharan African context, colonial and postcolonial historical provenance result in automobile dependency. Thus, to protect the interests of ordinary citizens and other road users, contemporary policy choices should prioritise non-motorised and public forms of transport. These will serve to ensure sustainable transport and mobility futures towards enhanced quality of life.

As already noted, this chapter is based on a comparative analysis of Ghana and Namibia. The choice of Ghana and Namibia can be justified on a number of grounds. To begin with, the differential historical and political economic contexts of the two countries provide for a better understanding of how their respective transport systems emerged. The dissimilarities also serve to unpack potential divergent ramifications of urban transport for quality of life. Moreover, appreciable differences between the two countries in terms of population and socio-spatial dynamics make for an interesting comparative analysis. For instance, Ghana had a 2020 population estimate of 31,073,000 and a population density of 136.56 given the fact that its land area is approximately 227,540 km^2 (United Nations: Department of Economic and Social Affairs 2019; The World Bank Group 2021). Moreover, the nation's socio-spatial tapestry is largely devoid of a racial binary. Namibia on the other hand had a 2020 population of approximately 2,541,000 as a well as population density of 3.086 given its land area of 823,290 km^2 (United Nations: Department of Economic and Social Affairs 2019; The World Bank Group 2021). It is characterised by a spatial binary premised on the dynamics of race and class. Ultimately, my relative familiarity with the transport systems in the two countries serves to ensure a more nuanced and in-depth analysis. The proposed model will now be applied to comparatively analyse the respective urban transport situations in Ghana and Namibia (Table 10.2).

10.3 Ghana

The implications of urban transport for the environment, health and equity in Ghana will be discussed first before focusing on Namibia.

10.3.1 Environment

The historical origins of urban transport in modern-day Ghana may largely be traced to the 1800s when the British colonial authorities emphasised the development of rail transit. The goal was to facilitate the exploitation and transportation of raw materials from the then Gold Coast to Europe (Poku-Boansi 2020). As time went on, attention began to shift towards the provision of roads in hinterlands so they could serve as feeders for the rail sector (Jedwab and Moradi 2011; Gould 1960a, b; Poku-Boansi 2020). Since Ghana's independence in 1957, the overwhelming policy emphasis has

Table 10.2 Urban transport framework applied to Ghana and Namibia

Urban transport	Theoretical or conceptual foundations	Quality of life		
		Environment	Health	Equity
Historical origins and contexts	**Ghana**: Colonialism and postcolonialism **Namibia**: Colonialism, apartheid and postcolonialism	**Ghana**: Colonial era planning (separation of places of work and residence); prioritisation of road transit after independence **Namibia**: Colonial era planning (low-density developments and monofunctional cities); apartheid-era planning (spatial segregation)	**Ghana**: Rapid spatial expansion from urbanisation and motorisation **Namibia**: Low physical access from monofunctional urbanism and apartheid spatial planning	**Ghana**: Historically less prioritisation of the needs of pedestrians, cyclists and physically challenged persons **Namibia**: Poor access by low-income residents or neighbourhoods
Dominant modes	**Both Ghana and Namibia**: Automobile dependency	**Ghana**: Road transport; motorised transit (mini-buses, taxis and private vehicles) **Namibia:** Road transport; motorised transport (taxis and private vehicles)	**Ghana**: Transport-induced pollution; poor air quality; health problems such as respiratory ailments **Namibia:** Low health access associated with transport planning; health problems such as respiratory ailments; high road fatality rates	**Ghana**: Few or no crosswalks, sidewalks and spaces dedicated for use by the physically challenged; poor safety zones for children **Namibia:** Less emphasis on public and non-motorised forms of transport

(continued)

Table 10.2 (continued)

Urban transport	Theoretical or conceptual foundations	Quality of life		
		Environment	Health	Equity
Actors and interests	**Ghana**: Citizen-oriented public interest **Namibia**: Equity-oriented public interest	**Ghana**: Ordinary residents, community-based groups and residents' associations; operators of mini-buses, taxis and private cars **Namibia**: Ordinary residents, community-based groups and residents' associations; operators of taxi services and private car owners	**Ghana**: Low- and high-income residents **Namibia**: Low-income residents; marginalised groups	**Ghana**: Pedestrians, cyclists, physically challenged persons and children **Namibia**: Low-income residents, pedestrians, cyclists, marginalised groups
Contemporary policy choices	**Both Ghana and Namibia**: Multi-modality; non-motorised and public transport	**Ghana**: Reversal of colonial era planning practices; prioritisation of non-motorised and public transport; adoption of electric vehicles **Namibia**: Reversal of apartheid-era planning policies; prioritisation of alternative modes of transport; integration of land use and transport planning	**Ghana**: Emphasis on non-motorised transport; integration of land use and transport planning; sustainable public transport **Namibia**: Emphasis on non-motorised transit; prioritisation of physical access to health facilities; focus on improving road safety	**Ghana**: Emphasis on safety, gender balance, universal access, efficient spatial design and participatory planning **Namibia**: Emphasis on sustainable and non-motorised transport forms

(continued)

Table 10.2 (continued)

Urban transport	Theoretical or conceptual foundations	Quality of life		
		Environment	Health	Equity
Sustainable transport and mobility futures	**Both Ghana and Namibia**: Ecological and social sustainability	**Ghana**: Reduced negative ecological impacts associated with pollution, congestion and reliance on fossil fuels; adoption of alternative modes of transport **Namibia**: Reduced pollution; upending the existing preponderance of road and motorised transportation	**Ghana**: Higher life expectancy from improved air quality; bus rapid transit; rail-based mass transit and an integrated light rail system **Namibia**: Higher life expectancy from improved access to health facilities and road safety; bus rapid transit; cycling and walking	**Ghana**: Enhanced accessibility by pedestrians, cyclists and the disabled; safer travel options for children **Namibia**: Enhanced accessibility for low-income residents, pedestrians, cyclists, marginalised groups

Source Author's construct based on document and discourse analyses, as well as interviews been on the development of road transport in order to increase accessibility to all parts of the country (Poku-Boansi 2020; Jedwab and Moradi 2011). Moreover, due to the poor public transport system, the nation has also experienced the emergence of an informal transport sector dominated by mini-buses or *trotros* and taxis (Poku-Boansi 2020).

According to Essel and Spadaro (2020), road transport in Ghana currently comprises approximately 95% of all internal transit in Ghana, with the remaining 5% being made up of rail, air and water transport. The specific breakdown of road transport is as follows: trotros and large buses (48.2%); non-motorised transport, i.e. cycling and walking (37.6%); taxis, private cars and motorcycles (13.7%); and passenger rail (0.5%). In Accra, an estimated 1 million trips are made daily to and from the central business district by mini-buses and taxis. These vehicles are inefficient in terms of the amount of road space used, implying that congestion is a major challenge especially on arterial routes (Armah et al. 2010; Quarshie 2006).

It is also important to note that colonial era urban planning practices were based on the separation of places of residence from work. This practice has continued even after independence, resulting in increased travel distances and wasteful commuting (Obeng-Odoom 2010, 2015). Moreover, government's economic liberalisation policies since the 1980s have fuelled the importation of vehicles into the country at an alarming rate. The situation has been exacerbated by the poor public transport system and increased pace of road construction which have fostered a growing preference for the use of the private automobile. The result has been increased traffic congestion with concomitant challenges such as noise and air pollution (Obeng-Odoom 2010).

From an environmental perspective, the main actors and interests in Ghana's urban transport system are ordinary residents in cities, as well as community-based groups and residents' associations. These persons or organisations directly bear the brunt of the negative environmental impacts such as congestion and pollution associated with transport choices. As a result, it is important to obtain their inputs when formulating policy responses to deal with these issues. Moreover, operators of mini-buses, taxis and private cars whose activities contribute to environmental challenges in the transport sector may also be regarded as important actors and interests.

To address the noted environmental challenges as a way of creating sustainable transport and mobility futures in Ghana, colonial era urban planning practices based on the separation of places of residence from work need to be revisited with the view of cutting down on travel distances and wasteful commuting. Moreover, policies which prioritise public and non-motorised transport may be envisaged (Obeng-Odoom 2010). It is however important to note that the empirical feasibility of cycling and other non-motorised forms of transport may be dependent on individual factors. These include personal attitudes, normative values, as well as perceptions regarding the convenience of cycling, traffic lights, congestion and individual ownership of bicycles (Acheampong 2017). Also, residents who have low levels of income or work from home are more likely to opt for walking as a mode of transport as opposed to high-income residents who commute to work (Acheampong 2020).

The adoption of electric vehicles such as buses for public transport can also be an approach to the creation of sustainable transport and mobility futures. Specifically, this mechanism will serve to reduce the negative ecological impacts associated with current buses which primarily rely on fossil fuels. Admittedly, the successful roll-out of such an initiative may be constrained by the unreliable power supply and the paucity of charging infrastructure. However, the existence of a favourable policy environment can mitigate these problems by facilitating investments in renewable energy infrastructure. This will serve to boost the electricity supply and enhance the charging system (Teko 2018).

10.3.2 Health

In terms of health, the historical emphasis on road transportation in Ghana has contributed to the rapid spatial expansion of cities and towns. In major cities such as Accra, increased urbanisation and motorisation have led to appreciable levels of transport-induced pollution. The result has been a concomitant decline in air quality (Nerquaye-Tetteh 2009; Essel and Spadaro 2020). Between March 2005 and December 2008, a comprehensive scheme for measuring air quality in Ghana was implemented by the Environmental Protection Agency of Ghana, the United Nations Environment Programme and the United States Environmental Protection Agency (Armah et al. 2010). A total of 745 roadside samples were collected for various pollutants such as particulate matter 10 (PM_{10}), carbon monoxide (CO), nitrogen oxide (NO_2), sulphur dioxide (SO_2), lead (Pb) and manganese (Mn). For instance, 75% of the roadside samples obtained in Accra surpass the national daily average PM_{10} limit of 70 $\mu g/m^3$ while 87% went beyond the World Health Organisation (WHO) level of 50 $\mu g/m^3$ (Nerquaye-Tetteh 2009; Armah et al. 2010; Essel and Spadaro 2020). Moreover, 40% of NO_2 samples were in excess of the 40 $\mu g/m^3$ yearly figure stipulated by the WHO (WHO 2006; Essel and Spadaro 2020).

Evidently, the reactive (rather than proactive) nature of the urban planning system has contributed immensely to the current status quo in which health-related impacts are relatively less prioritised with respect to the transport sector. This is evident in terms of the rapid population growth which persistently outpaces the ability of local authorities to effectively plan for the needs of residents. Moreover, the historical emphasis on motorised forms of transport with little or no mitigating measures for their environmental impacts lends credence to this assertion.

In Accra, residents associate environmental pollution with health problems such as respiratory ailments. Thus, they exhibit a greater propensity to participate in environmental policy design initiatives with the goal of mitigating these challenges (Odonkor and Adom 2020). The main actors and interests in terms of the implications of urban transport for health in Ghana are the residents (both high- and low-income) whose life expectancies are impacted either directly or indirectly by air pollution and other transport-related emissions. In one vein, low-income residents may be more negatively affected since they are the ones who normally patronise mini-buses and other informal transport forms associated with higher levels of pollution. On the other hand, however, wealthy people in Ghana are more likely to own and drive private cars, implying less physical exercise and thus potential health issues such as obesity.

To address the health-related challenges of urban transport towards creating sustainable transit and mobility futures in Ghana, a policy shift towards more ecologically friendly urban planning options such as non-motorised transport can be envisaged. There should also be proper integration of land use and transport planning initiatives to ensure a more orderly development of cities. Moreover, a sustainable public transport approach coupled with ride-sharing and pooling initiatives may be envisaged (Essel and Spadaro 2020). Towards attaining these and other goals, Ghana's transport policy of 2008 articulates the promotion of non-motorised forms of transit such as cycling and walking. It also emphasises the creation of a sustainable and road-based public transport system with options such as bus rapid transit. Moreover, it prioritises rail-based mass transit and an integrated light rail system to connect urban business nodes with suburban residential areas in cities (Essel and Spadaro 2020).

10.3.3 Equity

In terms of equity, the historic emphasis on road transportation in Ghana has led to the dominance of motorised forms of transit. The inequity dimensions of Ghana's urban transport system are thus evident in terms of relatively less prioritisation of the needs of pedestrians, cyclists and physically challenged persons. Specifically, the presence of few or no crosswalks, sidewalks and spaces dedicated for use by wheelchairs and other needs of the disabled lend credence to this assertion (Abane et al. 2019). It is estimated that pedestrian-related incidents account for 42% of Ghana's road traffic deaths. Moreover, 68% of the pedestrian deaths from accidents are determined by the state of crossing infrastructure, as well as the attitudes of pedestrians (Obeng-Atuah et al. 2017).

Additionally, the low availability of safe zones for children in many vicinities poses accident risks for school-going kids, particularly by speeding vehicles (Abane et al. 2019). The safety of children is also compromised by the fact that many of them commute to school unaccompanied by adults, with resultant accident risks. Moreover, even in instances where buses are availed to convey children to and from school, these are usually determined by the needs of parents for convenience, rather than the safety of children per se (Poku-Boansi et al. 2019).

It is therefore evident that the main actors and interests for transport equity in Ghana are pedestrians, cyclists, physically challenged persons and children. Ghana's non-motorised transport strategy for 2019–2028 thus places a premium on values such as safety, gender balance, universal access, efficient spatial design and participatory planning. It promotes the adoption of non-motorised forms of transport through systems which prioritise the needs of pedestrians, bicycles and green spaces. Thus, the provision of parking facilities on streets should occur after cycling, pedestrian and public transit needs have been attended to (United Nations Environment Programme et al. n. d.). It also articulates the creation of safe zones for children, particularly within the vicinity of elementary schools. The strategy stipulates the clear delineation of vendor spaces to prevent encroachment on places reserved for pedestrians. It seeks to minimise accidents by improving the design of road intersections (United Nations Environment Programme et al. n. d.).

Towards creating more equitable urban transport and mobility futures, local authorities in Ghana need to move away from the historical emphasis on modernism towards a postmodernist approach which is more sensitive to social context. This would entail planning with the needs of pedestrians, cyclists, the disabled and children in mind. Specifically, there should be a greater focus on the provision of physical infrastructure and social spaces to serve their needs. Moreover, there must be a better allocation of resources towards addressing the long-term mobility needs of these demographic groups. This should prioritise better access and spatial distribution of services or facilities.

10.4 Namibia

Having unpacked the situation in Ghana, the discussion on the environmental, health and equity aspects of urban transport now focuses on Namibia.

10.4.1 Environment

In terms of the environment, the history of urban transportation in Namibia is inherently linked with its colonial and apartheid heritage. Specifically, the colonial era planning system was based on low-density developments undergirded by monofunctional urban forms. This was subsequently reinforced by apartheid spatial planning which prioritised segmentation of the urban landscape on the bases of race and class. As a result, poor black residents were mostly confined to peripheral townships marked by poverty and poor service delivery (Müeller-Friedman 2006). Moreover, these places had relatively poor infrastructure such as road networks, as compared to the affluent suburbs.

In Windhoek for instance, many blacks to date reside in low-income north-western townships such as Katutura, Hakahana and Wanaheda. Even after independence, the socio-spatial polarisation has been compounded by the influx of people, especially from rural to urban areas. Given that cities are less equipped to meet the service and other needs of these new residents, there has been a proliferation of informal settlements at the urban periphery (Müeller-Friedman 2006). These places are especially marked by poor-quality infrastructure, portending negatively for transport accessibility.

Namibia has also been historically dominated by road transportation with the automobile as the preferred mode of travel. Specifically, the relatively poor public transport system implies that taxis and private cars are the preponderant modes of transport in major cities (Robinson and Fisher 2018–2019). In Windhoek for instance, the breakdown of transport modal choices by residents is as follows: car (43%); taxi or bus (36%); walking (21%); and cycling (1%). The preponderance of a car-oriented mode of urban transport in Namibia (i.e. mostly four-seater sedans) which occupy more space on roads and create traffic congestion portends negatively on the environment. This problem is also compounded by the relatively less emphasis on non-motorised forms of transport such as cycling and walking which have less ecological impacts as opposed to motorised transport (Robinson and Fisher 2018–2019).

Greenhouse gas emissions in Namibia have thus increased steadily over the years. This may be attributed in part to air pollution from exhaust emissions associated with road transportation in the country (Nwagbara and Iyama 2019). Moreover, the construction of roads is associated with land encroachment, deforestation, as well as the destruction of natural habitats which portend negatively for the protection of the environment (Nwagbara and Iyama 2019).

From an environmental perspective, the main actors and interests in Namibia's urban transport system are also ordinary residents, community-based organisations and residents' associations. These persons or groups who are directly affected by the deleterious environmental impacts of urban transport have the most to gain from policy reforms or initiatives which seek to reverse the negative legacy emanating from the colonial and apartheid eras. Moreover, operators of taxi services and private car owners have important interests since their activities directly or indirectly contribute to the environmental challenges associated with the transport system.

Namibia's new road transport policy was launched in 2018. The new document sought to make up for the limitations in previous policy approaches by moving away from the narrow focus on road transportation. Thus, it stressed the need to prioritise other modes of transport such as air, water and rail. Moreover, it underscored the importance of sustainable transport options through the focus on people and goods, rather than just vehicles (Rasmeni 2018). Also, Namibia's road transport sustainability plan seeks to enhance transport access and mobility by improving public transportation. Moreover, it recommends the integration of land use and transport planning to achieve this goal. It also aims at protecting Namibia's ecology and physical environment by minimising the negative environmental impacts of transport. Additionally, it promotes the use of green construction materials and techniques to ensure ecological sustainability (Roads Authority & Texas A and M Transportation Institute, n. d).

The creation of sustainable transport and mobility futures in Namibia would occur when interventions such as the new urban transport policy result in the upending of the existing preponderance of road transportation as well as the concomitant reliance on taxis and private vehicles. Achieving this feat would however not be easy, given the fact that the fundamental building blocks which created the current status quo remain unchanged even after independence. It is therefore important to actively engage municipalities and residents in order to ensure the embracing of non-motorised forms of transport such as walking and cycling. Ultimately, however, the main solution lies in dismantling the legacy of segregation through the creation of more mixed neighbourhoods which are equally served by facilities such as transport networks.

10.4.2 Health

As regards the health implications of urban transport for the Namibian context, the historical provenance of monofunctional urban forms coupled with apartheid spatial planning implies that physical access to health facilities could be a major challenge, especially in low-income communities. This situation may be especially exacerbated by urban sprawl which can result in the emergence of geographically dispersed settlements and facilities. Transport in Namibia is thus pivotal to accessing health facilities, especially by vulnerable groups such as persons infected with the Human Immuno Deficiency Virus (HIV) or the Acquired Immune Deficiency Syndrome (AIDS). Thus, in instances where residents are located far away from basic services, their health needs could be seriously compromised (Bicycle Empowerment Network Namibia et al. 2008).

The health aspect of urban transport in Namibia is also evident in terms of the negative health conditions such as respiratory infections and stress associated with the dominant car-oriented modes of transport. While the argument can be made that these conditions are not likely to be as dire given the relatively low population density in the country, the increased rate of urbanisation portends that residents are steadily being exposed to the same risks as those in other African countries with high population densities.

The main actors and interests when it comes to the health implications of urban transport in Namibia appear to be low-income residents who bear the brunt of the ailments associated with the dominant transport modes. Moreover, as already noted, marginalised groups who experience challenges in accessing health care are also affected. Towards dealing with the negative health aspects of urban transport in Namibia to ensure sustainable transport and mobility futures, contemporary policy emphasis should shift towards the adoption of non-motorised transit forms such as cycling and walking. Moreover, there should be the prioritisation of physical access to health facilities by residents of low-income neighbourhoods or communities.

Namibia has one of the highest levels of road traffic accident-related deaths per annum in the world. According to the WHO, this is estimated at 30.4 per 100,000 persons. The figure is higher than the average of 26.6 persons for sub-Saharan Africa. Across the continent, it is exceeded only by eight countries, namely Burundi, Democratic Republic of Congo, Central African Republic, Burkina Faso, Guinea Bissau, Liberia, Malawi and Zimbabwe. Moreover, Namibia is among the 17 countries in Africa which are actually experiencing increasing levels of death from road traffic accidents (WHO 2018; Jones et al. 2020: 297–298). The implications for quality of life are enormous, including reduced life expectancy and accident-related health challenges. The urban dimensions of Namibia's high level of road accidents are especially evident given the relatively high concentration of these incidents within cities and towns, as well as the poor conditions of road infrastructure in some municipalities.

Namibia's road transport sustainability plan thus seeks to enhance the nation's road safety by protecting users and ensuring better public education. The ultimate goal is to reduce the number of fatalities from road accidents. The plan also seeks to maintain the country's road infrastructure through better funding mechanisms as well as cost-efficient techniques. Moreover, there should be better regulation of the number of vehicles and passengers who ply the roads (Roads Authority, Texas A and M Transportation Institute, n. d; Government of the Republic of Namibia: Ministry of Works, Transport and Communication 2000).

Urban planning in Namibia should also seek to ensure healthy futures through transport by enhancing access to health facilities, minimising pollution and ensuring better road safety. These can occur by shifting away from the dominant monofunctional urban forms characterised by sprawl towards mixed land uses and a more compact form of urban development. Moreover, an emphasis towards multi-modal transport will help to reduce pollution levels. Also, urban strategies should seek to localise national-level plans for enhanced road safety through better law enforcement, especially in terms of speed limits and driving under the influence of alcohol.

10.4.3 Equity

Besides health, the equity dimensions of urban transport in Namibia are evident in terms of the legacy of apartheid-era spatial planning which has resulted in low-income neighbourhoods being located away from employment and other economic opportunities. Many residents in these areas spend a significant part of their disposable incomes on transportation, portending negative ramifications for access to basic services (Robinson and Fisher 2018–2019). Moreover, the dominance of a car-oriented mode of travel (which primarily consists of privately owned four-seater sedan vehicles) has resulted in a situation whereby the public transport system is poorly developed (Robinson and Fisher 2018–2019).

Another equity issue in Namibian urban transport is the relatively less emphasis on non-motorised forms of transport. As a result, sidewalks and bicycle lanes are either non-existent or poorly developed. This is especially a major challenge in low-income neighbourhoods where residents would likely find these alternative forms of travel more affordable (Robinson and Fisher 2018–2019).

Overall, the urban equity dimensions of transport in Namibia are evident in terms of differences in levels of access, affordability, infrastructure and quality of service delivery. Generally speaking, residents in poor urban communities such as informal settlements have less access to transport services due to their locations away from major areas of economic activity. Moreover, the relatively high costs of transport (in relation to their disposable incomes), coupled with poor infrastructure and low quality of service delivery, put them at a disadvantage.

The main actors and interests regarding equity in Namibian urban transport thus appear to be low-income residents, pedestrians and cyclists. Moreover, as noted earlier, marginalised groups such as persons living with HIV/AIDS may also be included in this category. In order to create sustainable transport and mobility futures towards enhancing the residential quality of life, it is therefore important to prioritise the needs of these persons or groups. This requires a policy reorientation at both the national and municipal levels.

Windhoek appears to be a trailblazer when it comes to municipal transport policy-making in Namibia. Thus, the city boasts of both the Sustainable Urban Transport Master Plan (SUTMP) and the Non-Motorised Transport Strategy (NMTS). The SUTMP has the broad goal of promoting public and non-motorised transport to reduce overreliance on taxis and private cars (City of Windhoek 2013). Moreover, the NMTS seeks to address existing safety, infrastructure and equity gaps associated with non-motorised transport in the city. It thus prioritises the creation of sidewalks, bicycle lanes and other infrastructure which promote walking and cycling. Additionally, it focuses on the needs of women to promote gender equity in trip modes or choices (City of Windhoek 2018).

Generally, urban planning should prioritise the needs of both high- and low-income communities towards attaining equitable transport futures. This should entail strategic investments in infrastructure as well as the creation of an enabling physical environment for travel. Moreover, emphasis on cost-effective travel options such as walking and cycling would benefit low-income residents while also affording the wealthy opportunities for exercise.

10.5 Syntheses and Observations

Overall, the chapter observes a number of differences between Ghana and Namibia when it comes to urban transport and quality of life. To begin with, the respective historical contexts of the two countries may be distilled in terms of divergent population growth, colonial and postcolonial constructions, as well as government policies. In the case of Ghana, tremendous population growth has increasingly led to a more compact urban form associated with high levels of demand for transport services. The colonial era was initially associated with the prioritisation of rail transport, even though attempts were later made to switch to road transport. In the postcolonial epoch, there has been an overwhelming emphasis on roads as the preponderant mode of transport. However, government policies in the recent past have sought to emphasise multi-modality and non-motorised forms of transport in order to deal with challenges such as congestion and pollution.

In the case of Namibia, while the urban population has been growing steadily over the years, the relatively low level of growth in absolute terms implies that urban sprawl is common. This results in the need for increased accessibility across geographically dispersed locations. Moreover, the colonial and apartheid eras with their emphases on spatial polarisation resulted in the emergence of a highly unequal urban landscape which was relatively unfavourable to the majority black population. Following independence in 1990, there was also a significant influx of blacks from rural to urban areas, resulting in the emergence of informal settlements, particularly at the urban periphery. These areas are characterised by poor infrastructure such as untarred roads leading to poor transport access by residents. Government policies have therefore sought to redress the ills of spatial inequality while simultaneously promoting sustainability through multi-modal and non-motorised forms of transport.

As regards the environment, urban transport in Ghana is characterised by challenges such as traffic congestion and pollution. In the case of Namibia, there is the problem of sprawl and poor physical access to communal facilities. For health, transport pollutants result in respiratory and other ailments in Ghana. The situation in Namibia is primarily associated with increased road accidents which have led to a tremendous cost in terms of human lives. As regards equity, the Ghanaian context reveals less prioritisation of the needs of pedestrians, cyclists, the physically challenged and children thus posing safety hazards for these demographic groups. And for Namibia, a spatial binary characterises transport use and access. The implication is that wealthy residential neighbourhoods tend to have better roads and physical access to communal facilities such as health care. However, low-income areas such as informal settlements are associated with poor road networks and less access to public facilities.

Overall, the creation of sustainable transport and mobility futures in sub-Saharan Africa requires strategic urban planning interventions. This is due to the fact that urban areas by virtue of their population densities and the extent of the built-up land areas are associated with many of the negative effects of transport such as pollution and congestion. Moreover, city authorities are best placed to initiate innovative plans or implement many of the national-level policies as a way of effectively responding to the highly localised nature of transport impacts. In the case of Ghana, urban planning interventions must promote multi-modality, reduce pollution, as well as ensure better access for children, the aged and the disabled. These can occur in terms of enhanced public transport, as well as the provision of infrastructure for non-motorised forms such as sidewalks and cycling. Moreover, the situation in Namibia requires a new mode of urban planning which prioritises the creation of mixed-income neighbourhoods in order to minimise sprawl and enhance transport accessibility. There should also be a stricter enactment and implementation of road safety by-laws to reduce transport-related fatalities.

10.6 Implications and Conclusion

Broadly speaking, an urban transport framework for enhanced quality of life is manifested in terms of positive environmental, health and equity outcomes (Fig. 10.1). This chapter articulates that a model for sustainable transport and mobility futures in African cities may be conceptualised in terms of multi-modalism, prioritisation of the public interest, as well as policy reforms. Specifically, emphasis should be shifted away from the overreliance on road transport to embrace alternative forms such as walking and cycling which are more environmentally friendly. To make these alternative transport modes attractive to residents, greater awareness of their benefits must be created through public education. Moreover, the inconvenience associated with these modes of transport should be redressed through increased investments in infrastructure such as pavements and sidewalks.

The prioritisation of the public interest in urban transport must occur through greater resource commitments towards improvements in public transport. For instance, local authorities in both Ghana and Namibia should establish more reliable and attractive bus rapid transit systems widely available to the majority of the population. Part of the strategy should also encompass the creation of designated bus routes along major thoroughfares to reduce traffic congestion and make these services appealing to the masses. In terms of policy reforms, governments in both

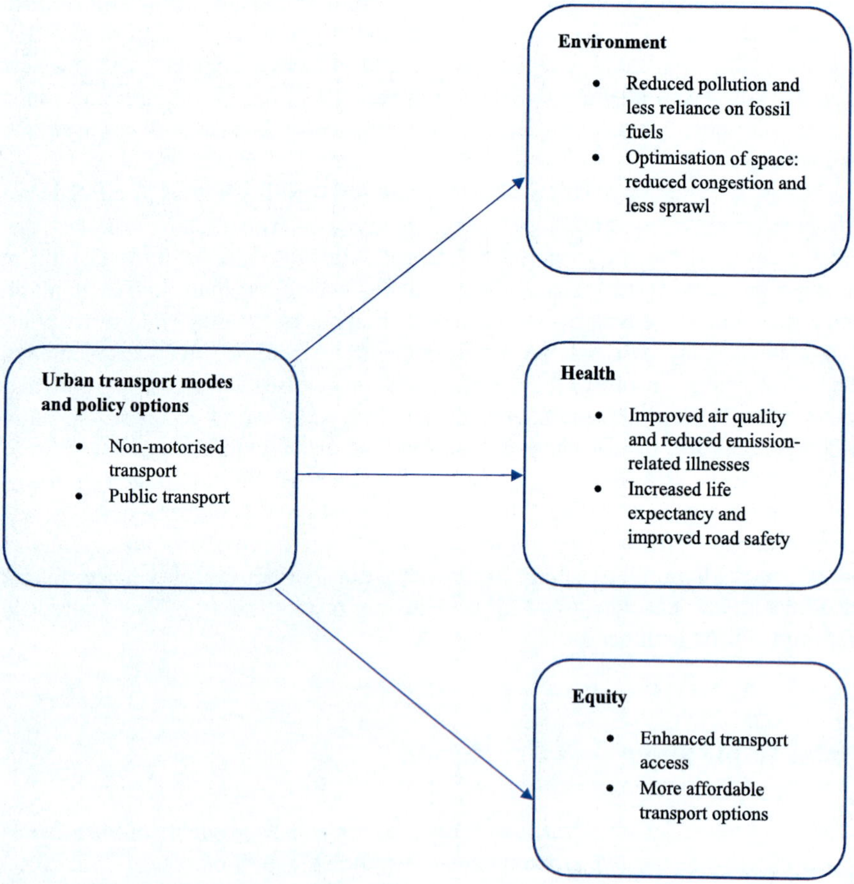

Fig. 10.1 Positive quality of life implications of urban transport framework *Source* Author's construct based on document and discourse analyses, as well as interviews

countries must adopt bottom-up approaches premised on regular engagements with the community and public interest groups to obtain their perspectives. These will form the bases for changes or innovations to policy initiatives adopted to improve the urban transport system.

This chapter contributes to the existing scholarly discourse in three main ways. To begin with, it unfurls conceptual relationships between urban transport and quality of life in the African context. Specifically, it distills these based on the broad themes of environment, health and equity. The advantage of this approach is that it unpacks transport as an evolving policy choice whose spatial and social impact must be considered in the urban planning process. This analysis also reveals the roles of unique historical and political-economic factors in shaping transport choices and their concomitant ramifications for the quality of life. Thus, it enriches understanding of how the national context affects the evolution of urban morphological patterns and resulting influences on residents. Finally, the chapter articulates a more sustainable approach to the design of urban transport systems. This occurs through the adoption of ecologically friendly modes of transit. The goal is to meet the ever-changing travel needs of residents while simultaneously protecting the urban landscape from deleterious impacts such as pollution and congestion.

Acknowledgements The author is grateful to the Volvo Research and Educational Foundations for providing grant funding for this research project and creating the platform for cross-institutional collaboration. Special thanks also go to the team leaders (Prof Charles Adams and Dr Gift Dumedah from the Kwame Nkrumah University of Science and Technology), Dr Eric Yankson (from the Namibia University of Science and Technology) and Prof Steven Jones (from University of Alabama).

References

Abane AM, Amoako-Sakyi RO, Owusu SA, Agyemang KK Odame PK (2019) Transport and social exclusion in Ghana. International network for transport and accesibility in low income countries supported by the Volvo Research and Educational Foundations

Acheampong RA (2017) Towards sustainable urban transportation in Ghana: exploring adults' intention to adopt cycling to work using theory of planned behaviour and structural equation modelling. Transp Dev Econ 3(18):1–11

Acheampong RA (2020) Spatial structure, intra-urban commuting patterns and travel mode choice: analyses of relationships in the Kumasi Metropolis, Ghana. Cities 96(102432):1–20

Aftabuzzaman M, Mazloumi E (2011) Achieving sustainable urban transport mobility in post peak oil era. Transp Policy 18:695–702

Armah FA, Yawson DO, Pappoe AANM (2010) A systems dynamics approach to explore traffic congestion and air pollution link in the city of Accra, Ghana. Sustainability 2:252–265

Biagi B, Ladu MG, Meleddu M (2018) Urban quality of life and capabilities: an experimental study. Ecol Econ 150:137–152

Bicycling Empowerment Network Namibia (BEN Namibia); Yelula/U-khâi; AIDS Law Unit/Legal Assistance Centre (ALU/LAC); International Community of Women Living with HIV/AIDS (ICW-Namibia) (2008) Impact of transport on access to health services for PLWHA in Namibia. (Aug 2008)

Carvalho L, Mingardo G, Van Haaren J (2012) Green urban transport policies and cleantech innovations: evidence from Curitiba, Göteborg and Hamburg. Eur Plan Stud 20(3):375–396
City of Windhoek (2013) Sustainable urban transport master plan: final-main report
City of Windhoek (2018) City of Windhoek: non-motorized transport strategy-final report. (31 May 2018)
Cleophas C, Cottrill C, Ehmke JF, Tierney K (2019) Collaborative urban transportation: recent advances in theory and practice. Eur J Oper Res 273:801–816
Creutzig F, Mühlhoff R, Römer J (2012) Decarbonizing urban transport in European cities: four cases show possibly high co-benefits. Environ Res Lett 7:1–9
Doi K, Kii M (2012) Looking at sustainable urban mobility through a cross-assessment model within the framework of land-use and transport integration. IATSS Res 35(2):62–70
El Din HS, Shalaby A, Farouh HE, Elariane SA (2013) Principles of urban quality of life for a neighborhood. HBRC J 9:86–92
Essel D, Spadaro JV (2020) Health and economic impacts of transport interventions in Accra, Ghana. Word Health Organization
Gössling S (2016) Urban transport justice. J Transp Geogr 54:1–9
Gould P (1960a) The development of the transportation pattern in Ghana. Northwestern University Press, Chicago
Gould PR (1960b) Transportation in Ghana. Northwestern University, Evanston, IL
Government of the Republic of Namibia: Ministry of Works, Transport and Communication (2000) The Namibian road sector reform: background, structure and philosophy. (May 2000)
Haq G, Matin N, Perschon J, Schwela D, Whitelegg J, Vanderschuren M, Jackson M, Tamale KA, Feresu S Lungu C (2012) Transport and environment in sub-Saharan Africa. In: Haq G, Schwela D (eds) Transport and environment science technology network. (Oct 2012)
Jedwab R, Moradi A (2011) Transportation infrastructure and development in Ghana. PSE Working Papers Number 2011–24. https://halshs.archives-ouvertes.fr/halshs-00607207
Jones S, Odero K, Adanu EK (2020) Road crashes in Namibia: challenges and opportunities for sustainable development. Dev South Afr 37(2):295–311
Jones S, Tefe M, Appiah-Opoku S (2013) Proposed framework for sustainability screening of urban transport projects in developing countries: a case study of Accra, Ghana. Transp Res Part A Policy Practice 49:21–34
Kane M, Whitehead J (2017) How to ride transport disruption –a sustainable framework for future urban mobility. Aust Plan 54(3):177–185
Khalil HA (2012) Enhancing quality of life through strategic urban planning. Sustain Cities Soc 5:77–86
Khreis H, May AD, Nieuwenhuijsen MJ (2017) Health impacts of urban transport policy measures: a guidance note for practice. J Transp Health 6:209–227
Lucas K (2011) Making the connections between transport disadvantage and the social exclusion of low income populations in the Tshwane Region of South Africa. J Transp Geogr 19(6):1320–1334
Müeller-Friedman F (2006) Beyond the post-apartheid city: de/segregation and suburbanization in Windhoek, Namibia. Afr Geograph Rev 25(1):33–61
Nerquaye-Tetteh E (2009) Urban air quality monitoring in Accra: case study. Accra, Ghana: environmental protection agency—Ghana. http://www.unep.org/urban_environment/PDFs/BAQ09_ghanacasestudy.pdf
Nwagbara VU, Iyama WA (2019) Assessment of the environmental impact of road infrastructure in countries: a study of the Namibia scenario. J Geosci Environ Prot 7:86–101
Obeng-Atuah D, Poku-Boansi M, Cobbinah PB (2017) Pedestrian crossing in urban Ghana: safety implications. J Transp Health 5:55–69
Obeng-Odoom F (2010) Drive left, look right: the political economy of urban transport in Ghana. Int J Urban Sustain Dev 1(1–2):33–48
Obeng-Odoom F (2015) Sustainable urban development in Africa? The case of urban transport in Sekondi-Takoradi, Ghana. Am Behav Sci 59(3):424–437

Odonkor ST, Adom PK (2020) Environment and health nexus in Ghana: a study on perceived relationship and willingness-to-participate (WTP) in environmental policy design. Urban Climate 34(100689):1–12

Olvera LD, Plat D, Pochet P (2013) The puzzle of mobility and access to the city in Sub-Saharan Africa. J Transp Geogr 32:56–64

Poku-Boansi M (2020) Path dependency in transport: a historical analysis of transport service delivery in Ghana. Case Stud Transp Policy 8:1137–1147

Poku-Boansi M, Amoako C, Obeng-Atuah D (2019) Urban travel patterns and safety among school children around Accra, Ghana. J Transp Health 15(100660)

Quarshie M (2006) Integrating cycling in bus rapid transit system in Accra. In: Morrison GM, Rauch S (eds) Highway and urban environment: proceedings of the 8th highway and urban environment symposium. Nicosia, Cyprus. (12–14 Jun 2006)

Rasmeni M (2018) Works Ministry launches new transport policy. https://economist.com.na/40654/wheels-wings/works-ministry-launches-new-transport-policy/. (12 Dec 2018)

Roads Authority; Texas A and M Transportation Institute (n. d.) Road transport sustainability plan for Namibia

Robinson B, Fisher R (2018–2019) Windhoek, Namibia: moving from transport planning to action. In: TUMI initiative's transformative stories. Transformative Urban Mobility Initiative (TUMI)

Teko E (2018) Ghana projct scoping: e-mobility options. In: Lah O (ed) Wuppertal Institut, UN-Habitat and UN-Environment

The World Bank Group (2021) Land area: square kilometres. https://data.worldbank.org/indicator/AG.LND.TOTL.K2

United Nations Environment Programme, FIA Foundation & Institute for Transportation and Development Policy (n. d.) Ghana's NMT strategy: 2019–2028

United Nations: Department of Economic and Social Affairs (2019) World population prospects 2019. https://population.un.org/wpp/Download/Standard/Population/

World Health Organization (WHO) (2006) WHO air quality guidelines for particulate matter, ozone, nitrogen dioxide and sulfur dioxide assessed: global update 2005. WHO. http://whqlibdoc.who.int/hq/2006/WHO_SDE_PHE_OEH_06.02_eng.pdf

World Health Organization (WHO) (2018) Global status report on road safety 2016. https://www.who.int/violence_injury_prevention/road_safety_status/2018/en/

Eric Yankson is the Associate Dean for Research and Innovation of the Faculty of Engineering and Spatial Sciences of the Namibia University of Science and Technology. His research interests include land use and transport planning, sustainable urbanism, political economy and policy analysis.

Part III
Public Transport Policy and Governance

Chapter 11
Public Transport Policy and Governance in African Cities: An Introduction

Michael Poku-Boansi

The growing incidence of urbanisation, increasing motorization and the need for safe, reliable, inclusive and sustainable public transport systems have received considerable attention among transport planners, policy makers and development partners. Central to ongoing efforts are the reform policies and governance regimes. These initiatives are seen as key to the realization of a sustainable public transport future for African cities and are discussed in the third Part of this book. In Chap. 12, Asimeng examines the problems with BRT implementation in African cities with the aim of providing a plausible approach based on the reverse product life cycle (RPLC) concept. His approach captures BRT implementation problems in African cities for improved mass transportation in the future. Within the context of several studies and reports identifying the problems with BRT implementation in Africa, he argues for a different feasible approach which is likely to rekindle government officials' interests in mass transportation and situate BRT as a plausible public transportation option for the future of African cities.

In Chap. 13, Poku-Boansi and Asibey examine the politics and socio-spatial conflicts between transport operators and the state using Ghana as a case. They take a political economy approach to discuss and unpack the politics around the pricing of public transportation services in Ghana. They explore the socio-political environment for operating and pricing public transportation in urban Ghana. They argue that there is a complex interaction between actors in the pricing of public transportation resulting in tension and mistrust among state actors on the one hand and the operators on the other hand.

Finally, in Chap. 14, Jones, Adanu, Adams and Ambunda explore how the provision of safe public transport is paramount to harnessing rapid urbanisation as well as connecting the opportunities of urban development to the rest of the continent.

M. Poku-Boansi (✉)
Department of Planning, Kwame Nkrumah University of Science and Technology, Kumasi, Ghana
e-mail: mpoku-boansi.cap@knust.edu.gh

R. A. Acheampong et al. (eds.), *Transport and Mobility Futures in Urban Africa*, The Urban Book Series, https://doi.org/10.1007/978-3-031-17327-1_11

They argue about how technological advancements can and will make Africa's public transport system safer and more secure in the future.

To begin delving into the detailed chapters presented in this section, we offer a brief overview of the conceptual understanding of public transportation policy and governance. Specifically, we briefly examine efforts in reforming public transport systems and the governance regimes in Africa.

11.1 Public Transport Services and Governance: An Overview

The importance of an efficient public transport system in developing countries has long been acknowledged, debated and discussed among academics, urban and transport planners and policy makers. Recent debate on this issue has become necessary in many African countries where urbanisation is rife and transport infrastructure is weak and services are poor (Poku-Boansi 2021; Cohen 2006; Turok and McGranahan 2013; Darkwah and Cobbinah 2014). Globally, the success of cities as engines of economic growth has been primarily driven by transportation. For example, Dube et al. (2011) argue that an efficient and inclusive urban mobility is essential for economic and social development, as it enables citizens to access goods, services, jobs, markets, educational opportunities and social contacts. Other scholars (see Pereira and Andraz 2012; Aschauer 1989; Njoh 2000) have argued that government expenditure on public transportation can potentially increase productivity or reduce the cost of production and increase economic growth, in terms of reduced cost of travel. In this case, it is understandable that a transport system which provides high quality services has become the sine qua non in transport studies (see Dell'Olio et al. 2010; Randheer et al. 2011).

11.2 Public Transport Service Challenges

In many African countries, research indicates that rapid urbanisation has led to several transport challenges including rising travel demand, especially in major cities (Rahman et al. 2012), increasing vehicle ownership (Cervero 1996, 2013), safety issues (Obeng-Atuah et al. 2016; Poku-Boansi and Adarkwa 2013; Poku-Boansi et al. 2019; Obeng-Atuah et al. 2016), congestion (Adarkwa and Poku-Boansi 2011), unreliable public transport services (Poku-Boansi and Marsden 2018), poor and inadequate pedestrian facilities (Amoako et al. 2014; Poku-Boansi 2020) and unscheduled services, as well as indiscipline among public transport operators in terms of non-adherence to traffic regulations (Behrens et al. 2016; Agyemang 2015). May and Marsden (2010) have also argued that uncontrolled growth in African urbanisation and motorization has contributed to urban land use and transportation system that is socially, economically, and environmentally unsustainable.

Research (e.g., Cervero 2000; Cervero and Golub 2007; Cohen 2006; Adarkwa and Poku-Boansi 2011) indicates that public transport services in African cities are mostly classified as 'informal'[1] with minibuses, taxis, motorcycles and vans dominating the operations (Klopp et al. 2019). In contrast, however, studies have shown that the services offered play a crucial role in meeting the mobility needs of the urban population (Poku-Boansi and Marsden 2018) as in some jurisdictions (e.g., Ghana, Kenya, Senegal) they provide over 50% of transport services (World Bank 2000). Kumar and Agarwal (2012) have estimated that in Lagos, 15% of the population earns a living through the informal transport sector. This thus makes it difficult for the policy makers to amend the system due to the political risk associated with such reforms.

11.3 Public Transport Reforms in African Cities

Unfortunately, regardless of the reported and known cases of the importance of transport to the socio-economic development, the current state of public transport systems in most African cities generates more questions than answers, in terms of their inclusiveness, efficiency, safety and sustainability. However, in response to these challenges, most governments in Africa have initiated steps to reform their public transport systems with support from multilateral organizations such as the World Bank and the African Development Bank (ADB). For example, since 1987, the implementation of the Africa Transport Policy Program (SSATP)—which is an international partnership of 40 African countries, Regional Economic Communities, continental institutions, the United Nations Economic Commission for Africa, public and private sector organizations, and international development agencies and organizations—aims at strengthening policies and strategies to promote efficient, safe and sustainable transport for the people in Africa. Through the activities of SSATP, attempts have been made to develop a robust and transparent regulatory and institutional framework which is an essential condition for improving the performance of transport operators and efficient use of the transport infrastructure. This has become essential because, in most African countries, the regulatory and institutional framework for public transport is inadequate. National transport *policy* statements have generally failed to address the fundamental issues facing the sector such as pricing, the extent to which competitive markets in transport operations should be encouraged, the purpose and scope of regulatory and licensing controls and the respective roles of government and the private sector.

Klopp et al. (2019) suggest that Africa's urban transport sector have received local and global capital and in the process has created change with the focus on creating

[1] This description is misleading primarily because the transport itself is not informal but its provision is embedded in informal processes that also involve state and other actors (Klopp et al., 2019).

more people-centered public transport-oriented development. This, according to Klopp et al. (2019), calls for reorganizing the relationship of the popular transport sector, land use and government that leads to stronger outcomes for the public interest including improved service and access. This change also requires improved metropolitan governance including transparent and integrated regulation over transportation and land use. Consequently, governments in Africa are working to address the existing institutional and political constraints towards models of transportation investment and planning and implementing reforms around public transport and related land use improvements (Poku-Boansi and Marsden 2018; Lindau et al. 2014; Klopp 2012; Klopp and Mitullah 2016). The growing investments in mass transit systems in Africa's major cities (Poku-Boansi and Marsden 2018) are in commuter rails and bus rapid transit (BRT) systems. Currently, few mass rapid transport systems exist in Africa even though many projects have been planned. Though laudable, the implementation of BRTs has encountered many difficulties after their launch (Poku-Boansi and Marsden 2018; Asimeng 2021). The object of promoting mass transit is to ensure efficient, safe, reliable and inclusive public transport systems that promote ness.

From the foregoing, one can argue that the key to Africa's public transport reforms is clear policy statement. This is because, without clear policy statements to give direction to decision-makers and operators, countries in Africa will lack a solid basis for countering non-competitive practices by transport operators and for encouraging private investments in the transport sector.

The three chapters in this section of the book are set within the context of examining the problems with the implementation of public transport reforms, especially with BRTs in African cities, unpacking the politics around the pricing of public transportation services and exploring how technological advancements can and will make public transport system safe and secure in African cities. Each contribution relies on various sources of data and complementary methodologies to provide in-depth analysis and understanding of the politics associated with public transport service pricing, reform implementation and governance regimes.

References

Adarkwa KK, Poku-Boansi M (2011) Rising vehicle ownership, road way challenges, and traffic congestion in Kumasi. In: Adarkwa KK (ed) the future of the tree: managing the growth and development of Kumasi. University Press, Kumasi, pp 128–152

Agyemang E (2015) The bus rapid transit system in the greater accra metropolitan area, Ghana: looking back to look forward. Nor Geogr Tidsskr Nor J Geogr 69:28–37. https://doi.org/10.1080/00291951.2014.992808

Amoako C, Cobbinah PB, Niminga-Beka R (2014) Urban infrastructure design and pedestrian safety in the Kumasi Central business district, Ghana. J Transp Saf Sec 6(3):235–256

Aschauer DA (1989) Is public expenditure productive? J Monet Econ 23(2):177–200. https://doi.org/10.1016/0304-3932(89)90047-0

Asimeng ET (2021) Bus rapid transit implementation with the inclusion of incumbent paratransit operators in African cities: lessons from Accra, Urban. Plan Transp Res 9(1):534–548. https://doi.org/10.1080/21650020.2021.2000485
Behrens R, McCormick D, Mfinanga D (2016) An introduction to paratransit in Sub-Saharan African cities. In: Paratransit in African cities: operations, regulation and reform. Routledge, Taylor & Francis, London, New York, pp 1–25
Cervero R (1996) Mixed land-uses and commuting: evidence from the American housing survey. Transp Res A 39(5):361–377
Cervero R, Golub A (2007) Informal transport: a global perspective. Transp Policy 14:445–457. https://doi.org/10.1016/j.tranpol.2007.04.011
Cervero R (2000) Informal transport in the developing world. United Nations Habitat. http://unhabitat.org/sites/default/files/download-managerfiles/Informal%20Transport%20in%20the%20Developing%20World.pdf
Cervero R (2013) Bus rapid transit (BRT): an efficient and competitive mode of public transport. In: URD working paper 2013–01. 20th ACEA scientific advisory group report. escholarship.org/uc/item/4sn2f5wc.pdf. Accessed 2 Sep 2015
Cohen B (2006) Urbanization in developing countries: current trends, future projections, and key challenges for sustainability. Technol Soc 28(2006):63–80
Darkwah RM, Cobbinah PB (2014) Stewardship of urban greenery in an era of global urbanisation. Int J Environ Ecol Geol Geophys Eng 8(10):671–674
Dell'Olio L, Ibeas A, Cecín P (2010) Modelling user perception of bus transit quality. Transp Policy 17(6):388–397
Dubé J, Des Rosiers F, Theriault M, Dib P (2011) Economic impact of a supply chain in mass transit in urban areas: a Canadian example. Transp Res Part A Policy Pract 45(1):46–62
Klopp JM (2012) Towards a political economy of transportation policy and practice in Nairobi. Urban Forum 23:1–21. https://doi.org/10.1007/s12132-011-9116-y
Klopp J, Mitullah W (2016) Politics, policy and paratransit: a view from Nairobi. In: Behrens R, McCormick D, Mfinanga D (eds) Paratransit in African cities: operations, regulation and reform
Klopp JM, Harber J, Quarshie M (2019) A review of BRT as public transport reform in African cities. In: VREF research synthesis project governance of metropolitan transport background paper
Kumar A, Zimmerman S, Agarwal OP (2012) International experience in bus rapid transit (BRT) implementation: synthesis of lessons learned from Lagos, Johannesburg, Jakarta, Delhi and Ahmedabad case studies. The World Bank Institute, Washington, DC
Lindau LA, Hidalgo D, de Almeida LA (2014) Barriers to planning and implementing bus rapid transit systems. Res Transp Econ 48:9–15
May A, Marsden G (2010) Urban transport and mobility: transport and innovation unleashing the potential (online). International Transport Forum. https://www.itfoecd.org/sites/default/files/docs/10highlights.pdf
Njoh AJ (2000) Transportation infrastructure and economic development in Sub-Saharan Africa. Publ Works Manag Policy 4(4):286–296. https://doi.org/10.1177/1087724X0044003
Obeng-Atuah D, Poku-Boansi M, Cobbinah PB (2016) Pedestrian crossing in urban Ghana: safety implications. J Transp Health. https://doi.org/10.1016/j.jth.2016.06.007i
Pereira AM, Andraz JM (2012) On the regional incidence of highway investments in the USA. Ann Reg Sci 48:819–838. https://doi.org/10.1007/s00168-010-0409-1
Poku-Boansi M (2020) Path dependency in transport: a historical analysis of transport service delivery in Ghana. Case Stud Transp Policy. https://doi.org/10.1016/j.cstp.2020.07.003
Poku-Boansi M (2021) Contextualizing urban growth, urbanisation and travel behaviour in Ghanaian cities. Cities 110:103083. https://doi.org/10.1016/j.cities.2020.103083
Poku-Boansi M, Adarkwa KK (2013) The determinants of demand for public transport services in Kumasi, Ghana. J Sci Technol 33(3)
Poku-Boansi M, Marsden G (2018) Bus rapid transit systems as a governance reform project. J Transp Geograp 70:193–202. https://doi.org/10.1016/j.jtrangeo.2018.06.005

Poku-Boansi M, Amoako C, Obeng Atuah D (2019) Urban travel patterns and safety among school children around Accra, Ghana. J Transp Health https://doi.org/10.1016/j.jth.2019.100660
Rahman MS, Timms P, Montgomey F (2012) Integrating BRT systems with rickshaws in developing cities to promote energy efficient travel. Procedia-Soc Behav Sci 54:261–274
Randheer K, Al-Motawa AA, Vijay PJ (2011) Measuring commuters' perception on service quality using SERVQUAL in public transportation. Int J Market Stud 3(1):21
Turok I, McGranahan G (2013) Urbanization and economic growth: the arguments and evidence for Africa and Asia. Environ Urban 25(2):1–18
World Bank (2000) Cities in transition: world bank urban and local government strategy. World Bank Policy Paper, Washington, DC

Michael Poku-Boansi is a Professor of Planning at the Department of Planning, and the Deputy Director, School of Graduate Studies, Kwame Nkrumah University of Science and Technology, Kumasi, Ghana. Michael has over fifteen (15) years' of transportation and land use planning, resettlement planning, climate change and resilience planning experience spanning both academic and private sector appointments. Professor Poku-Boansi is a Fellow of the Rockefeller Foundation Professional Fellowship Programme and a Commonwealth Academic Fellow. He has authored/co-authored several peer-reviewed research articles published in reputable international journals and numerous project reports. Prof. Michael Poku-Boansi is also a Fellow of the Ghana Institute of Planners (FGIP).

Chapter 12
Bus Rapid Transit Implementation in African Cities: The Case for a More Incremental Reform Approach

Emmanuel Theodore Asimeng

Abstract Self-regulated paratransit services have become the main means of public transport in African cities. The rapid implementation of bus rapid transit (BRT) in Bogotá and other Latin American cities has attracted the interest of African cities to reform the ubiquitous paratransit services to regulate the public transport sector and ensure mass transportation. The approach in African cities has mainly supported the incumbent paratransit operators to become operators of the new service in an incremental corridor-by-corridor manner. While this approach avoids incumbent operators' aggressive resistance, it has encountered low interest from the incumbent paratransit operators. Despite the need for a mass transport service like BRT, implementation in African cities has not achieved the expected outcomes. This article argues for a more incremental approach that follows the "reverse product life cycle" concept. The arguments are substantiated by evidence from paratransit reform studies in African cities. This new approach takes into consideration (1) the time required to improve the capacity and competence of national and sub-national governments and incumbent paratransit operators and (2) spreading the financial capital required for governments and incumbent operators. This new approach makes a case for public transport reform implementation institutions to avoid the difficult task of developing a new service in the form of BRT from the onset. Instead, the approach advocates for gradually improving the existing service until a new mass public transport service like BRT is eventually realised.

Keywords African cities · Bus rapid transit · Incremental planning · Paratransit operators · Reverse product life cycle

---

E. T. Asimeng (✉)
German Institute Of Development and Sustainability (IDOS), Bonn, Germany
e-mail: theodore.asimeng@idos-research.de; theodore.asimeng@gmail.com

Institute of Urban and Regional Planning, Technische Universität Berlin, Berlin, Germany

R. A. Acheampong et al. (eds.), *Transport and Mobility Futures in Urban Africa*,
The Urban Book Series, https://doi.org/10.1007/978-3-031-17327-1_12

12.1 Introduction

Public transport in African cities is characterised by a mix of formal bus services, informal small and medium bus services, rail lines, motorcycles, and auto rickshaws. Small and medium-sized self-regulated buses operating in an unscheduled manner have become the main means of public transport in most African cities. The operations of these mini and midi buses, often called paratransit, have grown over the years due to their flexibility and demand responsiveness, low-cost services without subsidy from governments, ability to serve areas neglected by formal services, and collapse of formal bus services (Behrens et al. 2016; Cervero 2000). These paratransit services locally called *daladala* in Dar es Salaam, *matatu* in Nairobi, *danfo* and *mule* in Lagos, *trotro* in Accra, and *minibus-taxi* in Johannesburg provide more than 50% of public transport in most African cities (Behrens et al. 2016). Despite the benefits of paratransit services, their focus on monetary gain at the expense of traffic regulations, coupled with weak to no regulation from public officials, has resulted in aggressive competition for passengers, resulting in fatal accidents. In recent times, public officials have sought to regulate the ubiquitous paratransit operations to address the adverse effects of their operations and implement an efficient mass transport service for the growing urban population. An efficient mass public transportation system would likely reduce present and future demand for private cars and unsafe public transport services such as motorcycles and auto rickshaws.

Public institutions in some cities have considered bus rapid transit (BRT) as the appropriate response to reform paratransit services through regulation of the existing operators and implementing a bus-based mass transport system without the disadvantages of paratransit, but with similar advantages of rail services. Since BRT introduction in Curitiba and the quick implementation in Bogotá, many cities have adopted it to an extent where it has been implemented in 182 cities worldwide (Global BRT Data 2022). Africa has BRT in five cities with 131 km, whereas Latin America has BRT in 61 cities with 1,960 km. Following the implementation success of BRT in Latin American cities, several African cities have sought to do the same. There are ongoing reforms with BRT at various stages in Addis Ababa, Dar es Salaam, Kampala, Nairobi, Lagos, Accra, and many cities in South Africa (Schalekamp 2017).

As experienced with the introduction of new technologies, BRT implementation in the Global South has faced institutional, financial, legal and political problems (Lindau et al. 2014). While African BRTs have sought to follow BRT examples in Latin American cities, they have not achieved the same implementation success. African BRTs have faced opposition and low interest from incumbent paratransit operators, low financial capital investment from both paratransit operators and public institutions, and governance issues (Asimeng and Heinrichs 2020; Klopp et al. 2019; Poku-Boansi and Marsden 2018). As a result of the problems, some scholars have questioned the intentions of international bodies that advocate for BRT in African cities (Rizzo 2015; Wood 2015). Yet, BRT remains the cheapest mass transportation option available to some African cities compared to metro rail and tram (Deng and Nelson 2011).

This chapter aims to analyse the problems with BRT implementation in African cities to provide a plausible approach based on the reverse product life cycle (RPLC) concept. This approach captures BRT implementation problems in African cities for improved mass transportation in the future. Many studies and reports have identified the problems with BRT implementation in Africa, and this chapter uses this literature. However, very few have sought to identify a different feasible approach for BRT implementation. Such an approach is likely to rekindle public transportation institutions interests in mass transportation and situate BRT as a plausible public transportation option for the future of African cities.

This chapter does not focus on the political context in which BRT implementation is done but highlights BRT features and components that have made it challenging to implement in African cities. The following section briefly describes BRT and the components that differentiate it from conventional bus services. The chapter then discusses BRT implementation approaches that have served as examples for African cities. BRT implementation in three African cities and the challenges contributing to a stall in implementation are then described. This is followed by considering a different approach for BRT implementation in African cities that incorporates the identified problems. Finally, the chapter concludes with an outlook on how an efficient mass transport system like BRT would shape future transportation in African cities.

12.2 The Components of BRT

BRT has been defined by Wright and Hook (2007) as a high-quality bus-based transit system that delivers fast, comfortable, and cost-effective urban mobility through a segregated right of way infrastructure and rapid and frequent operations. It can also be considered a new public transport system that combines the speed and reliability associated with rail transport but with conventional buses' flexibility and lower cost (Deng and Nelson 2011). BRTs have, therefore, become popular globally due to the ability of city authorities to implement quickly at lower to moderate cost and the possibility to include existing paratransit operators (Deng and Nelson 2011; ITDP 2017; Wright and Hook 2007). Essentially, the idea of BRT is for people to think of trains but see buses instead.

BRT has certain components that set it apart from other bus-based transit systems. As outlined in Table 12.1 are the characteristics that make BRT efficient as rail transport but cheaper to implement. For BRTs to be as efficient as rail, it needs to have dedicated lanes disallowed to other motorists. However, because of limited space, cost of infrastructure, legal and other problems, some BRTs share a part of the route with other vehicles.

BRT, through its components, provide a means to improve bus-based mobility system performance. It thus provides advantages in the form of:

Table 12.1 Components and characteristics of BRT systems

Component	Characteristic
Running ways	Vehicles operate in exclusive bus ways, dedicated bus lanes or may operate in mixed traffic
Stations/stops	Sufficient shelter from inclement weather ranging from enhanced shelters to large transit centres
Vehicle type	High-capacity buses with a level of passenger comfort and non-polluting
Services	A variety of service alternatives, including all stop route(s), limited stop service, feeder services
Fare collection	Multi-door boarding for customers with prepaid fare
ITS	Adoption of technology to enhance the safety, comfort and reliability of the BRT

Deng and Nelson (2011)

- Reduced travel time: time spent waiting for a vehicle, in the vehicle for passengers to transfer from one vehicle to the other, and time spent at each stop or transfer is enhanced.
- Reliability: passengers have reliable bus arrival and departure periods.
- Image and identity: BRTs are well branded to the public with well-crafted communication. Therefore, the public knows the company responsible for service and can directly report safety, security, and misconduct issues for improvement of the service.
- Cashless pre-boarding payment: while electronic fare collection allows for prepayment and cashless transactions, it also provides transparency on the revenue generated.
- Mass transport: large new buses with routine maintenance used for BRTs allow mass transportation comfortably at lower environmental pollution levels than paratransit.
- Technology: the use of intelligent transport systems (ITS) in BRT enhances safety, comfort and reliability through surveillance and tracking of the vehicles. In most cases, to avoid reckless driving, the BRT companies are paid by distance covered as reported by a tracking device.

Since it is not possible for some cities to meet all the components and characteristics of BRT, a BRT standard has been developed to provide a common definition, a scoring system, and a planning tool for BRTs worldwide (ITDP 2014). The BRT standard certifies BRT corridors as Basic, Bronze, Silver or Gold, placing it within a hierarchy of international best practices. BRT Basic is regarded as the minimum precondition to receiving Bronze, Silver or the Gold ranking. Some corridors may be awarded Basic as part of meeting some of the core elements of BRT. Bronze standard BRT has characteristics that elevate it above the Basic standard as it achieves better operational efficiencies or quality of service. The Silver standard BRT meets most of the international requirements and is likely to be cost-effective with sufficient demand to meet the relatively higher cost of implementation. The Gold standard meets all the international requirements. They achieve the highest operational performance and quality of service but may cost a little more than the silver standard. The Gold-rated BRT is often recommended for cities.

12.3 BRT Implementation Approaches

Two main timescales have been adopted for BRT implementation—the incremental corridor-by-corridor and abrupt city-wide implementation. With the city-wide approach, the BRT is designed to commence operations at one go for the entire city. This has been referred to as the "big bang" approach. The incremental approach is based on a corridor-by-corridor implementation and operation until enough corridors connect the entire city. Another critical but delicate decision to BRT implementation is including or excluding existing paratransit operators. Non-inclusion of incumbent operators may result in faster implementation, allow for competitive bidding and selection of operators with the requisite investment for efficient services at a lower cost. However, the non-inclusion of incumbent operators may result in strong opposition to the BRT and may cause violence, as observed in Quito (Chrustie et al. 2010).

Despite having different historical, spatial and economic resources for governing public transportation, African city institutions responsible for BRT implementation have been inspired by Latin American BRTs. This is because BRT was pioneered in Latin America and has achieved extensive implementation in Latin American cities. African institutions and stakeholders visit Latin American cities to learn the implementation process and operational mechanisms to adapt to the local conditions in their respective cities (McCaul and Ntuli 2011; Mobereola 2009; Wood 2015).

But three distinct methods can be elicited from BRT implementation in Latin American cities based on the incremental or big bang approach and the inclusion or exclusion of incumbent paratransit operators. The first method is an extensive incremental approach with the inclusion of existing operators. This method was adopted in Curitiba; the city is regarded to have pioneered BRT in the 1970s. The city implemented a bus system that did not have all the components that would make it a BRT, but other components were gradually included and improved over the period and could be called a full BRT in the 2000s (Lindau et al. 2010). For instance, the system started with a cash-based payment mechanism per route. This was improved to an integrated payment system where a single payment system was possible for transfers between trunk and feeder. After some years, an electronic card-based system was installed in the 2000s to replace the paper and coin cash system. Today, Curitiba's BRT system has most of the components in Table 12.1 with 7 corridors rated silver and gold standards and operates on a 74 km trunk network moving about 721,000 passengers per day (Global BRT Data 2022).

The second method is what has become popular among BRT implementing agencies. It is an incremental method with the inclusion of existing operators. Unlike Curitiba, Bogotá implemented BRT with all the components depicted in Table 12.1

in about 4 years (Ardila-Gomez 2004; Ardila 2005; Hidalgo and Graftieaux 2007b). This popularised the technology for its adoption globally. In contrast to the BRT in Curitiba, the BRT in Bogotá showed that BRTs could be implemented quickly to suit politicians' interest and demonstrate to electorates that a new bus system can be planned and implemented within a government's term of office.

The third method is the "big bang" approach with an open bidding process for existing operators and potential new operators. While the plan for the BRT in both Curitiba and Bogotá was to implement BRT city-wide but commence operations on an incremental corridor-by-corridor basis, in Santiago de Chile, the plan was to develop a city-wide BRT but to start operations on a single day. After 6 years of designing, constructing and installing technology for the BRT, it was scheduled to commence operations across the city and integrated with the light rail on 10th February 2007 (Hidalgo and Graftieaux 2007a; Muñoz and Gschwender 2008). At the start of operations, there were issues with the required number of buses, electronic payment, and GPS systems affecting smooth operations of the BRT. These problems were gradually addressed over the period to realise improvements in performance standards as planned by the city authorities.

As depicted in the cases, BRT implementation can take a long or short period, depending on the adopted approach and local conditions. The option of a city-wide "big bang" approach is more capital intensive and may encounter many unforeseen challenges in contrast to incremental corridor-by-corridor implementation. As a result, the corridor-by-corridor approach is method that has been used in African cities.

12.4 BRT Implementation in African Cities

BRT implementation has not had the same success compared to those in Latin American cities. Generally, the approach adopted by African cities has sought to follow the Bogotá approach of short-term implementation using incremental corridor-by-corridor and inclusion of existing operators. The inclusion of existing paratransit operators has become mandatory due to the large number of people who find employment in that sector. Therefore, any attempt to exclude them would make the government unpopular. The following section focuses on BRT implementation in Lagos, Johannesburg and Accra, emphasising why there has not been a commensurate success as their Latin American counterparts.

12.4.1 Lagos

The Lagos BRT-Lite which is the first BRT in Africa commenced service provision in 2008. It initially operated on a 22 km trunk only corridor with about 60% segregated bus lane (Mobereola 2009; Otunola et al. 2019) until 2020, when it was extended by 13.6 km (Abiodun 2020). The World Bank financially supported the BRT implementation through the Lagos urban transport project. The BRT commenced with the incumbent paratransit association, National Union of Road Transport Workers (NURTW) and existing formal bus company LAGBUS operating buses regulated by the Lagos Metropolitan Area Transport Authority (LAMATA). This BRT operates as a trunk-only system without removal of the existing paratransit fleet from the corridor. Payment is done per number of passengers with both paper tickets and an electronic payment system. Since its implementation, it has not met the lower tier BRT Basic standard and therefore not technically considered as BRT. Table 12.2 shows the components of the Lagos BRT. During the early years of its service provision, it was said to move about 200,000 passengers per day (Fashola 2009), but a later study shows that it commutes about 90,000 passengers per day (Klopp et al. 2019).

A significant challenge for current and future expansion and more patronage of the BRT Lite has been the role of incumbent operators and public institutions. Firstly, public officials realised that any public transport reform without the participation of incumbent paratransit operators would not work, and therefore, their participation was mandatory (Mobereola 2009). However, the paratransit operators have shown

Table 12.2 Components of BRT in Lagos, Johannesburg, and Accra

Components of BRT	BRT-Lite in Lagos	Rea Vaya in Johannesburg	Aayalolo in Accra
Running way	65% exclusive dedicated bus lanes with no scheduled operations	Exclusive dedicated bus lanes and scheduled operations	No dedicated bus lane
Stations/stops	Enhanced shelters	Large transit shelter	Enhanced shelters
Vehicle type and ownership	Large buses owned and leased by the paratransit franchise and LAGBUS	Large buses owned by the bus operating companies	Large buses owned by government and leased to paratransit BOCs
Services	Express and all stop services; trunk only	Express and all stop services; trunk and feeder	Express and all stop services; trunk only in 2016. All stop service only after restart
Fare collection	Prepaid paper ticket and electronic	Prepaid electronic only	Prepaid electronic in 2016, paper tickets after restart in 2019
ITS	Yes, but revenue depends on number of passengers	Yes, with revenue based on distance travelled per bus	Yes, but revenue depends on number of passengers
Removal of incumbent paratransit vehicles	No removal of old paratransit fleet	Existing paratransit fleet removed or relocated from BRT corridor	No removal of old paratransit fleet

limited interest in the BRT. Moreover, the paratransit operators do not have the requisite skills for managing and driving large buses in a formal manner (Orekoya 2010). As a result, though the drivers were trained, the NURTW could not manage their operations as required by formal services leading to the cancellation of their contract after many complaints from commuters (Klopp et al. 2019). Since the paratransit operators were not removed from the corridor and the NURTW was simultaneously responsible for their paratransit operations, they could go back as paratransit workers. Secondly, the NURTW and the paratransit operators do not keep financial records to access loans, neither do they save towards purchasing buses. The lack of capital and savings resulted in inability to procure more buses for the BRT.

Public institutions lack of financial capital for infrastructure development has affected expansion and service quality (Mason-Jones et al. 2012; Orekoya 2010). Though public officials have prioritised BRT among other transport modes to improve mobility in the city, low capital for repairs and improving the existing infrastructure and for expansion has resulted in a snail pace of the BRT. With further financial support from the World Bank, the city was able to extend the BRT lane by 13.6 km. Another challenge facing the expansion of the BRT is weak institutional cooperation and capacity for managing formal transport services while regulating the existing paratransit service (Klopp et al. 2019; Oshodi et al. 2016). LAMATA must cooperate with 16 other local institutions in the city with diverse interests and low formal transportation capacity. Finally, LAMATA has several roles within the urban transport project and the BRT in the form of service provider, road transport industry coordinator and regulator.

12.4.2 Johannesburg

Johannesburg's BRT became operational in 2009 as the first full BRT in Africa (Allen 2013). Rea Vaya is modelled after the Bogotá BRT, and the BRT standard rates it as Bronze with trunk lines and feeder routes. Rea Vaya currently operates on a 43.5 km segregated lane with scheduled operations, and about 42,000 passengers per day use it. Unlike the mini and midi buses used by individuals for paratransit services, large buses operated by newly formed bus companies of incumbent paratransit operators are used. The revenue is based on the distance covered and not the number of passengers to curb competition among drivers. Paratransit vehicles were removed and replaced with BRT buses on the corridor through incorporation into the BRT or withdrawal of services with compensation. Table 12.2 depicts the BRT components adopted in Johannesburg. The source of funding for Rea Vaya is mainly from the South African national treasury. The bus companies must ensure that drivers have the requisite skills for operating in a formal manner. Some of the paratransit operators have been trained for managerial and technical skills.

Negotiations between public officials and existing paratransit operators for the first BRT took about 3 years compared to the projected one month of negotiations (McCaul and Ntuli 2011). Subsequent negotiations for other corridors have again taken longer than expected. Despite the long negotiation period, the compensation paid to incumbent paratransit operators to leave the BRT corridors has substantially contributed to the ongoing need for operating subsidy.

Generally, the BRT in Johannesburg has come at a high implementation cost to the government. High ridership was expected to compensate for the cost, but that has not been the case. The urban spatial plan of Johannesburg from the apartheid regime has led to the situation where the city has a lower density than South American cities. Therefore ridership is not as high as Latin American BRTs (Scorcia and Munoz-Raskin 2019). This means that the BRT must be subsidised for an extended period contrary to initial plans. Another problem affecting the BRT is the governance of transportation services in the Gauteng region of which Johannesburg is a part. The Provincial government has different goals for the provision and governance of public transport, which is at variance with that of the city government of Johannesburg (Klopp et al. 2019).

12.4.3 Accra

One of the main aims of the Ghana urban transport project that commenced in 2008 through support of the World Bank was the implementation of a BRT. A BRT could not be implemented mainly due to cost overruns and financing gaps (The World Bank 2017). Other factors contributing to the shift in BRT implementation are institutional capacity and stiff opposition from the incumbent paratransit operators. In place of the BRT, a conventional bus transit service called *Aayalolo* was implemented with improved bus stops and terminals infrastructure in 2016 (Poku-Boansi and Marsden 2018; The World Bank 2017). The existing paratransit associations formed three bus companies to operate on a 22 km trunk route in mixed traffic with other motorists. The components of BRT that were implemented in Accra are shown in Table 12.2.

The government bought buses and leased them to the operators for the service. An electronic pre-board fare system was used for the service and was planned to have a scheduled start from the terminals but compete with other motorists on the road. The paratransit fleet aggressively competed with the bus service especially at bus stops where passengers wait to board either the Aayalolo or *trotro*. The plan was not to completely remove the paratransit services (Finn et al. 2009). But the absence of dedicated lanes meant the Aayalolo spent much time in traffic congestion which affected the planned schedule and commuters patronage in favour of the existing paratransit services due to their smaller fleet size. Ultimately, the Aayalolo stopped operations in September 2018 due to low patronage and high operational cost (Asimeng 2021). Later in April 2019, it restarted operations on a different route without scheduled transit but on a "fill and go" basis. The Aayalolo currently operates during morning and evening peak hours.

Although the paratransit associations formed companies to operate the Aayalolo, limited interest and trust in government-led reforms resulted in a decision to keep their services in competition with the Aayalolo. Because the paratransit operators and associations do not keep records of their income to access loans and simultaneously lack the financial capacity to purchase new buses for mass transport, they cannot contribute to the investment capital required for formal bus services. Another problem has to do with their lack of competence for formal operations. As a result, the Greater Accra Passenger Transport Executives (GAPTE), the institution formed to regulate the Aayalolo services hired personnel to manage the bus companies for the paratransit associations that had formed them.

GAPTE and other institutions responsible for the BRT implementation in Ghana have encountered problems of capital investment for infrastructure, institutional and governance issues, and competence for regulating incumbent paratransit operators. The pilot's planned corridor was to have a flyover, which was deemed too expensive to build after estimations were done, prompting the selection of a different corridor, but without dedicated lanes (The World Bank 2017). The BRT implementation resulted in the creation of new laws and institutions including the Greater Accra Passenger Transport Executive (GAPTE). The existing institutions required re-organisation to deliver on the BRT. Since governance processes that flow out of change emerge over time, reforming an existing paratransit system with low interest in the reforms would likely overwhelm the process. As part of the BRT implementation, 13 metropolitan and municipal assemblies (MMDAs) with varying capacities were to regulate the paratransit operators while GAPTE manages the cross-jurisdictional issues. Some of these MMDAs do not have the personnel for public transport and therefore cannot regulate the services of the paratransit operators.

12.5 A New BRT Implementation Approach for African Cities

Utilising the product life cycle concept, Venter (2013) described the paratransit sector in Johannesburg as having reached a period of stasis and maturity requiring new ideas for growth, for which BRT is the answer. This is true for Johannesburg and most African cities where mini and midi buses have become the main means of public transportation. The introduction of BRT technology provides the means for addressing the problems of paratransit while meeting the mass transportation requirements of the growing urban population efficiently and effectively. However, the BRT implementation approach modelled after the Bogotá BRT, where government institutions develop new institutions, infrastructure, regulations and re-organise incumbent paratransit operators within a few years to commence transit services, has proven challenging to implement in African cities. Essentially, the Bogotá approach has become a "big bang" approach making it burdensome to implement. Moreover, it has exposed national and sub-national governments' inadequate capital investments, capacity, and competencies to undertake such reforms in African cities. BRT implementation in African cities has also revealed the challenge of transitioning incumbent paratransit operators to become operators of formal transport services.

Despite these challenges, a bus-based mass transport system is the cheapest service that can be implemented with the inclusion of informal operators while reducing dependence on paratransit services. Though efficient and less expensive mass transportation is increasingly becoming an inevitable requirement in African cities, conventional bus services in mixed traffic as observed in Accra cannot meet this need. BRT and its components are relevant to meet the current and future mass transportation needs. A new approach for BRT implementation in African cities is required. Because of the problems of BRT implementation in African cities, this new approach must consider (1) the time required to improve the capacity and competence of national, sub-national governments and incumbent paratransit operators and (2) spreading the financial capital required for governments and incumbent operators over a more extended period. Thus, a more incremental approach to BRT implementation with the inclusion of existing operators is required.

A slight detour on the product life cycle concept is necessary to develop new ideas for public transport improvement. As shown in Fig. 12.1a, the product life cycle concept stipulates that products and services go through four stages, i.e. introduction, growth, maturity and finally decline (Day 1981; Rink and Swan 1979). Introducing new products or services generally requires high capital investment to get to the growth stage where the product attracts more consumers, increasing sales and revenue. At the maturity stage, the product is well known and achieves saturation due to having no new customers and competition from other providers. The product finally declines and is then removed from the market or replaced by others. However, firms prefer to avoid the decline stage and institute innovative measures to prolong the maturity stage (Hisrich et al. 2017; Levitt 1965). Measures to prolong the life cycle include promoting more usage among current users, developing varied uses among current users, creating new users by expanding the market, and finding new users. An alternative way to avoid the decline stage or ensure that the growth stage is continued is what Barras (1986, 1990) calls the reverse product life cycle (RPLC). The RPLC concept describes how incremental innovations aimed at improving the efficiency and effectiveness of a service can prolong and eventually lead to an entirely new service. The first stage of the reverse product life cycle involves improving the efficiency of delivering an existing service (Fig. 12.1b). The second stage involves using technology to improve the quality of the service, and at the third stage, technology assists in generating a wholly transformed or new service (Barras 1986).

BRT implementation following the Bogotá approach can be likened to initiating a new public transport service with operators of an old service at the decline stage of the product life cycle. As already demonstrated, the introduction of the new service in the form of BRT is burdened with many problems. A plausible approach that considers the problems of paratransit and BRT implementation challenges is the RPLC concept. By applying the reverse product cycle concept, innovative ideas for improving the declining paratransit service can be adopted to change the service delivery incrementally until a new service in the form of BRT is realised. Table 12.3 shows how the reverse product cycle concept can be adopted to improve the existing paratransit service until dedicated lanes are developed for BRT purposes.

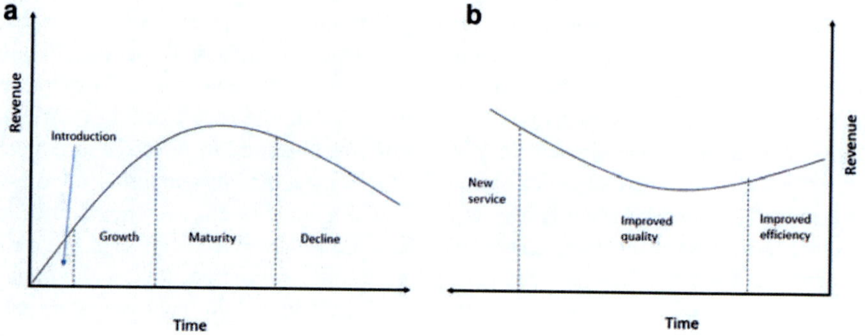

Fig. 12.1 **a** The product/service life cycle **b** the reverse product/service life cycle

Table 12.3 Components of the reverse product cycle approach for BRT reforms

	Initial phase	Second phase	Final phase
Innovation of existing services	Institutional capacity development; regulation of paratransit operations; Government supports paratransit operators to purchase larger buses for peak hours mass transportation	Mini-bus operators are supported to connect to the large buses; counter flow of buses; improvement in working conditions like formal workers	Electronic payment; dedicated bus lanes; development of more routes
Aim	Paratransit to be conversant with regulation of their service and improving efficiency of the current service through large buses	To improve the quality of the service and improve labour conditions to attract more paratransit operators	To have a new transport service in the form of BRT to differentiate it from paratransit services

The incremental reform could be envisaged through the following process: Since institutional capacity is required to regulate the paratransit services, the first phase ensures a clearly defined public transport framework with delineated institutional functions at the city or urban level. An institution(s) empowered to regulate the paratransit operators, communicate government policies effectively with leaders of the paratransit operators and coordinate the overall reforms should be created if non-existing already. This institution(s) should communicate with the government on paratransit operators' financial capacity and develop a scheme where the government supports paratransit operators to purchase new larger buses for trunk services in an unscheduled manner. The trunk services could be done at morning and evening peak hours only for a start. Skills development of paratransit operators is essential at this phase. Drivers should be trained on driving larger buses. At the same time, paratransit regulators should be trained on managing public transportation as required in formal services and identified vehicle owners trained on financial management and loan acquisition requirements and procedures.

At the second phase, existing paratransit services along the trunk lines of the large buses are encouraged to provide feeder services. At the same time, options should be considered to ensure the trunk lines provide faster services. This could be in the form of counterflow to the traffic where possible as traffic direction in morning and evening peak hours are often in one direction whereas the opposite route often has less vehicle traffic. The conditions of service should be improved like that of formal workers to ensure that this service is attractive to other paratransit operators while sustaining the operators. In the third phase, the civil and digital infrastructure required to make the bus transit system efficient in the form of dedicated bus lanes and digital payment can be added. Whereas it may be prudent to consider dedicated lanes for only the trunk lines, digital payment can begin with the trunk lines and extended to the feeder when there is a clear trunk to feeder arrangement.

The incremental process described above should be designed according to the local conditions of the city that intends to adopt it. Applying the reverse product cycle to BRT implementation avoids the high initial capital required for infrastructure while ensuring institutional and capacity development. High capital investment is not necessary from the onset, but rather innovations in the existing paratransit operators can be implemented incrementally until a full BRT is realised. The ideas in this approach are not new but can be considered modified from the BRT implementation in Curitiba. When Curitiba commenced its bus reforms in the 1970s, it took about 20 years to attain the efficiency of BRTs. African cities can emulate that and not the quick fix Bogotá approach, which has proven challenging to implement and may never get to the desired status to address the transportation problems for which BRTs are designed to change.

BRT implementation following the reverse product cycle concept does not consider BRT as a quick fix to the problems of incumbent paratransit operators but rather a gradual means of improving the existing service until the desired goal of BRT is achieved. Table 12.3 shows how the components of such a BRT could be implemented. This can be adapted to suit the capacity and resources of African cities while considering the difficulty of incorporating incumbent paratransit operators for formal services.

12.6 Conclusion

The future of transport and mobility in African cities depends mainly on the present decision of public institutions. The option of no action on public transport would likely result in more private car use, motorcycles, auto-rickshaw, and stagnation of the paratransit sector. The decision to reform the paratransit sector to a formal mass transport service like BRT would revolutionise transportation and mobility in African cities, ensuring stringent regulation of unsafe transport modes such as motorcycles and auto-rickshaws. It would also ensure a reduction in the patronage of paratransit services on trunk lines compelling them to provide last mile and feeder services to the BRT. There is also the likelihood of a reduction in private car use in favour of mass public transport like BRT due to the benefits of reduced travel time, safety and comfort, lower cost of travel, and access to transportation by the elderly and disabled persons often excluded from paratransit services.

Because of the benefits of mass transport services like BRT in African cities, this chapter advocates for a shift from the quick fix Bogotá approach, which has not achieved much success in BRT implementation in African cities, to the more incremental approach based on the RPLC concept. The shift in approach is necessary due to the challenges of incorporating existing paratransit service providers, lack of management skills for formal services, and lack of financial capital for investing in BRTs. Secondly, institutions that have been set up to implement BRT have encountered financial, and governance issues are often overwhelmed with multiple functions beyond their capacity. Both Accra and Nigeria could not implement BRT like Bogotá due to these challenges. Though BRT implementation in Johannesburg can be likened to Bogotá in terms of conception to implementation and following the BRT standard, it has come at a high cost that most African countries' public transportation budget cannot support. The incremental approach based on the RPLC concept offers a way for realising the much-needed relatively cheaper and easy to implement mass transport service without the burden associated with the Bogotá approach. Moreover, Yussuf (2011) has argued against applying theories developed based on Latin American experiences to the African informal economy due to differences in origin, causes and reasons behind persistence of informality in Africa.

Despite the challenges African cities have faced in implementing BRT, this chapter has suggested a different approach that is more plausible to implement. This approach considers the limited capital, capacity and institutional challenges faced by both incumbent paratransit and public institutions that initiate transport reforms. This approach acknowledges the significant employment opportunities many Africans have found within the paratransit sector and argues that drastic reforms are unnecessary. Additionally, this approach considers that the incumbent paratransit operators have low interest in formal operations and even resist their participation but continuous engagement for co-production of incremental solutions is likely to soften their position. Finally, the time required for institutional development and re-organisation for proper governance of transportation in African cities is captured by this approach. The main drawback of this approach is that political leaders may not have the opportunity for implementation within their term of government because more time is required to fully realise the BRT.

References

Abiodun K (2020) BRT and public transportation in Lagos. This Day Newspaper. Retrieved from https://www.thisdaylive.com/index.php/2020/09/25/brt-and-public-transportation-in-lagos/

Allen H (2013) Africa's first full rapid bus system: the Rea Vaya bus system in Johannesburg, Republic of South Africa. Case study prepared for the Global Report of Human Settlements. Retrieved from http://www.unhabitat.org/grhs/2013%0Ahttp://unhabitat.org/wp-content/uploads/2013/06/GRHS.2013.Case_.Study_.Johannesburg.South_.Africa.pdf

Ardila-Gomez A (2004) Transit planning in Curitiba and Bogota. Roles in interaction, risk, and change. Massachusetts Institute of Technology. Retrieved from https://dspace.mit.edu/handle/1721.1/28791

Ardila A (2005) Study of urban public transport conditions in Bogota. Retrieved from https://ppiaf.org/sites/ppiaf.org/files/documents/toolkits/UrbanBusToolkit/assets/CaseStudies/full_case/Bogata.doc

Asimeng ET (2021) Bus rapid transit implementation with the inclusion of incumbent paratransit operators in African cities: lessons from Accra. Urban Plann Transp Res 9(1):534–548. https://doi.org/10.1080/21650020.2021.2000485

Asimeng ET, Heinrichs D (2020) Why do paratransit operators resist participation in bus rapid transit? Case evidence from Bogota, Mexico City. Johannesburg Lagos Transp Rev 41(1):115–135. https://doi.org/10.1080/01441647.2020.1818872

Barras R (1986) Towards a theory of innovation in services. Res Policy 15:161–173

Barras R (1990) Interactive innovation in financial and business services: the vanguard of the service revolution. Res Policy 19:215–237

Behrens R, McCormick D, Mfinanga D (2016) An introduction to paratransit in Sub-Saharan African cities. In: Behrens R, McCormick D, Mfinanga D (eds), Paratransit in African cities: operations, regulation and reform. London and New York: Routledge.Taylor & Francis, pp 1–25

Cervero R (2000) Informal transport in the developing world. UN Habitat. Retrieved from http://mirror.unhabitat.org/pmss/getElectronicVersion.aspx?nr=1534&alt=1

Chrustie C, Docherty JS, Lira L, Mahuad J, Gadlin H, Honeyman C (2010) Negotiating Wicked Problems: Five Stories, (Aba)

Day GS (1981) The product life cycle: analysis and applications issues. J Mark 45:60–67

Deng T, Nelson JD (2011) Recent developments in bus rapid transit:aA review of the literature. Transp Rev 31(1):69–96. https://doi.org/10.1080/01441647.2010.492455

Fashola BR (2009) Transformational impact of transport projects in Lagos. In: World Bank Transport Forum

Finn B, Arthur BA, Gyamera S (2009) New regulatory framework for urban passenger transport in Ghanian cities. In: 11th conference on competition and ownership in land passenger transport. Delft University of Technology (The Netherlands) 20–25 September 2009, pp 20–25

Global BRT Data (2022) Global BRT data. Retrieved from https://brtdata.org/

Hidalgo D, Graftieaux P (2007a) A critical look at major bus improvements in Latin America and Asia: case study of Transantiago, Santiago, Chile, 1–39. Retrieved from http://siteresources.worldbank.org/INTURBANTRANSPORT/Resources/Hidalgo-Graftieaux-SANTIAGO-CaseStudy.pdf

Hidalgo D, Graftieaux P (2007b) A critical look at major bus improvements in Latin America and Asia: case study of Transmilenio, Bogota, Colombia. Retrieved from http://siteresources.worldbank.org/INTURBANTRANSPORT/Resources/Hidalgo-Graftieaux-BOGOTA.pdf

Hisrich RD, Peters MP, Shepherd DA (2017) Entrepreneurship, 10th edn. McGraw-Hill Education, New York, NY

ITDP (2014) The Brt standard. Retrieved from https://itdpdotorg.wpengine.com/wp-content/uploads/2014/07/BRT-Standard-20141.pdf

ITDP (2017) The BRT planning guide. Retrieved from https://www.sutp.org/files/contents/documents/resources/I_BRT-Planning-Guide/GIZ_SUTP_BRT-Planning-Guide_Complete_4th_EN.pdf

Klopp JM, Harber J, Quarshie M (2019) A review of BRT as public transport reform in African Cities. https://doi.org/10.13140/RG.2.2.29342.79686
Levitt T (1965) Exploit the product life cycle. Harv Bus Rev. Retrieved from https://hbr.org/1965/11/exploit-the-product-life-cycle
Lindau LA, Hidalgo D, Facchini D (2010) Curitiba, the cradle of bus rapid transit. Built Environ 36(3):274–282. https://doi.org/10.2148/benv.36.3.274
Lindau LA, Hidalgo D, Lobo ADA (2014) Barriers to planning and implementing bus rapid transit systems. Res Transp Econ, 1–7. https://doi.org/10.1016/j.retrec.2014.09.026
Mason-Jones K, Cohen B (2012) Case study: Lagos BRT-Lite
McCaul C, Ntuli S (2011) Negotiating the deal to enable the first Rea Vaya bus operating company: Agreements, Experiences and Lessons
Mobereola D (2009) Lagos bus rapid transit: Africa's first BRT scheme. The World Bank, Washington DC
Muñoz JC, Gschwender A (2008) Transantiago: a tale of two cities. Res Transp Econ 22:45–53. https://doi.org/10.1016/j.retrec.2008.05.010
Orekoya T (2010) The bus rapid transit system of Lagos, Nigeria. In: United nations forum on climate change mitigation, fuel efficiency & sustainable Urban Transport, Seoul Korea. Seoul
Oshodi L, Salau T, Seun OM, Emeka UR (2016) Urban mobility and transportation. In: Hoelzel F (ed) Urban Planning Process in Lagos. Heinrich Böll Stiftubg Nigeria and Fabulous Urban, pp 61–132
Otunola B, Kriticos S, Harman O (2019) The BRT and the Danfo: a case study of Lagos ' transport reforms from 1999–2019. IGC, Cities that Work, Case Study. Retrieved from https://www.theigc.org/wp-content/uploads/2019/10/lagos-BRT-case-study.pdf
Poku-Boansi M, Marsden G (2018) Bus rapid transit systems as a governance reform project. J Transp Geogr 70(June):193–202. https://doi.org/10.1016/j.jtrangeo.2018.06.005
Rink DR, Swan JE (1979) Product life cycle research: a literature review. J Bus Res, 219–242
Rizzo M (2015) The political economy of an urban megaproject: the bus rapid transit project in Tanzania. Afr Aff 114(455):249–270. https://doi.org/10.1093/afraf/adu084
Schalekamp H (2017) Lessons from building paratransit operators' capacity to be partners in Cape Town's public transport reform process. Transp Res Part A Policy Pract 104:58–66. https://doi.org/10.1016/j.tra.2017.08.002
Scorcia H, Munoz-Raskin R (2019) Why South African cities are different? Comparing Johannesburg's Rea Vaya bus rapid transit system with its Latin American siblings. Case Studies Transp Policy 7(2):395–403. https://doi.org/10.1016/j.cstp.2019.01.010
The World Bank (2017) Implementation completion and results report on an urban transport project in Ghana. Retrieved from http://documents.worldbank.org/curated/en/937111488510098083/pdf/ICR-Ghana-Urban-Transport-Project-P100619-FINAL-01262017-01312017.pdf
Venter C (2013) The lurch towards formalisation: lessons from the implementation of BRT in Johannesburg. South Africa. Rese Transp Econ 39(1):114–120. https://doi.org/10.1016/j.retrec.2012.06.003
Wood A (2015) The politics of policy circulation: unpacking the relationship between south African and South American cities in the adoption of bus rapid transit. Antipode 47(4):1062–1079. https://doi.org/10.1111/anti.12135
Wright L, Hook W (2007) Bus rapid transit planning guide. Institute for Transportation & Development Policy
Yussuf OS (2011) A theoretical analysis of the concept of informal economy and informality in developing countries. Europ J Soc Sci 20(4):624–636. Retrieved from http://libproxy.albany.edu/login?url=http://search.ebscohost.com/login.aspx?direct=true&db=a9h&AN=67654419&site=eds-live&scope=site

Emmanuel Theodore Asimeng is a Researcher at the German Institute of Development and Sustainability and the Technical University of Berlin. His research interest is sustainable urban transportation. He adopts behavioural and change management concepts to understand informal and paratransit transportation systems and the workers. Asimeng is passionate about measures for improving the quality of urban transportation services. He is also interested in how governments adopt industrial policy to improve transportation systems.

Chapter 13
Pricing Transport Services in Urban Ghana—The Politics and Socio-Spatial Conflicts Between Transport Operators and the State

Michael Poku-Boansi and Michael Osei Asibey

Abstract Dominated by the private sector and several informal operators, the contours of urban transport operation and pricing in many African countries are shaped by complex socio-economic factors, political alliances and local actions. In most cases, national and local governments' attempt to regulate urban transport operations and pricing has failed due to the emerging agencies of local associations of transport operators. This chapter explores the socio-political structures for operating and pricing public transportation services in urban Ghana. The chapter traces the emergence of transport operators' associations and their interests, alliances and conflicts in shaping urban transport pricing. In doing this, the chapter examines the interaction of political and economic processes—the distribution of power and agencies among diverse groups and individuals in the determination of public transportation pricing in Ghana. Using Kumasi, Ghana's fastest growing city, as its primary case, the study adopts qualitative data collection methods including: interview of transport operators' associations, drivers, passengers, and officials of public transport regulation institutions. The chapter reveals a complex relationship between urban transport operation and pricing, shaped by the formation of various informal and semi-formal operators' associations with common interest and values on the one hand; and government direct and indirect intervention on the other. The chapter recommends that actions are needed to create governance arrangements that foster co-production and reduce the tensions among the public sector actors on the one hand and the operators on the other hand.

Keywords Political economy · Pricing · Public transportation · Ghana · Informal institutions

---

M. Poku-Boansi (✉) · M. O. Asibey
Department of Planning, Kwame Nkrumah University of Science and Technology, Kumasi, Ghana
e-mail: mpoku-boansi.cap@knust.edu.gh

R. A. Acheampong et al. (eds.), *Transport and Mobility Futures in Urban Africa*, The Urban Book Series, https://doi.org/10.1007/978-3-031-17327-1_13

13.1 Introduction

Transportation plays a key role in the economic development of every society. It enhances the mobility of people and goods from one geographical area to the other, stimulating urban growth and productivity (Horn and Van Eeden 2018; Marais et al. 2020; Shao et al. 2020). Public transportation has been the backbone of urban mobility in developing countries as about 95% of the population, primarily rely on it for their everyday mobility needs (Wilson 2006).

Additionally, rapid urbanisation without the corresponding provision of adequate transport infrastructure has resulted in numerous transport challenges in urban Africa, including Ghana. These challenges include heavy traffic congestion in urban centres, informal nature of the sector, low public transport services, the use of few high-occupancy vehicles for mass transportation, increased road accidents and air pollution (Agyemang 2009; Oteng-Ababio and Agyemang 2012; Obeng-Atuah et al. 2016; Poku-Boansi et al. 2019). To this end, mobility within urban areas has become less enjoyable and, sometimes, extremely difficult. The urban poor and low-income households are the most affected by the problems of the public transport system (UN-Habitat 2013). In many parts of the Global South, including urban Africa, public transport systems are largely informal, where operators operate outside the officially sanctioned public transport sector. They are thus regarded to be a nuisance requiring public intervention and occasionally, eradication (Poku-Boansi 2020; Poku-Boansi and Cobbinah 2017; Cervero and Golub 2007; Sietchiping et al. 2012; Cirolia and Harber 2021). The informal transport sector has relied primarily on minibuses, taxis, motorcycles and vans (Del Mistro and Behrens 2015; Finn et al. 2011; Olvera et al. 2020) with several benefits and challenges.

The private sector and its transport operations have become a mainstay of Ghana's public transportation system, as it has provided valuable transport services to the citizenry in most cases where the state's intervention to organize, develop and maintain effective public transport services have failed. Public transportation, which is in the form of minibuses, shared taxis and midi-buses, are operated and owned by the private sector. The sector, who to exert their influence, has formed various transport unions, which control their mode of operation and practices aimed at protecting their interest and shared values (Fouracre et al. 1994). Their control of terminals and service quality has given them power in influencing the direction and growth of public transportation in Ghana (Cobbinah 2016).

The emergence of private transport operators and their associations following the Government of Ghana deregulation and privatization policies (Yobo 2014) have given total control of the operations of public transportation to the private sector. Even though the object of the deregulation policy was to ensure efficiency, it has resulted in low quality of service to the public, high cost of patronage, delayed journey times, increased urban traffic and the use of obsolete and uncomfortable vehicles (Oteng-Ababio and Agyemang 2012; Poku-Boansi and Cobbinah 2017; Poku-Boansi 2021). In the wake of these challenges is the issue of public transport pricing that has re-echoed political conflicts between the government on one hand

and the private transport associations on the other hand (e.g., Adams et al. 2014; Amoako et al. 2014; Peprah et al. 2014; Yorgri et al. 2016).

Public transport pricing measures play a key role in acceptability by the citizenry, as has been seen to be a restrictive measure in protecting the public interest by state intervention (Poku-Boansi and Marsden 2018; Okoye et al. 2010). However, the emergence of transport operators' associations has renewed the politics of responsiveness by the state in the welfare of its citizens in the political economy discourse. There have also been renewed interests to protect the interest of transport operators against the harsh economic environment and the direct control of the state in the pricing of transport services in urban Ghana.

This chapter explores the socio-political environment for operating and pricing public transportation in urban Ghana. Available evidence shows complex interaction between actors in the pricing of public transportation. Specifically, there are reported tensions among the public sector actors on the one hand and the operators on the other hand. From the above, it can be deduced that public transport systems in Ghana are the result of complex interactions of levels of government, driven by their agency, ability to raise taxes and overall funding levels and an understanding of the relative roles in public transport delivery. Public transport systems are thus not simply about mobility; however, they provide access to activities creating value to individuals or society. In this paper, therefore, we take a political economy approach to discuss and unpack the politics around the pricing of public transportation services in Ghana.

13.2 The Political Economy of Public Transport Service Pricing

The discourse of political economy has a renewed focus in contemporary times as the issue of the state's concern for public welfare is becoming a political and social responsibility within the socio-economic structures of decision-making. Political economy has been viewed as the "study *of social relations, particularly the power relations that mutually constitute the production, distribution and consumption of resources*" (Mosco 1996; 24). From this perspective of political economy to urban public transportation, citizen satisfaction, reduced journey times and affordable public transport services remain the primary resources of urban transport services.

The responsiveness of government in controlling market failures to favour its citizens is a key issue in the political economic literature (Poku-Boansi 2020; Poku-Boansi and Marsden 2018). The welfare of its citizens, particularly of the vulnerable and the low-income groups who are most likely to be impacted by public policy, is the prime interest of government. It should thus be the commitment and role of the government to protect its citizens from harsh economic environment. In the developing world, market failures and non-existent of market opportunities make most citizens rely on the state intervention for their subsistence, thereby consolidating political alliances in the socio-economic processes (Cobbinah et al. 2019; Takyi et al.

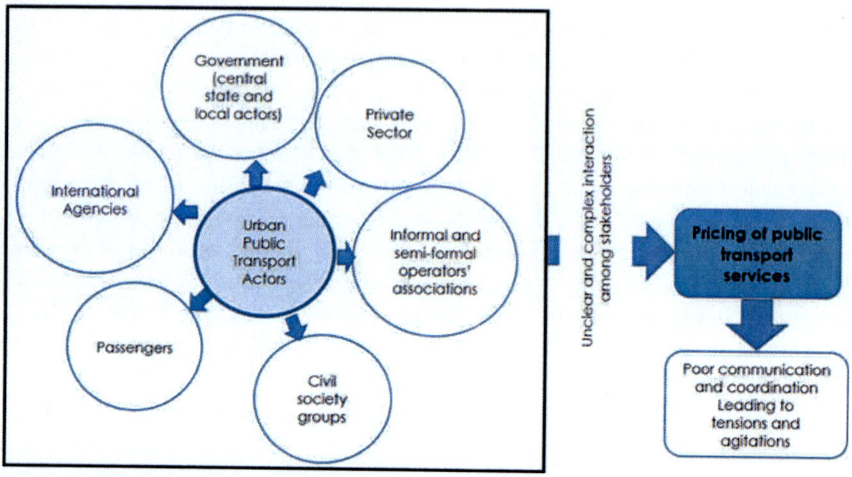

Fig. 13.1 Governance system for pricing public transport services

2013). There is, therefore, the need for political, social and economic state and non-state institutions to be apt, innovative and institute effective interventions to address the needs of all actors.

The interaction between the state and the private sector on the pricing of transportation has, therefore, arisen the conflict of ensuring the satisfaction of the citizenry as a tool for political accountability, welfare and social protection on one hand by the state and the maximization of economic gains for the rendering of a service by private sector on the other. In the political economy approach to transportation pricing, the production and distribution of transport services place the onus on the state to ensure a fair public transport pricing whereas the private sector hopes to ensure that the consumption of the services provided is able to replicate itself in the reproduction chain. There is, therefore, the need for continual dialogue through policy mainstreaming to satisfy socio-political interactions and interests of both the state and the private transport operators as well as all stakeholders in the political economy (see Fig. 13.1).

13.3 Exploring the Politics of Public Transport Pricing in Kumasi, Ghana—Overview of Research Approach

The research was undertaken in Kumasi (see Fig. 13.1), located 270 kms northwest of Ghana's national capital—Accra. It is the second largest city in Ghana (Ghana Statistical Service, GSS 2012). Since its establishment as the heart of the Ashanti Empire around the eighteenth century, Kumasi has been the cross-roads between the northern and the southern sectors of the country. The city's population was about

1,730,249 (GSS 2014) and is the fastest-growing metropolis in Ghana. The city covers approximately 214.3 km^2, which is about 0.9% of the region's land (GSS 2014).

The city's public transport system, similar to many others across the country, has road as the dominant transport mode of travel (Poku-Boansi and Adarkwa 2011). As indicated earlier, the public transport sector is dominated mainly by privately owned minibuses ("trotro" services) due to the failure of the state to provide adequate, reliable and comfortable intra-urban services (Agyemang 2015). Several challenges, confronting the sector, including congestion, inadequate road infrastructure, increasing demand for public transport and general inefficiencies have necessitated the emergence of several stakeholders (both public and private) to shape the transport sector. These actors formulate policies to provide infrastructure and improve mobility and most importantly, regulate the pricing of the services provided by operators. Additionally, these stakeholders all shape the development and management of the sector, contributing to the complex governance and pricing system of the transport sector.

The explanatory sequential mixed methodology where data were first undertaken through review of relevant documents from secondary sources, followed by primary data collection, was adopted for this study. The research began with literature review of local and international literature on the pricing of public transport services. The literature primarily centred on research work on the concepts and theories on urbanisation and mobility, the transportation sector as well as the politics and socio-spatial conflicts regarding the pricing of transport services in cities across the globe (Li et al. 2019) particularly, within the sub-Sahara African region. The local literature centred on governance systems at regularizing and managing the transport sector in Ghanaian cities (e.g., Adams et al. 2014; Amoako et al. 2014; Cobbinah 2016; Peprah et al. 2014; Yorgri et al. 2016; Adom-Asamoah et al. 2021). The documents again provided useful information for understanding the politics and socio-spatial conflicts regarding the pricing of transport services in Ghana. Most of these documents were collected prior to conducting semi-structured interviews, which allowed the researchers to relate interview responses to the information in the documents.

Using expert interviews, both state and private transport sector institutions were purposively engaged in discussions on pricing of transport services in the city and across the country. Institutions such as (i) the Development Planning Unit (DPU); (ii) the Urban Roads Department; (iii) the Ghana Road Transport Coordinating Council (GRTCC); and (iv) transport unions or associations, specifically, the Ghana Private Road Transport Union (GPRTU), Progressive Transport Owners' Association (PROTOA) and HERITAGE. These stakeholders were purposively engaged in expert interviews because of their direct responsibilities in ensuring efficient delivery and management of public transport. Three officials each were interviewed to better understand the phenomenon as presented in Table 13.1.

Due to the absence of a sampling frame and based on a similar approach adopted by Adom-Asamoah et al. (2021), 15 public intra-city transport operators (trotro) and some users of public intra-city transport services (trotro) were conveniently interviewed for relevant data across the city. The aim was to obtain their opinions on

Table 13.1 Interview respondents

Division/Unit	Number of respondents
Development Planning Unit	3
Urban Roads Department	3
Ghana Road Transport Coordinating Council (GRTCC)	3
Ghana Private Road Transport Union (GPRTU)	3
Progressive Transport Owners' Association (PROTOA)	3
HERITAGE	3
Total	**18**

Source Field Survey, December 2021

mechanisms that go into the pricing of transport services. The content and thematic analyses were employed to analyse the interview transcripts from the secondary and primary data from the respondents. The content analyses focused on the assessment of respondents' views on the politics and socio-spatial conflicts regarding the pricing of transport services.

13.4 Public Transport Governance in Kumasi—Evolution, Institutions and Actors

13.4.1 Evolution of Public Transport Governance

Ghana's public transport system, similar to many others within the global south, has road as the dominant transport mode of travel, which is used by all classes of people (Poku-Boansi and Adarkwa 2011; Vanderschuren and Phayane 2015; Vasconcellos 2013). The evolution of Ghana's public transport governance system is presented in Table 13.2.

It was revealed during the interviews that between the colonial period and until the early 1970s, the sector was part of the local government function. In the late 1960s, urban public transport in Ghanaian cities was consolidated in the Omnibus Services Authority (OSA) Limited, a publicly owned company that provided large and medium-size bus services in the main cities such as Accra, Kumasi and Tamale. It however also operated services in smaller cities as well as inter-city bus operations. Due to the focus in large cities, the deficit in public transport service especially in the medium and small cities was filled by minibuses (known as "trotro"; typically, a 14-to18-seater bus) and shared taxis, which became firmly established throughout the country (Finn 2008). The services provided by the trotro and shared taxis were readily available, accessible, convenient, and above all affordable (Agyemang 2015).

Table 13.2 Summary on the evolution of Ghana's public transport governance

Period	Details
Colonial period and until the early 1970s	Public transport was part of the local government function
Late 1960s	Urban public transport in Ghanaian cities was consolidated in the Omnibus Services Authority (OSA Ltd), a publicly owned company that provided large and medium-size bus services in the main cities such as Accra, Kumasi and Tamale but also operated services in smaller cities as well as inter-city bus operations
Mid-1970s	OSA entered a period of financial difficulty, loss of business and collapse of both capacity and market, ceasing operations by 2000
Early 1990s	Because of declining services and a growing population, the Government of Ghana "partly deregulated the sector, encouraging the private sector to participate in the provision of bus services"
Late 1990s to mid-2000s	Deficit in public transport service was filled by minibuses (known as "trotro") and shared taxis, which became firmly established. Their services provided are readily available, accessible, convenient, and above all affordable
	Over time, the trotro operators have unionized, with the GPRTU as the dominant entity that organizes more than 80% of the trotro sector
	Formation of various informal and semi-formal operators' associations with common interest and values on the one hand; and government direct and indirect intervention on the other
2007 and beyond	Ghana Urban Transport Project (GUTP) was subsequently launched to reform Ghana's public transport system by introducing a BRT system

Source Authors' construct, 2022

They mainly serve fixed routes between terminals, and buses depart only when they are full, which lead to very long waiting times in the off-peak times and difficult to board for the first 2–3 km along the route (Finn 2008).

Starting in the late 1970s, the regulatory and enforcement capacity of the local authorities dissipated, while the private transport operators enjoyed decades of monopoly (Agyemang 2015). In 2003, Metro Mass Transit Limited was created as a private–public initiative. The Government of Ghana had 45% stake in the ownership with the remaining 55% being held by private investors (Metro Mass Transit (MMT) Ltd. 2016). Some authors have, however, argued whether MMT is a private or public enterprise, as the "private" shareholders are in the main government-linked institutions (IBIS 2005; Finn et al. 2011; Yobo 2013).

A Ghana Urban Transport Project (GUTP) was subsequently launched in 2007 to reform Ghana's public transport system by introducing a BRT system in Accra and Kumasi, the two major cities. The implementation of the BRT also formed part of the Ghana National Urban Policy Action Plan, which sought to address the need for improvement of the urban transport system in the country's major cities (Poku-Boansi and Marsden 2018).

According to all the agency officials interviewed, the emergence and growth in the "trotro" services across many Ghanaian cities were due to the inability of the state's public transport system to meet the increasing demand, provide frequent, comfortable intra-urban and safe services to users/passengers. This confirms a similar assertion made by Abane (2010) and Agyemang (2015). According to the DUR officials, the transport sector of the city and Ghana as a whole.

> ....provides the major means of intra-city travel in the country, especially in the urban areas where about 95% of the service provided, is by the informal sector.........

13.4.2 Public Sector Failures and Dominance of Private Sector Operations

There are primarily four public transport operating entities providing services to the people of Kumasi. The operating unions and associations have the GPRTU dominating with over 80% of trotros and shared-taxis operators belong to its fold. There is also the PROTOA and the Ghana Cooperative Transport Association (GCTA), which are also national-level transport operators' associations that represent the interest of their members in the industry. According to all the officials of the transport associations interviewed, there are other private transport operators such as the Agate Transport, Kingdom Transport and Pergah Transport, which offer a range of bus, urban and intercity services. The MMT Company also provides urban transport services to the public in the cities of Accra and Kumasi. Furthermore, the Ghana Road Transport Coordinating Council (GRTCC) which is an umbrella body of all road transport unions and associations in Ghana also play a key role in the public transportation sector in Ghana. The GRTCC represents the interest of transport owners and their associations in the determination of transport pricing with the Government.

With reference to the above, the GPRTU and PROTOA officials reiterated that the public transport sector is primarily characterised by free entry and exit, with minimal restrictions. In addition, we largely control routes allocations, but other non-unionize drivers also operate within the cities which makes it difficult to ensure standards. The establishment of the unions such as the GPRTU and PROTOA was to help manage and regularize the informal public transport service operations, but we haven't been very successful.

An official of the GRTCC reported that the private transport operators have, over the last two decades, played key roles in Ghana's public transportation sector. Specifically, the GPRTU, PROTOA and a couple of other informal sector transport operators have become the reliable transport service provider since the 1980s, and have been involved in the pricing same. This has made these unions a strong political force in Ghana. According to an official of the GPRTU, "*the political alignment of GPRTU with the prevailing political faction in the 1980s, allowed the union to organise terminals and services*". They operated at three levels: national, local and branch, with many branches in Accra and Kumasi. According to Finn (2008), with the powers at their disposal, they organized their operations, departures from terminals, internal discipline, levying of fees, and pricing of services, among other. Finn (2008) further argued that fares were in practice and currently so, negotiated between the government and the operators represented by GPRTU and the Ghana Road Transport Coordinating Council. By default, it appears that this union has become the de facto principal regulating body for informal public transport provision, although main policy formulation is by the Ministry of Transport.

From the foregoing, it appears there are no strategic mechanisms to ensure efficient service provision in the city of Kumasi. To one of the transport operators:

> ….we have several associations and government institutions in this city. All they do is to take money without providing better services to improve upon service delivery and better pricing……..

The above has partly contributed to the increased number of "trotro" and taxis causing pollution, traffic congestion, and unnecessary delays in most Ghanaian cities (Poku-Boansi et al. 2019; Adom-Asamoah et al. 2021). These have led to the low quality of public transport services, which make the sector unattractive to the high-class groups (see Yeboah and Asibey 2019; Adom-Asamoah et al. 2021). Nonetheless, the relatively low-class population, who dominates the country's populace, seem to have limited options than to patronize the services of these trotro operators, thus making them captive to their services.

Relatedly, the officials of the DUR and DPU indicated that individual Metropolitan and Municipal Assemblies (MMDAs) can come together for Joint Jurisdictional Development purposes as enshrined in the Local Government Law (Act 462), 1993, LI 2232 (Joint Development Planning Board), and LI 1961 (Decentralised Departments of the Local Government system). These provisions allow the MMDAs to plan, regulate and manage public transport operations through their Departments of Transport by issuing license, promulgating regulations and by-laws. The interview findings, however, showed that implementing these provisions has proved difficult primarily because of capacity challenges and other factors. Even though the Transport Department is reported to be responsible for the Metropolitan Assembly's strategic policy on public transport and for guidelines to achieve efficient movement within the city, the department is not represented at the Assembly committee and core management meetings, which limits its influence on decision-making. Interference from other departments and other entities in transport-related issues is seen to be a major hindrance on work progress, compounding the political conflicts to make decisions towards managing the transport sector. It suffers from inadequate enforcement capacity and a lack of incentives to carry out the enforcement activities affecting the issue of licenses.

Given the powers of the informal operators under the various unions, and the current challenges of the Transport Department of the Kumasi Metropolis, the dynamics of pricing of transport services will tilt in favour of the private sector, which is already noted for providing inefficient transport services at higher cost to passengers.

13.4.3 Public Transportation and Pricing in Kumasi: Respondents' Perspective

Heavy traffic congestion, poor traffic management, inadequate facilities for pedestrians and bicycles, and poor road safety procedures were reported to define the transportation environment in Kumasi and most cities in Ghana. Again, the official indicated that urban transportation in Ghana faces significant governance challenges, as well as basic issues of transport management and planning and, above all, of pricing of transport services that are deemed complex. To the representative, these challenges do influence public transport pricing. For instance, it was revealed that the unclear and complex governance systems (tensions between public and private transport sector actors) result in disparities in price determination. Transport operators in several instances, fix prices without recourse to policy direction from state actors.

The above resonates the assertion of Gutman et al. (2015) and has necessitated the emergence of several stakeholders (both public and private) to shape the urban transport sector—formulate policies to provide infrastructure and improve mobility as well as regulate the pricing of services. These stakeholders all shape the development and management of the sector, contributing to the complex governance system of the transport sector. According to all representatives of the selected institutions, in most developing countries including Ghana, efforts by national governments to develop a public transit system have been unsuccessful and have been marred with financial impropriety and lack of public confidence as a result of delayed journey times and long traffic in its patronage as well as weak participatory approaches in determining the pricing of transport services. To an official of PROTOA,

> …….In Ghana, the situation is no different as efforts by the government to develop a public transit system over the last two decades have yielded no positive impact. The private sector has dominated the provision of public transit in most of the major cities including Accra and Kumasi…….

Owing to the above, transport pricing has, therefore, become a critical issue of public interest since leaving the determination of transport fares in the hands of the state or private transport operators and its powerful and growing associations will not inure to the benefit of the citizenry. An official of the DPU further stated that "*the socio-spatial politics and conflict in the determination of transport pricing has been largely centred on the deregulation of petroleum products on one side and the cost of operating transport services by the private sector*". Hence, to all respondents, transport pricing has been the transport and social policy goals of the Government of Ghana as it is considered and seen as one of the public welfare responsibilities of every state as similarly indicated by Yobo (2018).

According to all the officials interviewed, the relevant state institutions in retrospect have been fixing transport fares as a way of ensuring fairness and equity across the country and protecting the low-income groups as well as the vulnerable in the society.

> ……. Some years past, the fixing of transport prices was done mainly by the state through its relevant Ministries, which was largely non-participatory. Increases/hikes in fares often resulted in agitations from transport users. In recent years however, pricing of transport services has been participatory, involving both state and non-state actors……. … [Official of the GPRTU]

> ………. The fixing of transport prices, until recently, did not involve private transport unions, who dominate the transport sector… Anytime there are price increases, passengers complain a lot, but does not help much because they are not involved in the entire process…….. …. [Official of the DUR]

However, the interviews showed that the determination of transport prices in recent times since the deregulation and liberalization of the road transportation sector has adopted a more participatory approach in order to satisfy the interest of all stakeholders involved. The government of Ghana, transport operators' associations and the GRTCC through negotiations determine transport fares in contemporary times. This was confirmed during discussions with all agency officials. The consensus reached was that:

> ……. The determination of transport prices has been quite participatory in recent years, involving all relevant state and private sector institutions/associations/unions …..

Despite the more participatory approach adopted by the Government of Ghana, there is still dissatisfaction among the private transport operators' associations and other players in the sector (including passengers) primarily because of the autonomy they enjoyed in the past. According to an operator interviewed,

> ….. Transport fares are often unfair and unfavourable to us. All that you hear is that there has been an increase in transport fare …..

On the above, a passenger remarked that,

> ……..There is a lot of dissatisfaction and quarrel among passengers and drivers whenever there is an increase in transport prices. This happens because we [passengers] do not get to know about that…… …. Passenger B.

Furthermore, the drivers generally reported that:

> ……The Government does not ensure a strict application of the formula and therefore the transport fares that comes out is less significant than what consumers are actually supposed to pay……..

The above dissatisfaction notwithstanding, the agency officials further stated that the changes in transport prices often occur as a result of changes in fuel prices, and so have very little to do with price increases, in particular. Officials of the selected transport unions, confirming a statement by the GRTCC, on an increase in transport prices in 2021 stated that:

> ..... The increase has become necessary in view of the number of taxes on fuel announced by the government in the 2021 Budget and was subsequently approved by Parliament. With these developments, if we delay any further in increasing lorry fares by at least 20%, our transport business will collapse. We, therefore, wish to impress upon our customers and the general public to bear with us as we have also factored their financial situation into our calculations before arriving at the 20% ......
>
> ......However, we the private road transport operators have been at the receiving end of this decision for far too long. While we diligently obey the decision, prices of petroleum products keep increasing without corresponding increases in lorry fares.......

On the above, some officials of HERITAGE and PROTOA indicated that:

> ... this has become unfair business practice with a dire consequence on their businesses, and have therefore, resolved that any time fuel prices, spare parts, and other levies relating road transport go up, we will also adjust our lorry fares to reflect the increement.........
>
> .....We wish to state that we are not part of the GRTCC which together with the government fix percentage increase in lorry fares. It must interest the general public to note that the GPRTU for some time now has pulled out of the council and now joined the TUC and has since been negotiating with the government for increase in lorry fares on its own......

Some of the interviewed drivers were however happy about increases in prices of transport services. To them,

> ...... an upward adjustment in transport fares was crucial to ameliorating the current hardships drivers faced due to increase in prices of petroleum products and other taxes in the transport sector... whenever the prices of fuel went up, we are unable to make their daily sales we always charge less than was required.......

In all of the foregoing, officials of the transport associations argued that the transport fares that are always agreed on at negotiations with the Government do not actually reflect the true situation, and have therefore rendered most private operators incapable of breaking even as Government deliberately keep the transport prices low in order to protect the interest of poor and low-income earners in the country as similarly reported by Yobo (2018). This shows the direct and indirect intervention of Government to regulate transport fares on the basis of public welfare and political responsiveness against the interest of private operators whose principal aim is to maximize profit and reduce operational cost, has generated socio-political conflict between the two parties. There is, therefore, the need for policy mainstreaming to bring on board interests of all stakeholders to ensure a sustainable public transportation system and pricing regime that will inure to the benefit of the state, the private operators and the general public.

13.5 Discussion

Taking a political economy approach, this chapter discusses and unpacks the politics around studying the pricing of public transportation services in Ghana. The dominant narrative shows that public transportation plays an indispensable role in the development and socio-economic agenda of every nation. In spite of this, the pricing of public transportation services in Ghana and many other African countries is characterised by complexities, weak coordination and communication between relevant stakeholders, including passengers, which often result in tensions and agitations (Un-Habitat 2020; Veeneman and Mulley 2018; Marcucci and Stathopoulos 2012).

The findings showed that urban transport systems in low-income countries, which are essentially dominated by private groups and individuals, have been largely characterised by free entry and exit, with minimal restrictions. The emergence of private transport operators and their associations, following the deregulation and privatization policy, has given total control of the operations and practices of public transportation to the private unions resulting in several challenges (Poku-Boansi 2020; Poku-Boansi et al. 2019). In the wake of these challenges is the issue of public transport pricing that has re-echoed political conflicts between the government on one hand and the private transport associations on the other. Most importantly, public transport pricing measures play a key role in citizen acceptability, as has been seen to be a restrictive measure in protecting the public interest (Adom-Asamoah et al. 2021; Yeboah and Asibey 2019).

However, as revealed by the study, the emergence of transport operators' associations has renewed the politics of responsiveness by the state to the welfare of its citizens in the political economy discourse. It has also renewed interests in protecting the interest of transport operators against the harsh economic environment and the direct control of state in pricing transport services (Adom-Asamoah et al. 2021; Oteng-Ababio and Agyemang 2012; Adams et al. 2014; Amoako et al. 2014). The quest by the state to promote efficient and sustainable public transport systems as well as favourable pricing regime has led to the formation of two schools of thought. The first suggests the need for public intervention and, occasionally, calls for having dominant control in the determination of transportation fares. The second, however, supports public reform-driven intervention as it emphasizes the critical role these private operators play in meeting the mobility needs of the urban population, providing over 50% of transport services in some cities, and contribute significantly towards fixing favourable transport fares (Poku-Boansi and Marsden 2018; Okoye et al. 2010). Governments' efforts, over the last few decades, have essentially leaned towards the second thought with the implementation of strategies to reform the existing regimes—allowing a participatory approach towards transport price determination.

The results showed a complex relationship between urban transport operation and pricing, shaped by the formation of various informal and semi-formal operators' associations with common interest and values on the one hand; and government direct and indirect intervention mechanisms on the other. The determination of transport fares since the deregulation and liberalization of the sector was revealed to have adopted a more participatory approach in order to satisfy the interest of all stakeholders involved in the sector. The government and transport operators' associations through negotiations determine transport fares in Ghana in contemporary times. Despite the more participatory approach, there still exists dissatisfaction among the private transport operators' associations and other players in the sector (including passengers). These stakeholders, however, all shape the development and management of the sector, contributing to the complex governance system of the pricing of transport services. It is important to re-iterate that local authorities and informal and semi-formal operators' associations are responsible for taking policy-related decisions on several aspects including pricing of public transport services, which is relevant to the transport sector (Vasconcellos 2013; Stehle et al. 2020; Westman et al. 2019). These actors, however, do not much strategic plan analysing how they expect urban mobility and pricing to evolve (thus setting the necessary actions to be taken to accommodate future transport demand) or if they have such plans, they lack detail (Lah et al. 2018).

The foregoing resonates earlier assertions that the pricing of public transport services has been notably challenging in cities of many developing countries, which is partly inherent due to the complexity in the politics of governance systems (Stehle et al. 2020; Westman et al. 2019; Veeneman and Mulley 2018; Marcucci and Stathopoulos 2012). Poku-Boansi (2020) further notes that some of the main difficulties on the pricing of transport services are related to the complex and unclear nature of coordination/governance roles between the national transport actors and the various local transport authorities. Particularly in the developing region, including Ghana, policies on developing the transport sector (including fixing transport prices) are highly centralized with little involvement of other relevant state and non-state stakeholders at the local level who are the implementers of such policies.

13.6 Conclusion

The chapter explores the socio-political structures for operating and pricing transport services in Ghana, using Kumasi as a case. The results showed that public transport systems in Ghana are the result of complex interactions of levels of government, driven by their agency, ability to regulate transport prices and an understanding of the relative roles of link and place in public transport delivery. Public transport systems are thus not simply about mobility; however, they provide access to activities creating value to individuals or society. In spite of the complexity, the many stakeholders and seemingly participatory nature of the process have the potential of helping address the numerous transportation pricing challenges in Ghana and other developing countries that share similar decentralisation system. This has the capacity of helping to pool resources, enhance coordination and sharing of cost and benefits across various stakeholders. The chapter, based on the foregoing, suggests that the public urban transportation sector should be reformed to reduce this dichotomy through a policy mainstreaming of all stakeholders on both sides.

References

Abane MA (2010) Travel behaviour in Ghana: empirical observations from four metropolitan areas. J Transp Geogr 189–376

Adams CA, Opoku–Boahen R, Abdul Muhsin Zambang M (2014) Parking management in metropolitan cities in West Africa case study of the Kumasi paid parking scheme, Ghana. Int Ref J Eng Sci 3(6):1–8

Adom-Asamoah G, Asibey MO, Nyarko S (2021) Rethinking or affirming the "Affordability" debate on the efficiency of urban public transport usage: case of Kumasi, Ghana. Transp Dev Econ 7:22. https://doi.org/10.1007/s40890-021-00131-6

AFD & MEEDDM (2009) Who pays What for urban transport? Handbook of good practices. CODATU

Agyemang E (2009) Traffic congestion: the bane of a BRT system in Accra, Ghana. Published MPhil Thesis, Norwegian University of Science and Technology, Trondheim. http://ntnu.diva-portal.org/smash/record.jsf

Agyemang E (2015) The bus rapid transit system in the Greater Accra Metropolitan Area, Ghana: looking back to look forward. Nor Geogr Tidsskr—nor J Geogr 69:28–37

Amoako C, Cobbinah PB, Niminga-Beka R (2014) Urban infrastructure design and pedestrian safety in the Kumasi Central Business District, Ghana. J Transp Saf Secur 6(3):235–256

Cervero R, Golub A (2007) Informal transport: a global perspective. Transp Policy 14(6):445–457

Cirolia LR, Harber J (2021) Urban statecraft: the governance of transport infrastructures in African cities. Urban Stud. https://doi.org/10.1177/00420980211028106

Cobbinah PB (2016) Urbanization in Ghana: insights and implications for urban governance. https://doi.org/10.4018/978-1-5225-0187-9.ch005

Cobbinah PB, Amoako C, Asibey MO (2019) The changing face of Kumasi central, Ghana. Geoforum 101:49–61

Dansoh A, Frimpong S, Ampratwum G, Oppong D, Osei-kyei R (2020) Exploring the role of traditional authorities in managing the public as stakeholders on PPP projects: a case study. Int J Constr Manag 20(6):628–641

Del Mistro R, Behrens R (2015) Integrating the informal with the formal: an estimation of the impacts of a shift from paratransit line-haul to feeder service provision in Cape Town. Case Stud Transp Policy 3(2):271–277

Finn B (2008) Market role and regulation of extensive urban minibus services as large bus service capacity is restored—case studies from Ghana, Georgia and Kazakhstan. Res Transp Econ 22(1):118–125

Finn B, Kumarage A, Gyamera S (2011) Organisational structure, ownership and dynamics on control in the informal local road passenger transport sector. Presented at the 12th conference on competition and ownership in land passenger transport, South Africa, conference proceedings, pp 133–146

Fouracre PR, Kwakye EA, Okyere JN, Silcock DT (1994) Public transport in Ghanaian cities—a case of union power. Transp Rev 14(1):45–61

Gbadamosi KT (2010) An evaluation of the impact of bus rapid transit in urban intra-city passenger movement in Lagos State. A paper to be presented at WCTR 2010. Centre for Transport Studies, Lisbon, Nigeria

Ghana Statistical Service (2014) 2010 Population and Housing Census. District Analytical Report. Kumasi Metropolitan Area, Accra, Ghana Statistical Service

Ghana Statistical Service [GSS] (2012) 2010 population and housing census. Summary report of final results. GSS, Sakoa Press Limited, Accra, Ghana
Gutman J, Sy A, Chattopadhyay S (2015) Financing African infrastructure can the world deliver? Brookings, Washington, DC
Horn A, Van Eeden A (2018) Measuring sprawl in the Western Cape Province, South Africa: an urban sprawl index for comparative purposes. Reg Sci Policy Pract 10:15–23
IBIS (2005) Study of Urban Public Transport Condition in Accra, Ghana. IBIS Transport Consultants Ltd., Accra
Lah O, Alveano S, Arioli M, Chesterton V, Sdoukopoulos L (2018) Sustainable urban mobility solutions for Asia, Latin America and the Mediterranean region. In: Sustainable urban mobility pathways: policies, institutions, and coalitions for low carbon transportation in emerging countries. https://doi.org/10.1016/B978-0-12-814897-6.00002-8
Li Y, Miao L, Chen Y, Hu Y (2019) Exploration of sustainable urban transportation development in China through the forecast of private vehicle ownership. Sustainability (Switzerland) 11(16). https://doi.org/10.3390/su11164259
Marais L, Denoon-Stevens S, Cloete J (2020) Mining towns and urban sprawl in South Africa. Land Use Policy 93:103953. https://doi.org/10.1016/j.landusepol.2019.04.014
Marcucci E, Stathopoulos A (2012) Multi-level governance and transport policy: the case of local roads in Italy. Int J Transp Econ. XXXIX (1):15–38
Mbara TC (2002) Transport: How have African cities managed the sector? What are the possible options? Paper presented at the Urban and City Management Course for Africa, Kampala
Metro Mass Transit Limited (2016) The establishment of MMT. 2016 Annual Report. Metro Mass Transit Limited, Accra
Mosco V (1996) The political economy of communication: rethinking and renewal. Sage, London
Obeng-Atuah D, Poku-Boansi M, Cobbinah PB (2016) Pedestrian crossing in urban Ghana: safety implications. J Transp Health. https://doi.org/10.1016/j.jth.2016.06.007
Okoye V, Sands J, Debrah AC (2010) The Accra pilot bus-rapid transit project: transport-land use research study. Millennium Cities Initiative and Accra Metropolitan Assembly, Accra
Olvera LD, Plat D, Pochet P (2020) Looking for the obvious: motorcycle taxi services in Sub-Saharan African cities. J Transp Geogr 88:102476
Oteng-Ababio M, Agyemang E (2012) Virtue out of necessity? Urbanisation, urban growth and Okada services in Accra, Ghana. J Geogr Geol 4(1):148–162
Peprah C, Oduro CY, Ocloo KA (2014) On-street parking and pedestrian safety in the Kumasi metropolis: issues of culture and attitude. Dev Count Stud 4(20):85–95
Poku-Boansi M (2020) Path dependency in transport: a historical analysis of transport service delivery in Ghana. Case Stud Transp Policy 8(4):1137–1147
Poku-Boansi M (2021) Contextualizing urban growth, urbanisation and travel behaviour in Ghanaian cities. Cities 110:103083. https://doi.org/10.1016/j.cities.2020.103083
Poku-Boansi M, Adarkwa KK (2011) An analysis of the supply of urban public transport services in Kumasi, Ghana. J Sustain Dev Africa 13(2):28–40
Poku-Boansi M, Cobbinah PB (2017) Land use and urban travel in Kumasi, Ghana. GeoJ 83:563–581
Poku-Boansi M, Marsden G (2018) Bus rapid transit systems as a governance reform project. J Transp Geogr 70:193–202
Poku-Boansi M, Amoako C, Obeng Atuah D (2019) Urban travel patterns and safety among school children around Accra, Ghana. J Transp Health. https://doi.org/10.1016/j.jth.2019.100660
Pucher J, Hyungyong P, Kim MH (2005) Public transport reforms in Seoul: innovations motivated by funding crisis. J Pub Transp 8(5):41–62
Pucher J, Korattyswaroopam N, Ittyerah N (2004) The crisis of public transport in India: overwhelming needs but limited resources. J Pub Transp 7(4):1–20
Shao Z, Sumari NS, Portnov A, Ujoh F, Musakwa W, Mandela PJ (2020) Urban sprawl and its impact on sustainable urban development: a combination of remote sensing and social media data. Geo-Spat Inf. Sci. https://doi.org/10.1080/10095020.2020.1787800

Sietchiping R, Permezel MJ, Ngomsi C (2012) Transport and mobility in sub-Saharan African cities: an overview of practices, lessons and options for improvements. Cities 29(3):183–189
Stehle F, Hickmann T, Lederer M, Höhne C (2020) Urban climate politics in emerging economies: a multi-level governance perspective. Urbanisation. https://doi.org/10.1177/2455747120913185
Takyi H, Poku K, Emmanuel AK (2013) An assessment of traffic congestion and its effect on productivity. Int J Bus Soc Sci 4(3):225–234
UN-Habitat (2013) Planning and design for sustainable urban mobility: global report on human settlements. Routledge, New York
Un-Habitat (2020) The urban mobility challenge. In: Planning and design for sustainable urban mobility. https://doi.org/10.4324/9781315857152-10
Vanderschuren M, Phayane S (2015) Non-motorised transport facility guidelines accommodating people. Urban Rural Eng: Civil Eng Siviele Ingenieurswese 23(9):68–72
Vasconcellos E (2013) Urban transport, environment, and equity: the case for developing countries, 2nd ed. Earthscan, New York, USA
Veeneman W, Mulley C (2018) Multi-level governance in public transport: governmental layering and its influence on public transport service solutions. Res Transp Econ 69:430–437
Westman LK, Castán Broto V, Huang P (2019) Revisiting multi-level governance theory: politics and innovation in the urban climate transition in Rizhao, China. Polit Geogr. https://doi.org/10.1016/j.polgeo.2019.01.002
Wilson T (2006) Public transport operations in Ghana. modernghana: http://www.modernghana.com/news/102058/1/public-transportoperations-in-ghana.html. Accessed 18 Nov 2012
Yeboah V, Asibey MO (2019) Transport and historical changes in Kumasi's growth and form. Case Stud Transp Policy 7(2019):802–813
Yobo E (2013) The politics of public transportation in Ghana: the case of Metro Mass Transit Limited. University of Ghana, Ghana. http://ugspace.ug.edu.gh/handle/123456789/5284
Yobo E (2014) Role of the state in Ghana's road transport system. Res J Pub Policy 1(3):1–13
Yobo E (2018) State intervention in public transportation in Ghana. J Transp Logist 3(1):36–51
Yorgri E, Wen PC, Hong PL (2016) Parking planning and policy in the CBD of Accra, Ghana. Int J Adv Agricult Environ Eng 3(2):1–5

Michael Poku-Boansi is a Professor of Planning at the Department of Planning, and the Deputy Director, School of Graduate Studies, Kwame Nkrumah University of Science and Technology, Kumasi, Ghana. Michael has over fifteen (15) years' of transportation and land use planning, resettlement planning, climate change and resilience planning experience spanning both academic and private sector appointments. Professor Poku-Boansi is a Fellow of the Rockefeller Foundation Professional Fellowship Programme and a Commonwealth Academic Fellow. He has authored/co-authored several peer-reviewed research articles published in reputable international journals and numerous project reports. Prof. Michael Poku-Boansi is also a Fellow of the Ghana Institute of Planners (FGIP).

Michael Osei Asibey is a Lecturer in the Department of Planning, Kwame Nkrumah University of Science and Technology, Kumasi, Ghana. Dr. Asibey's research focus are in four broad areas: transportation studies; environmental policy and planning, sustainable cities and urban containment; and climate change adaptation and resilience and urban greenery conservation. Michael has authored/co-authored several peer-reviewed research articles published in reputable international journals

Chapter 14
Safe, Inclusive Public Transport in Africa—Challenges and Opportunities Identified in Ghana and Namibia

Steven Jones, Emmanuel Adanu, Charles Adams, and Robert Ambunda

Abstract Access to safe and inclusive public transportation is set out in the Sustainable Development Goals (SDGs) as foundational to improving the human condition. Public transport throughout Africa, however, is a complex endeavour. It spans many modes such as the largely informally (or semi-formally) operated vans and minibuses, traditional taxis, as well as motorcycle taxis common in cities throughout the continent. While new technologies (e.g., Intelligent Transportation Systems, ride-hailing apps) and construction projects to add infrastructure capacity may continue to improve the operational efficiency of these modes and shape the future mobility trends in many cities in Sub-Saharan Africa (SSA), there is still much work to be done to address the all too pervasive lack of safety and personal security which threaten the adoption of certain public transport modes across the continent. Current and future transport modes must be safe to promote inclusivity. This chapter examines the existing research published in the literature on challenges and opportunities facing public transport safety across the continent. This information is then combined with further research into public transport safety and security issues in Ghana and Namibia. In doing so, it presents evidenced-based strategies to meet these challenges, while also identifying and illustrating opportunities to address them through human factors focused efforts (education, outreach, training, and enforcement), as well as

S. Jones (✉)
University of Alabama, Alabama, USA
e-mail: steven.jones@ua.edu

S. Jones · E. Adanu
Alabama Transportation Institute, Alabama, USA
e-mail: ekadanu@ua.edu

C. Adams
Kwame Nkrumah University of Science and Technology, Kumasi, Ghana
e-mail: caadams.coe@knust.edu.gh

R. Ambunda
University of Namibia, Windhoek, Namibia
e-mail: rambunda@unam.na

Stellenbosch University, Stellenbosch, South Africa

R. A. Acheampong et al. (eds.), *Transport and Mobility Futures in Urban Africa*,
The Urban Book Series, https://doi.org/10.1007/978-3-031-17327-1_14

deployments of new technologies to collect and analyse relevant data on which to develop and deploy a *Safe System* based approach to mitigation.

Keywords Safe systems · Public transport · Transport safety · Intelligent transportation systems

14.1 Introduction

Three corollaries:

- The future of Africa is urban.
- Sustainable urban development relies on safe, inclusive transport.
- The future of safe, inclusive transport in Africa is technology.

Assuming the above corollaries are true,[1] they serve as both summary and challenge. Myriad academic articles, books, and reports generated by national and international development agencies acknowledge, even celebrate, the coming urban future of Africa (Njoh 2003; UN-Habitat 2015; Lufumpa and Yepes 2017). There is an even larger body of literature connecting the importance of transport to sustainable development across Africa (Too and Earl 2010; Njenga and Odero 2014). Indeed, access to safe and inclusive transport is set out in the Sustainable Development Goals (SDGs) as foundational to improving the human condition globally (United Nations 2015). With more specificity, the United Nations Economic and Social Council describe safe and inclusive transport as:

> …key to enabling participation in society, by providing access to socioeconomic and life-enhancing opportunities. It is broadly defined as the ability to safely and reliably access a preferred destination by navigating in an environment considerate of individual needs. Safe and inclusive transport and mobility can, thus, have a domino effect towards enhancing an inclusive society that leaves no one behind (UN ESCAP 2020).

Transport throughout Sub-Saharan Africa (SSA), however, is a complex endeavour. It comprises many modes and operating modalities (e.g., formal, semi-formal, and informal). It spans enormous geographical spaces, contrasting topographical features, cultural traditions, jurisdictional boundaries, and economic realities—financial resources constrain both the provision and consumption of transport throughout SSA.

Non-motorized transport (NMT) is, and will continue to be, an essential form of mobility and accessibility throughout SSA. But history and current events make it clear that humans will increasingly rely on some form of motorized transport to meet the expanding mobility and accessibility requirements of rapidly developing societies. While there has been much written on the increasing dependency on private automobiles, most Sub-Saharan Africans move around via specialized

---

[1] By 2050, some 66% of the global population will be considered urban with Africa contributing significantly to this growth—urbanization in Africa has been progressing at a rate more than 10 times that of Europe in recent years (UN-Habitat, 2016).

public transport called paratransit, characterized by small vehicles (passenger cars and minibuses), individual ownership, and unregulated operations (Behrens et al. 2015) while an increasing amount of transport is provided by two- and three-wheeled vehicles. It is precisely these modes that will continue to connect the activities of future African cities with each other as well as those that make up life in the peri-urban landscapes and hinterlands surrounding them. This chapter explores how the provision of safe public transport is paramount to harnessing rapid urbanization as well as connecting the opportunities of urban development to the rest of the continent. After introducing some safety issues facing African public transport through observations from Ghana and Namibia, a discussion follows on how technological advancements can (and will) make African public transport more safe and secure in the future and, by extension, more inclusive.

14.2 African Public Transport Safety

Throughout Africa, there are numerous examples of conventional, government-supported public transport in the form of urban bus systems, including newer Bus Rapid Transit (BRT), as well as recent deployments of light-rail transit and other forms of commuter trains. Most of the paratransit in Africa, however, is provided by privately-operated vans and minibuses—colloquially known as *trotros* in Ghana, *matatus* in Kenya, *car rapides* in Senegal, etc.—or the motorcycle taxis known as *boda bodas* in Tanzania, *okadas* in Nigeria and Ghana, and their three-wheeled relatives (e.g., *tuk tuks* or *motor kings*). There are unique circumstances like Namibia, where motorcycles are relatively scarce and the urban population largely relies on shared passenger car taxis in lieu of buses and minibuses, which are used more so for intercity and intra-regional travel (Ambunda and Nakale 2020). Like the rest of the world, SSA is experiencing the Mobility as a Service (MaaS) revolution whether through encroachment from abroad by companies such as Uber and Lyft, or locally developed platforms such as *LEFA* in Namibia, *Bolt* in Nigeria, and many more (Boutueil and Quillerier 2020).

According to the 2018 Global Status Report on Road Safety, in 2016, the African region recorded the highest rate of road deaths per 100,000 population compared to all other world regions (World Health Organisation 2018). Considering that the African region has a relatively lower share of the global vehicle population, this high transport fatality rate points towards higher death rates per vehicle across the region—an indication of the involvement of higher passenger capacity public transport modes being involved in fatal crashes. Despite the generally poor roadway infrastructure and older vehicle fleets, public transport driver behaviours, attitudes, and driving styles constitute a major contributing factor for road crashes (Pearce et al. 2000; Abegaz et al. 2014; Sam et al. 2018). Indeed, much has been written on the relative road safety of various public transport modes functioning throughout Africa. In some cases, the focus has been on a specific mode (e.g., Adanu et al. 2020) while other studies have focused on issues specific to urban or rural contexts (e.g., Jones et al.

2016a; Agyemang et al. 2021). Table 14.1 presents a snapshot of the diverse literature on African public transport safety organized by mode. The findings summarized in Table 14.1 are organized into three broad categories: *Conventional* public transport connoting urban bus systems, light-rail, etc.; *Paratransit* to represent any of the many vehicle configurations intended to serve up to 15–20 passengers or so and typically privately operated; *Two- or Three-wheeled* vehicles; and recent insight into the road safety evidence amassing on *MaaS* deployment in Africa.

The overarching theme from the studies summarized in Table 14.1 above is that public transport, irrespective of the mode, across the continent has been a major contributor to traffic fatalities and injuries. The general lack of protective equipment such as seatbelts and helmets, coupled with poor infrastructure, old vehicle fleets, and risky operator behaviours increases the severe injury and fatality rates. Further, the negative consequences of public transport operation threaten the human resource base of the continent considering that younger people constitute a higher share of road crash casualties (World Bank 2017).

Table 14.1 Snapshot of public transport safety

Mode	Key findings	Source
Conventional	Bus drivers need to be better educated and better trained when initially learning to drive	Pearce et al. (2000)
	Commercial bus drivers' surveyed exhibited poor knowledge of road signs and speed limits as well as other factors affecting basic road safety	Okafor et al. (2013)
Paratransit	Crashes involving minibuses have increased probability of higher severity outcomes	Abegaz et al. (2014)
	Weekends, absence of road median, night-time conditions, bad road terrain (curved, wet and rough roads), hit-pedestrian collisions, and drunk driving are associated with more severe accident outcomes	Sam et al. (2018)
Two/three wheel	Passengers are more at risk of sustaining injuries compared to drivers	Oluwadiya et al. (2016)
	Gender, age, marital status, and license status affect helmet use	Akaateba et al. (2015)
	Day of the week, weather condition, road geometry, location type and traffic condition affect motorcycle crashes	Aidoo and Amoh-Gyimah (2019)
MaaS	Proliferation of Uber in South Africa has not had a significant impact on drunk driving crash reduction	Majid (2019)

While the discussion above offers a snapshot of the road safety challenges facing public transport in SSA, issues such as exposure to air and noise pollution negatively impact public health and safety (GSRF 2014). Personal security, or lack thereof, is also an important component of public transport safety, and one that disproportionately impacts women (Porter et al. 2020). For example, a recent survey of university students in Windhoek, Namibia indicated that personal security concerns associated with using shared-taxis far outweighed the fear of road crashes, especially among women (Jones and Makamani 2021). Sam and Abane (2017) documented a survey of 237 intercity bus passengers on Ghana noting significant concerns for personal security and support for such technological interventions as improved passenger and baggage identification systems in addition to CCTV cameras at transit terminals. And, as the COVID-19 pandemic has illustrated, public transport is central to both the spread of communicable diseases and the ability for socioeconomic systems to recover from them (Zhen et al. 2020).

14.3 Observations from Ghana

The most recent road safety data from the Building and Road Research Institute (BRRI) indicate that there were some 11,000 reported crashes (2,203 fatalities and 13,285 injuries) in 2019 in Ghana. Among these, about 62% involved passenger cars while minibuses/buses and motorcycles comprised 18 and 16% of crashes, respectively. Additionally, about half of the crashes in 2019 were attributable to private vehicles compared to about a quarter reported as involving public transport vehicles. As previously noted, public transport in Ghana is predominantly provided by *trotros* and *okadas*, with some intercity and intra-regional travel provided by privately operated large capacity buses. Motorcycles have been a popular mode of transport, particularly in the northern regions (Kudebong et al. 2011) which happen to be the poorer regions in the country due mainly to sparse development, poor road network connectivity, and overall diminished accessibility. Despite the safety risks associated with their use, a high proportion of residents of these regions are captive users of motorcycles for essential socioeconomic activities. In the last several years, the proliferation of motorcycles (and three-wheelers) for passenger and goods transport in many rapidly urbanizing communities across Ghana has contributed to an increase in motorcycle crashes, injuries, and fatalities (National Road Safety Commission 2013; Agyemang et al. 2021). Motorcycle-related fatalities increased a staggering 200% between 2006 and 2015—a period otherwise exhibiting a 3% reduction in national traffic fatalities (BRRI 2016). Even though evidence suggests that motorcycle fatalities are considerably reduced using helmets, numerous previous studies have documented low helmet use across the country (Ackaah and Afukaar 2010; Dapilah et al. 2017; Nimako-Aidoo et al. 2018).

Ackaah and Adonteng (2011) found that high occupancy public transport vehicles (minibuses/buses) accounted for nearly 28% of all fatal crashes that occur in the country, making them the second highest sources of road fatalities after pedestrian crashes. This trend has persisted over the years due to speeding on mostly single carriageway major highways linking cities, towns, and villages (Sam et al. 2018), high proportion of goods vehicles especially on national highways which frequently form platoons making overtaking very risky. As such, most of these fatal crashes are catastrophic head-on crashes. Run-off-road crashes involving public transport vehicles also tend to be fatal due to collisions with fixed objects such as trees and roadside features such as bridge abutments. Many of these crashes were documented to be attributable to driving fatigue from working long hours due to pressures from vehicle owners or competition among drivers for passengers in order to meet daily "sales" limits (Sam and Abane 2017). However, driver behaviours and the reasons behind them are complex. For example, Ulzen et al. (2018) reported a significantly higher incidence of adult Attention Deficit Hyperactivity Disorder (ADHD) among *trotro* drivers in Accra as compared to the general population there and went on to link risky driving behaviours to ADHD. Furthermore, minibus owners/operators tend to overstretch the lifespan of the vehicles in order to maximize profits. Such older vehicles pose an increased risk as they are: (a) more likely to contribute to crashes due to mechanical failures; and (b) more likely to result in serious injuries due to diminished ability to withstand the crash impacts and lack of onboard safety devices. The problem is well known as some public transport users often rely on their own assessment of vehicles condition as proxy for the overall safety of the particular driver and/or route (Sam et al. 2018). Indeed, it has long been understood that public transport operators use subtle strategies such as inscriptions on the vehicles to assure users of their safety and security (Date-Bah 1980). And public transport users generally recognize safety implications of using these vehicles (Sam et al. 2018). Understandably many *trotro* users, like motorcycle users, are constrained to this mode of transport due to affordability—as such, safety is often forced to be a secondary concern.

The dynamic and evolving relationship among the various actors and components, perhaps, may be correlated with the legal frameworks under which the public transport operators work. For instance, in 2012, Ghana's Parliament passed the Road Traffic Regulation 2012 (Legislative Instrument 2180) that bans the commercial use of motorcycles and three wheelers. Despite the ban, the number of these vehicles has been increasing because of the critical mobility services they provide. Exercising regulatory oversight over their operation and enforcement of traffic laws present a complex legal and political conundrum due to their increasing modal share, and spatial disaggregation throughout urban and peri-urban road networks. So perhaps, Ghanaian authorities need to revisit the role of motorcycles within the fabric of the transport system and look for ways to safely integrate them as "feeders" serving the first/last mile segments of public transport trips as opposed to unregulated over-the-road operations.

14.4 Observations from Namibia

Public transport safety has not been widely studied in Namibia, yet a recent survey indicated that most Namibians rely on public transport (Ambunda and Sinclair 2020). The most recent road safety data published by the national Motor Vehicle Accident (MVA) Fund indicate that there were some 4,800 reported crashes (607 fatalities and 5,646 injuries) in 2019 across Namibia (MVA, 2019). Among these, roughly half involved passenger cars while buses and motorcycles comprised 3.0 and 1.4%, respectively—less than 1% were attributed to vehicles coded as vans. Additionally, more than two thirds of crashes in 2019 were attributable to private vehicles compared to 13% of crashes reported as involving public vehicles (e.g., buses). Without related traffic volume and vehicle registration data, it is not possible to determine the relative danger of these modes. And further complicating the issue of identifying public transport crashes is the aforementioned fact that much of it is provided by private vehicles (predominantly passenger car shared taxis). Detailed analysis of crash records from the National Road Safety Council (NRSC) analysed by Jones et al. (2019) revealed that crashes involving vehicle types such as light delivery vehicles and panel vans often recorded one or more passengers (in a few cases as many as 10). As noted above and elsewhere in the literature, there is ample evidence of goods vehicles being used to transport passengers, especially in rural areas (Jones et al. 2016a, b). For-hire-service vehicles that include taxis, buses, and passenger-carrying commercial vehicles account for nearly 50% of all traffic crashes in the country. The predominantly two-lane and unpaved roads in the country account for a high number of often fatal head-on collisions and run-off-road crashes, many of which result in rollovers. On the average, each single crash involves about three people. According to Adanu et al (2020) for-hire service vehicle, crashes were nearly 98% more likely to record severe injuries when they have passengers on board. And interestingly, drivers 30 years of age and younger were overrepresented in crashes involving for-hire vehicles. The drivers of these vehicles are mostly less than 30 years as the public transport sector provides employment to less educated youths.

Of the 5,646 injured in Namibian road crashes in 2019, 58% were vehicle passengers and 23% were drivers (the rest were mostly pedestrians). Slightly more than twice the number of vehicle passengers died in road crashes than did drivers. Acknowledging the role of drivers in public transport crashes, in 2019 the MVA Fund, in conjunction with national law enforcement, the Namibia Bus and Taxi Association (NABTA) and the Namibia Public Passenger Association (NPPTA), launched the *Green Dot* campaign. The campaign involved conducting pre-trip checks of vehicle and driver fitness as well as general inspection of vehicle loads at local transit stations—vehicles that passed the checks were fitted with green stickers (i.e., dots) to indicate that they were inspected and considered ready for service.

The high road fatality rate in Namibia has a detrimental effect on the socioeconomic growth of the country, with a proportion of fatal crashes involving the most economically active part of the population. The high fatality has been largely attributed, among others, to the absence of appropriate road safety policies and underperforming road safety management systems (Amweelo 2016). As a result, Namibia has undertaken the review of existing road safety laws (Road Safety Act No. 29 of 1992) and development of a new Road Safety Management Bill (RSMB) to bring about uniformity and effective cooperation among road safety stakeholders to ensure effective road safety management. The RSMB is intended to culminate in the formation of a Road Safety Agency of Namibia, charged with improving crash data collection and storage systems as well as supervising road safety policies and programmes across all stakeholders and government agencies.

14.5 Safe System Approach and the Role of Technological Innovation

In its most basic form, a *Safe Systems* approach to road safety is based on the concepts of Safe Roads, Safe Speeds, Safe Vehicles, and Safe People. While much has been written on conventional means of advancing Safe Systems (Larsson and Tingvall 2013; Naumann et al. 2020), this chapter focuses on technology as being key to a safer public transport future in SSA. The region is well aware of the coming 4th Industrial Revolution. With it comes the promise of new technologies that can be leveraged to make transport safer and more inclusive such as the Internet of Things (IoT), automation, machine learning, and artificial intelligence—even blockchain technology is under investigation for its potential in improving transport services in Africa (Herko 2019).

Mobile information and communication technologies (ICT) have already revolutionized how Africans interact with one another and participate in various elements of society from social to political to agricultural to transport (Asongu and Le Roux 2017). Indeed, projects such as the Digital Matatus effort in Nairobi and Mapa Dos Chapas work in Maputo have already illustrated how these technologies can be used to make public transport more efficient and accessible (Kloop and Cavoli 2019). As IoT continues to develop on the continent, it will meet the already rapid expansion of mobile communications technology (wireless networks, smartphones, etc.) that have fuelled progress in other sectors. As such, it can be expected that the overall connectivity of African life, and transport is important to that, will expand.

Connectivity will have a direct impact on public transport safety as it can empower users to share their perceptions of safety and security in ways that can influence behaviours of operators, owners, and even law enforcement and policymakers. In that way that mobile phones and burgeoning MaaS are making public transport more operationally demand responsive, it is also making it more safety responsive. For instance, apps such as the Kenya-based *Little Cab*, *SafeBoda*, and *UberBoda*, have features for customers to rate public transport operators, the quality of services provided, and even report suspicious activities. Additionally, social media platforms (that will continue to expand throughout Africa along with ICT), such as the Facebook-based

Trotro Diaries in Ghana, provide a platform for *trotro* users to share daily experiences and safety tips. This has the potential to influence driver and user behaviours, hence improving the overall safety of public transport and, in doing so, making it more inclusive and accessible to children, women, elderly, and persons with disabilities.

Ultimately, safe public transport systems should provide the foundation for a connected urban life. Peprah et al. (2019) described a holistic smart mobility approach to transport in Ghanaian cities where they found that although Ghanaian cities are not mobility-smart, they demonstrated how the concept can be operationalized to mitigate some of the adverse effects of urbanization in the cities. Slavova and Okwechime (2016) further explored the potential of the *Smart Cities* paradigm to improve transport efficiency and address public safety issues throughout Africa. A more connected Africa, through ICT and IoT, will transform existing informal and semi-formal public transport systems into a proliferation of MaaS applications that significantly improve the safety and inclusivity of all public transport stakeholders. Furthermore, these systems will create new opportunities to develop data-driven systems with cascading potential for improvements in how Africans move about.

Data is an essential element in identifying risk factors associated with public transport safety. The lack of comprehensive data has been an obstacle to performing in-depth analysis towards meaningful data-driven policies and mitigation measures. Underreporting and discrepancies in crash data introduce biases in data analysis results and that can lead to erroneous conclusions and policy decisions (Mannering et al. 2016). However, the development of advanced machine learning techniques provides the opportunity to overcome some of the major crash data issues (missing data, erroneous data, etc.), which stifle efforts to develop informed countermeasures. SSA countries also have the opportunity to leap-frog to technologically driven machine vision strategies to detect conflicts and unsafe locations to reduce the occurrence of traffic crashes and security issues (e.g., St-Aubin et al. 2015). An example is a recent incident in Nairobi wherein the death of a *boda boda* passenger late one night was captured on a municipal CCTV camera. Interestingly, related to the discussion above, the video of this tragedy made its way around social media in Kenya resulting in such a public outcry that *UberBoda* responded by banning operations after 11 pm (Kisekka 2019).

To the extent that the Safe Systems approach emphasizes the design and operation of safe roads (in addition to focusing on road user behaviours), such as the above example of using machine vision to identify potentially unsafe and insecure locations, other related technologies can be applied to better inform design and maintenance of the roadways on which public transport vehicles (including motorcycles) operate. For example, Silyanov et al. (2020) summarized a number of ways new technologies can be used (and synergized) to collect, visualize and analyse road data in developing countries. And Jones et al. (2021) documented the potential for using the machine learning-based technique *Tensor Flow*, to automate the scanning and analysis of publicly available photo images of a Namibian highway to identify potential safety deficiencies.

Additionally, corruption and misconduct on the part of law enforcement officers compromise their ability to confront risky public transport driver behaviours (Hua et al. 2010; Wells and Beynon 2011; Jonck and Swanepoel 2016; Tankebe et al. 2020). For example, bribery of law enforcement officers is an affront to road safety as public transport operators can ply the road networks with overloaded and rickety vehicles, with drivers engaging in various traffic law violations with impunity. The deployment of technology, such as traffic speed enforcement cameras or digital fare collection platforms, can minimize the windows of opportunity for corrupt practices involving law enforcement officers.

There is the need for concerted efforts on the part of governments across SSA to provide incentives for informal public transport operators to replace their ageing vehicle fleets, as the older vehicles pose significant safety risks to users. Advanced vehicle technologies provide the opportunity to ensure that the new generation of public transport vehicles that should be imported or developed for the African market meet certain minimum safety thresholds to be licensed to operate. For instance, policies should be enacted to ensure that newer public transport vehicles are equipped with seatbelts and technologies such as: location tracking devices, advanced collision warning systems, and in-vehicle cameras to improve safety and security en-route. Also, advanced technologies such as in-vehicle navigation (e.g., intelligent speed adaptation), route guidance, and real-time transport information systems are aimed to improve rider comfort and safety by ensuring driver compliance with traffic laws. And there are innovative solutions being explored such as the in-vehicle intelligent speed adaptation and warning systems (e.g., Akpa et al. 2016). In-vehicle systems and even the mobile phones of users can provide both real-time and archival data that can be used to better understand trends in public transport safety.

The design and provision of a safe public transport system, consisting of quality infrastructure, appropriate policies, and backed by the necessary enforcement strategies, can shape public transport driver behaviours and promote the necessary culture to improve user safety and security. Encouraging a culture of safe behaviours among public transport drivers and security for users within the safe system framework will rely heavily on advances in technological innovations.

In order to advance safe and inclusive public transport throughout future Africa, this chapter proposes relying on technological innovation to enhance the application of a Safe Systems approach. This African perspective could be based on the following goals facilitated by technology as mapped out in Fig. 14.1:

- Operationalizing safety and personal security as a value proposition;
- Changing the way transport safety data is collected and analysed;
- Enhancing the integrity and effectiveness of enforcement; and
- Augmenting design and maintenance activities.

Figure 14.1 is not intended to be an exhaustive list of applicable technology applications. Rather, it is meant to illustrate the potential for a range of available innovations to be leveraged towards achieving safer roads, speeds, vehicles, and people. And beyond such accessible technologies, Africa must position itself to be ready for the inevitability of connected infrastructure systems (Ackaah 2019) such as

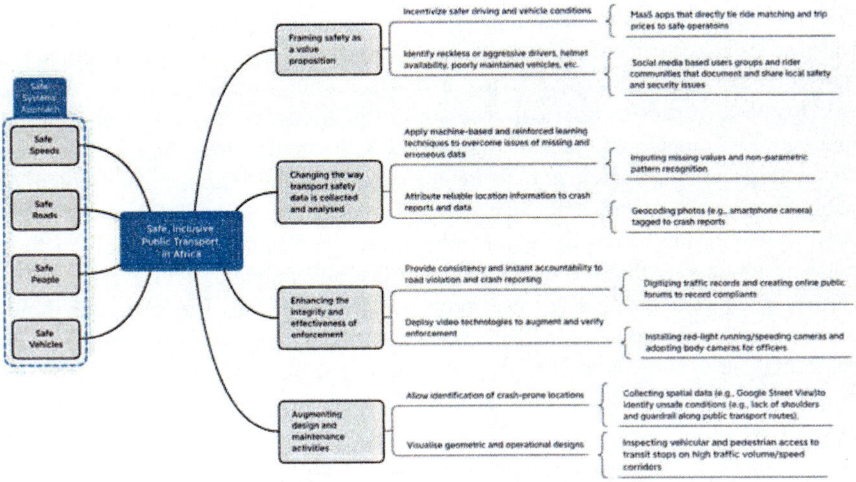

Fig. 14.1 Sample Technologies for a System Approach to Public Transport Safety

the Traffic Management Centres in Accra, Nairobi, etc. and the globally anticipated transition to autonomous vehicles (Morrison and van Belle 2020).

14.6 Steps Towards Realization

A safe public transportation system will be fundamental for sustaining the economic and social growth of the rapidly urbanizing Africa, and it will be essential to ensuring that this growth is inclusive. Presently, public transport services are predominantly used by disadvantaged and vulnerable population groups of society (Porter et al. 2020). A significant proportion of the growing middle class also rely on some form of public transport for their daily mobility needs—until a point where they can afford a private vehicle. Robust public transport is key to serving the full spectrum of users and discouraging the expansion of private vehicle use that, given the challenges (physical, financial, social, etc.) of adding roadway capacity, will only exacerbate existing road safety issues, further the crippling traffic congestion (and extend it further outside the cities), as well as render functional mobility and accessibility more exclusive rather than inclusive.

Overcrowding in mass intra-city public transport vehicles is a common phenomenon across the continent, particularly during peak hours on weekdays, and this is becoming an increasingly problematic issue for the safety, security, and comfort for customers as more and more people use these buses. Sam et al. (2014) have observed that passengers' perception of safety and security is cardinal to their choice of public transport operator.

Accessibility issues threaten inclusivity of public transport in SSA in two main ways: accessible services for the general public and accessible vehicles for vulnerable transport users such as people living with disabilities, women, children, and the elderly. With respect to physical accessibility of transport services, users tend to find the most problematic part of the journey to be getting from home to the stop and getting from the vehicle to their final destination—the first/last mile problem of public transport (Shaheen and Chan 2016). The future of successful, high-ridership systems may or may not involve modernized vehicle fleet but will certainly include route developments that provide reasonably fast and frequent service close to the various origins and destinations that frame the rhythm of daily life. Two- and three-wheeler motorcycle taxis are currently preferred modes to overcome the first/last mile problem in many sprawling cities across the continent. To overcome the first/last mile problem, many progressive cities in the developed countries are using innovative transport solutions such as shared micromobility (scooters, bicycles, and electric versions of both). Unfortunately for many African cities, the first/last mile journey often comprises less than well-maintained surfaces on lower classification roads and streets, and many lack basic infrastructure such as footpaths—these conditions can be particularly difficult at night where no lighting is present. It follows that mobility needs of vulnerable transport users must be factored into the planning policies from the offset, with the focus on inclusion set on adapting the environment and not the person, as retrofitting accessibility features later may prove to be a challenge.

In order to overcome the safety and security issues of public transport in SSA, investment in cleaner, safe vehicles that prevent crashes and protect road users, including occupants, pedestrians and cyclists, in the event of a crash is highly needed. The continent needs to move from the use of motorized two-and three-wheelers as primary transport on main roads and integrate them into an overall functionally hierarchal system wherein they (as well as micromobility technologies) serve as feeders for a more planned overall system. Such systems would involve concerted efforts at land use planning with particular focus towards transit-oriented development.

There will be a need for governments to exercise stronger influence on the activities and operation of public transport, beyond user price controls. A precursor to that will be the need to subsidize the cost of replacing older public transport vehicles with modern zero- and low-emission vehicles that provide greater access and promote safety and inclusiveness especially for people living with disabilities. Such progress is not without precedent. The South African government, having recognized the high volume of road-unworthy and non-compliant taxis, that compromise the safety of all road users, undertook a taxi recapitalization programme (TRP) in 2006. This programme is aimed, inter alia, to improve safety for public transport passengers and integrate the informal public transport sector into a more formal and regulated environment, allowing for the identification of taxis (Chiloane-Tsoka 2016; South African Government 2021). Though a costly exercise, it has helped to directly address the hostile challenges that come with unregistered taxis, endangering the lives of both passengers and other drivers, therefore, improving the safety satisfaction of users in an environment mostly reluctant to apply scheduling and route allocation controls.

In South Africa, access to public transport is considered a basic right for all citizens (Act 108 of 1996, page 1251). This basic right is intended to be complemented by a safe, timely, and affordable transport for all users. It is a bold step and has set a precedent for other countries—in fact, Mexico just recently passed national legislation declaring access to safe, inclusive mobility a right afforded to its citizens (Hidalgo 2020). Such an effort will require leadership and partnerships across a range of stakeholders. The open market business module of the informal public transport sector may need to be further re-structured to ensure that only well-trained drivers and riders are licensed to operate public transport vehicles. By assigning each operator to an identifiable public transport association, it will be easy to regulate their operations. Road safety outreach and education programmes can then be adequately targeted if public transport driver associations are well established. Additionally, adequate security at transit nodes (stations and stops), along transit corridors and routes, and the feeder roads that connect the transit network and the neighbourhoods and regions it serves, is a critical component that drives transit ridership, especially for women, children, elderly, and persons with disabilities.

Ultimately, the safety and security concerns of public transport are underpinned by the regulatory frameworks and governance structures that influence the operation of the various modes (Pirie 2014). Thus, although there are significant similarities across the continent, the various components of the public transport system are embedded, and intricately and uniquely linked to the broader local socio-political and cultural current. Very often, the multi-institutional umbrella under which public transport systems operate hinder an effective oversight role in realizing the sector goals. According to Cirolia et al (2020), understanding the "institutional and territorial dynamics of African cities requires an understanding of African local governments".

The public transport governance structure, therefore, needs to be reassessed at the local level to establish clear mandates for the various institutions to work synergistically towards the creation of an inclusive and safe public transport system that can help African cities develop sustainably. This has the potential to be a game changer for the rapidly urbanizing SSA. The ability of local governments to build an efficient public mobility system for the growing urban workforce will, therefore, be fundamental to achieving sustainable socioeconomic development and livability throughout the region. In addition to building safer road infrastructure and replacing ageing vehicle fleets, changing operator and user behaviours will be a critical component of this holistic systems approach to safe and inclusive public transportation as has been shown in this chapter. The chapter has also summarized some of the major challenges affecting public transportation across SSA, while also identifying and illustrating lessons and opportunities to address them through human factors focused efforts (education, outreach, training, and enforcement), as well as deployment of new technologies to improve the safety and operation of public transport throughout sub-Saharan Africa.

References

Abegaz T, Berhane Y, Worku A, Assrat A, Assefa A (2014) Effects of excessive speeding and falling asleep while driving on crash injury severity in Ethiopia: a generalized ordered logit model analysis. Accid Anal Prev 71:15–21

Ackaah W (2019) Exploring the use of advanced traffic information system to manage traffic congestion in developing countries. Sci Afri 4:e00079

Ackaah W, Adonteng DO (2011) Analysis of fatal road traffic crashes in Ghana. Int J Inj Contr Saf Promot 18(1):21–27

Ackaah W, Afukaar FK (2010) Prevalence of helmet use among motorcycle users in Tamale Metropolis, Ghana: an observational study. Traffic Inj Prev 11(5):522–525

Adanu EK, Jones S, Odero K (2020) Identification of factors associated with road crashes among functionally classified transport modes in Namibia. Sci Afri 7:e00312

Agyemang W, Adanu E, Jones S (2021) A safe systems based comparison of motorcycle crash severity in rural and urban areas of Ghana. J Adv Transport 6336517

Aidoo EN, Amoh-Gyimah R (2019) Modelling the risk factors for injury severity in motorcycle users in Ghana. J Public Health 1–11

Akaateba MA, Yakubu I, Akanbang BAA (2015) Correlates and barriers associated with motorcycle helmet use in Wa, China. Traffic Inj Prevent 16(8):809–817

Akpa N, Booysen M, Sinclair M (2016) Auditory intelligent speed adaptation for long-distance informal public transport in South Africa. IEEE Intell Transp Syst Mag 8(2):53–64

Ambunda R, Nakale E (2020) Towards safe, inclusive and sustainable public transport solutions in Namibia. In: 8th International Conference on Sustainable Development. New York, NY, 20–21

Ambunda and Sinclair, 2020 Ambunda R, Sinclair M (2020) Road safety perceptions and attitudes in Namibia. In: Namibia road safety project. Stellenbosch University

Amweelo M (2016) The road safety in Namibia: focus on road traffic accidents. Int Sci Technol J Namibia 7:103–119

Asongu S, Le Roux S (2017) Enhancing ICT for inclusive human development in Sub-Saharan Africa. Technol Forecast Soc Chang 118:44–54

Behrens R, McCormick D, Mfinanga D (2015) Paratransit in African cities: operations, regulation and reform. Routledge, London

Boutueil V, Quillerier T (2020) The rise of paratransit digital platforms in large African metropolises: competitive strategies between international and local platforms. In: Proceedings of the 99th annual meeting of the transportation research board. Washington DC

BRRI (2016) Road traffic crashes in Ghana, statistics 2015. Building & Road Research Institute (BRRI), Kumasi, Ghana

Chiloane-Tsoka G (2016) Factors affecting the implementation of the taxi recapitalization project: the department of transport. Probl Perspect Manag 14(4):25–32

Cirolia LR, Harber J, Croese S (2020) Governing mobility in Sub-Saharan African Cities. The state of knowledge and research. Volvo Research and Educational Foundations, Gothenburg, Sweden

Dapilah F, Guba BY, Owusu-Sekyere E (2017) Motorcyclist characteristics and traffic behaviour in urban Northern Ghana: implications for road traffic accidents. J Transp Health 4:237–245

Date-Bah E (1980) The Inscriptions on the vehicles of Ghanaian commercial drivers: A sociological analysis. J Mod Afr Stud 18(3):525–531

GRSF (Global Road Safety Facility, The World Bank; Institute for Health Metrics and Evaluation) (2014) Transport for health: the global burden of disease from motorized road transport. The World Bank Washington, DC, IHME, Seattle, WA

Herko S (2019) A blockchain infrastructure for transportation in low income country cities, and beyond. J Br Blockchain Assoc 2(2):1–7

Hidalgo D (2020) Mexico's 'Right to Mobility' amendment could shift road safety discourse and save thousands of lives. The City Fix. https://thecityfix.com/blog/mexicos-right-to-mobility-amendment-could-shift-road-safety-discourse-and-save-thousands-of-lives/. Accessed May 3, 2021

Hua L, Noland B, Evans A (2010) The direct and indirect effects of corruption on motor vehicle crashes. Accid Anal Prev 42:1934–1942

Jonck P, Swanepoel E (2016) The influence of corruption: a South African case. Policing Int J Police Strat Manag 39(1):159–174

Jones S, Tefe M, Zephaniah S, Tedla E, Appiah-Opoku S, Walsh J (2016a) Public transport and health outcomes in rural Sub-Saharan Africa—a synthesis of literature and professional opinion. J Transp Health 3(2):211–219

Jones S, Walsh J, Appiah-Opoku S (2016b) Rural transport health and safety in Sub-Saharan Africa research needs—a snapshot of expert opinion from an online survey. Dev South Afr 33(5):677–702

Jones S, Odero K, Adanu EK (2019) Road crashes in Namibia: challenges and opportunities for sustainable development. Dev South Afr 37(2):295–311

Jones S, Makamani R (2021) Assessing the impact of transport barriers on University student experience in Windhoek, Namibia. In: Working paper to be submitted to the Journal of African Economies

Jones S, Bullard J, Johnson E (2021) Application of machine learning to automate the identify road safety issues: case study of the B2 highway in Namibia. In: Working paper to be submitted to the African Journal of Science and Technology

Kisekka J (2019) You won't be able to Use UberBoda after 11 pm in new safety measure. Dignited. Kampala, Uganda. https://www.dignited.com/51698/you-wont-be-able-to-use-uberboda-11-pm-in-new-safety-measure/. Accessed March 3 2021

Kloop J, Cavoli C (2019) Mapping minibuses in Maputo and Nairobi: engaging paratransit in transportation planning in African cities. Transp Rev 39(5):657–676

Kudebong M, Wurapa F, Nonvignon J, Norman I, Awoonor-Williams JK, Aikins M (2011) Economic burden of motorcycle accidents in Northern Ghana. Ghana Med J 45(4)

Larsson P, Tingvall C (2013) The safe system approach–a road safety strategy based on human factors Principles. In: International conference on engineering psychology and cognitive ergonomics. Springer, Berlin, Heidelberg, pp 19–28

Lufumpa C, Yepes T (2017) Rapid urbanization and the growing demand for infrastructure in Africa. In: Lufumpa C, Ncube M (eds) Infrastructure in Africa: lessons for future development

Majid F (2019) Does the ride-sharing service Uber reduce drunk-driving deaths in South Africa? Health Policy Res 14(1)

Mannering FL, Shankar V, Bhat CR (2016) Unobserved heterogeneity and the statistical analysis of highway accident data. Anal Methods Accid Res 11:1–16

Morrison G, van Belle J (2020) Customer intentions towards autonomous vehicles in South Africa: an extended UTAUT Model. In: 2020 10th International Conference on Cloud Computing, Data Science & Engineering (Confluence). Noida, India, 29–31 January, 525–531

Motor Vehicle Accident Fund (2019) Road crash and claims report 2019

National Road Safety Commission (2013) Road traffic crashes in Ghana. Statistics, Accra, Ghana

Naumann R, Sandt L, Kumfer W, LaJeunesse HS, Lich K (2020) Systems thinking in the context of road safety: can systems tools help us realize a true "safe systems" approach? Curr Epidemiol Rep 7:343–351

Nimako Aidoo E, Bawa S, Amoako-Yirenkyi C (2018) Prevalence rate of helmet use among motorcycle riders in Kumasi, Ghana. Traffic Inj Prevent 19(8):856–859

Njenga P, Odero K (2014) African position paper on the role of transport in the sustainable development goals. AFCAP2/ASCAP: GEN2006A, UKAID, London, UK

Njoh A (2003) Urbanization and development in sub-Saharan Africa. Cities 20(3):167–174

Okafor I, Odeyemi K, Dolap D (2013) Knowledge of commercial bus drivers about road safety measures in Lagos, Nigeria. Annal Afric Med 12(1):34–39

Oluwadiya K, Ojo O, Adegbehingbe O, Mock C, Popoola OS (2016) Vulnerability of motorcycle riders and co-riders to injuries in multi-occupant crashes. Inter J Inj Control Saf Promot 23(2):189–196

Pearce T, Maunder DAC, Mbara TC, Babu DM, Rwebangira T (2000) Bus accidents in India, Nepal, Tanzania, and Zimbabwe. Transp Res Rec 1726(1):16–23
Peprah C, Amponsah O, Oduro C (2019) A system view of smart mobility and its implications for Ghanaian cities. Sustain Cities Soc 44:739–747
Pirie G (2014) Transport pressures in urban Arica: practices, polices, perspectives. In: Pieterse E, Parnell S (eds) Africa's urban revolution
Porter G, Abane A, Lucas K (2020) User diversity and mobility practices in Sub-Saharan African cities: understanding the needs of vulnerable populations. The state of knowledge and research. Volvo Research and Educational Foundations, Gothenburg, Sweden
Sam E, Abane A (2017) Enhancing passenger safety and security In Ghana: appraising public transport operators' recent interventions. J Sci Technol 37(1):101–112
Sam EF, Adu-Boahen K, Kissah Korsah K (2014) Assessing the factors that influence public transport mode preference and patronage: perspectives of students of University of Cape Coast (UCC), Ghana. Int J Dev Sci 3(2):323–336
Sam E, Daniels S, Brijs K, Brijs T, Wets G (2018) Modelling public bus/minibus transport accident severity in Ghana. Accid Anal Prev 119:114–121
Shaheen S, Chan N (2016) Mobility and the sharing economy: potential to facilitate the first-and last-mile public transit connections. Built Environ 42(4):573–588
Silyanov V, Sodikov J, Kiran R, Sadikov A (2020) An overview road data collection, visualization, and analysis from the perspective of developing countries. IOP Conf Ser Mater Sci Eng 832:012056
Slavova M, Okwechime E (2016) African smart cities strategies for Agenda 2063. Africa J Manag 2(2):210–229
South African Government (2021) Taxi recapitilization programme, government programmes
St-Aubin P, Saunier N, Miranda-Moreno L (2015) Large-scale automated proactive road safety analysis using video data. Transport Res Part C Emerg Technol 58(Pt B):363–379
Tankebe J, Boakye KE, Amagnya MA (2020) Traffic violations and cooperative intentions among drivers: the role of corruption and fairness. Polic Soc 30(9):1081–1096
Too L, Earl G (2010) Public transport service quality and sustainable development: a community stakeholder perspective. Sustain Dev 18:51–61
Ulzen T, Higginbotham JC, Donnir G, Jerome L, Segal A (2018) Undiagnosed attention deficit/hyperactivity disorder (ADHD) among unionized drivers in Ghana: public health and policy implications. Accid Anal Prev 114:12–16
UN ESCAP (2020) Safe and inclusive transport and mobility. Note by the secretariat. United Nations Economic and Social Council. ESCAP/CTR/2020/4
UN-Habitat (2015) Sustainable urban development in Africa. United Nations Human Settlement Programme, Nairobi, Kenya
UN-Habitat (2016) Urbanization and development: emerging futures, world cities report 2016. Human Settlement Programme, Nairobi, Kenya
United Nations (2015) Transforming our world: the 2030 Agenda for Sustainable Development, A/RES/70/1. United Nations General Assembly, New York, NY, USA
Wells P, Beynon M (2011) Corruption, automobility cultures, and road traffic deaths: the perfect storm in rapidly motorizing countries? Environ Plan A Econ Space 43(10):2492–2503
WHO (World Health Organization) (2018) Global status report on road safety 2018
World Bank (2017) The high toll of traffic injuries: unacceptable and preventable. World Bank, Washington, DC
Zhen J, Chan C, Schooness A, Apatu E, Thabane L, Young T (2020) Transmission of respiratory viruses when using public ground transport: a rapid review to inform public health recommendations during the COVID-19 pandemic. South African Med J 110(6)

Steven Jones is a Professor of Civil, Construction & Environmental Engineering at The University of Alabama. He is the Deputy Director of the Alabama Transportation Institute where also leads its Transportation Policy Research Center. Steven has more than 29 years of transportation engineering and planning experience spanning both academic and private sector appointments. He spent 2019 teaching and conducting transportation research as a Fulbright Scholar at the Namibia University of Science and Technology. Steven has authored/co-authored more than 200 refereed publications and numerous conferences papers, design manuals, and project reports. He serves on the editorial boards of the Journal of Transportation Safety and Security and the African Geographical Review.

Emmanuel (Kofi) Adanu is an Associate Research Engineer in the Transportation Policy Research Center at The University of Alabama. Kofi focuses his research on sustainable transportation solutions with a special emphasis on road safety and applications of advanced technologies. Kofi earned a B.S. in civil engineering from the Kwame Nkrumah University of Science and Technology (KNUST), Ghana. He also holds an MSc and PhD in civil engineering from The University of Alabama where he also obtained an MsC in Applied Statistics.

Charles Adams is Professor and the Programme manager of the Road and Transportation Programme at KNUST and the Lead of the Road and Transportation group at the KNUST. Charles is the Director of the Transportation Research and Education Centre Kumasi (TRECK). He has expertise in traffic and transportation infrastructure systems planning, design, monitoring and maintenance. Charles has also contributed extensively to road condition monitoring for the Ministry of Roads and Highways, Parking planning and management in cities in Ghana and served as technical auditor on numerous transportation projects. He has participated and organized several national and international conferences on transportation and civil engineering. Charles is currently the President of the Ghana Institution of Engineering.

Robert Ambunda is a researcher and academic in the departments of Civil Engineering at the University of Namibia (Namibia) and Stellenbosch University (South Africa). His areas of research interest encompass transport modelling, transportation planning and design, Intelligent Transport Systems (ITS), MAAS and road safety. He has worked on multiple transportation infrastructure projects with particular focus on data driven road safety and public transport policy initiatives. He holds a Masters (MSc) and Ph.D. in Civil Engineering from Stellenbosch University.

Part IV
Non-motorized Transport and Traffic Safety

Chapter 15
Non-motorized Transport and Traffic Safety in African Cities: An Introduction

Chinebuli Uzondu

This section is dedicated to case examples of studies that provide a unique contribution to understanding and improvement of Non-Motorized Transport (NMT) and traffic safety in Africa. It offers readers a critical look at some key emerging issues within and around NMT and traffic safety and how infrastructure and policy challenges associated with progress in these domains can be addressed. The study of NMT and traffic safety could be approached from different perspectives and requires various approaches to understand and appreciate the concept. Studies presented in this section used a wide range of methods to conduct research into this increasingly important concept. They represent research carried out in real life and show the current challenges and experiences of the selected population. They identify relationships among people and environments that affect NMT and traffic safety. Topics include the future of non-motorized transport in Urban Africa; reflecting on walkability in Johannesburg; prospects for on-campus cycling on university campuses: the case of KNUST and the provision of NMT in the city of Lusaka: an analysis of policy and practice.

This section constitutes four (4) chapters, three of which focus on the most common aspect of NMT and traffic safety—Infrastructure & Safety—whilst the remaining one is related to steps taken to develop friendly policies targeted at improving NMT projects. The studies were carried out in different cities in Africa (Nairobi, Kenya; Accra (2), Ghana; Lagos, Nigeria; Kampala, Uganda; Johannesburg, South Africa and Lusaka, Zambia) and provided research evidence-based interesting and insightful suggestions and practical implications for policymakers.

Wood in Chap. 16 focusing on walkability, highlighted practical solutions by showing the steps taken in Johannesburg to develop pedestrian-friendly policies and projects using examples from three different projects. The author identified three

C. Uzondu (✉)
Department of Transport Management Technology, Federal University of Technology Owerri, Owerri, Nigeria
e-mail: chinebuliuzo@yahoo.com; Chinebuli.uzondu@futo.edu.ng

R. A. Acheampong et al. (eds.), *Transport and Mobility Futures in Urban Africa*, The Urban Book Series, https://doi.org/10.1007/978-3-031-17327-1_15

overlapping approaches used to design and develop walkable infrastructure in Johannesburg which includes: high-profile, flagship projects to publicize the principal role of walking in transport planning; integrating infrastructure with city-wide principles and policies and developing infrastructure in partnership with the local community and users.

Obeng and Komla in Chap. 17 focus on examining cycling in Ghana and how it is perceived by the general population and the authorities. The authors showed that there are currently various negative cultural stereotypes limiting the adoption of cycling in Ghana. However, this study shows how this mode could be developed for sustainable transport by exploring the feasibility of introducing on-campus cycling in a university environment. Through an online survey, the authors showed that current bicycle ownership on-campus was very low. However, most of the respondents expressed their willingness to use a bicycle on-campus if infrastructure was provided to support this mode and guarantee the safety of riders.

Through direct observations and key informants' interviews, Nchito and Lukomona in Chap. 18 provide insights into infrastructure conditions, challenges and integration of policies related to NMT on selected roads in Lusaka. They highlighted the challenges NMT users were facing which include road safety, encroachment of roadsides, poor state of the roads with respect to infrastructure provision and inadequate road safety education. The authors further revealed that there are huge opportunities for integrating NMT infrastructure policies to improve these existing roads in the city of Lusaka.

In Chap. 19, Uzondu and Etika attempt to provide an understanding of the role of NMT in improving urban mobility in Africa while highlighting the current trends and conditions of NMT in selected cities. Through online interviews with stakeholders, the authors present case studies from four selected cities in Africa by looking at the current condition of NMT in these cities, opportunities for improving the infrastructure, challenges associated with developing these, design of NMT environment and policy development. Experiences shared from various cities seemed to be similar, especially regarding the current condition of NMT. However, the study showed that there are currently a lot of opportunities that could be leveraged to improve the present condition.

Before going into the detailed information provided in this section, we offer a general understanding of NMT and traffic safety and how NMT provision could contribute to the sustainability of urban transport systems in Africa.

15.1 Understanding NMT and Traffic Safety: An Overview

Non-Motorized Transportation includes all forms of travel that do not rely on an engine-powered motor for movement. According to Litman (2012), it refers to walking, cycling, wheelchair, scooter and handcart use. They are commonly used in making short trips, especially in developing countries. It could be while going to bus stations to board vehicles, to the mall or running errands within walking and cycling distances etc. These modes are readily accessible to the population and their overwhelming benefit for public health and environmental sustainability cannot be overemphasized. According to VTPI (2010), they provide recreation, are cost-effective and bring about large health, economic and social benefits, particularly for those who do not have access to transport at all and whose transport needs cannot be satisfied by motorized vehicles. NMTs have significant environmental benefits, helping to reduce pollution levels and promoting physical activity. Studies have shown that 5–10% of car trips can be replaced by NMT provided good policies are in place to support the use of these modes (VTPI 2010). Properly designed walking and cycling routes can increase people's access to jobs, schools, and healthcare. However, despite the enormous benefits of NMT, key challenges related to poor infrastructure, high cost of bicycles, public perception, and safety of users negatively affect its use and these also carry large economic and social costs. Among these challenges, the most reported is safety- Safety of the users.

Safety means being protected from harm. Therefore, traffic safety focuses on any action undertaken to protect road users from harm in traffic. It is aimed at promoting and maintaining safety in traffic. NMT modes without adequate infrastructure may be risky for their mainly vulnerable users and therefore may lead to increase in traffic crashes. An example is the WHO report which shows that about 26% of road traffic crashes (RTCs) happening annually in the world involves vulnerable road users (VRUs- pedestrians, cyclists and motorcyclists) who have little or no physical protection. The rate of RTCs involving VRUs is disproportionately borne by countries as the African region has the highest proportion of deaths at 44% (Fig. 15.1). Economic development and the rapid growth of motorized transport have contributed to this burden in Africa. Crashes involving vulnerable road users are often avoidable, and efficient interventions exist, yet in most African cities, their safety does not attract the attention it deserves (World Health Organization 2013). VRUs constitute those who use the NMT modes.

In Africa, most NMT users walk and cycle. The challenges they face are often safety-related which is due to the lack of adequate infrastructure to support the modes. Mokitimi and Vanderschuren (2017) has shown that the majority of people do not use NMT modes because of the risk of getting into crashes. Could this be related to the seeming lack of or badly designed infrastructure? Current transportation infrastructure in most African cities does not encourage the use of NMT. This is because the traffic system has been designed for motorized transport. Transport policies and budgets clearly show that motorized transport modes are prioritised above NMTs. Lack of institutional support has been a hindrance to the development of NMTs in African cities. It is interesting to note that in most parts of Africa, roads are currently being designed and constructed without considering the needs and safety of NMT modes and users (Uzondu et al., 2018). NMTs share the same road space with motorized vehicles. They are constantly being exposed to unregulated traffic which compromises their safety. As a result, when there is an interaction involving an NMT, they are mostly affected because they are particularly vulnerable to injuries and crashes and are not protected the same way as the other road users (cars, buses, and tricycles). Kim and de Jong (2011), argue that there is little or no adequate

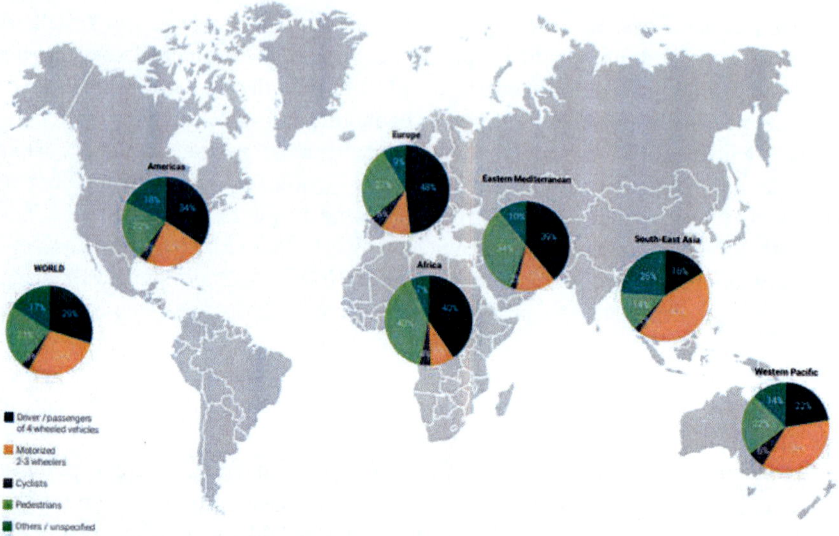

Fig. 15.1 Distribution of deaths by road user type by WHO Region (WHO 2018)

road infrastructure for sustainable transport in Africa, especially for walking and cycling. Additionally, the little NMT spaces available (footpaths and cycle lanes) have been converted to illegal car parks while some are now largely occupied by street traders/vendors, worsening an already bad situation (Transport Research Laboratory 2002). Transportation networks could be designed and constructed to accommodate all transport modes. For NMTs, this would mean a provision of more open spaces, footpaths and cycle paths and restricting motorized vehicles from accessing these areas (Tseu 2006). Institutional systems that promote and encourage NMT would ensure their inclusion and integration into city planning processes.

15.2 NMT and Sustainability of Urban Transport Systems in Africa

In recent times, there has been more interest in providing sustainable transport systems as it has become the centre of most transport decision-making. According to Yuri (2002), sustainable transport systems must meet the mobility and accessibility needs of people by providing safe and environmentally friendly modes of transportation while improving the quality of life. Sustainable transport systems are fundamental to meeting the 2030 agenda for sustainable development and in achieving the 17 SDGs (Global Mobility Report 2017).

The approaches to achieving these differ between cities. However, there is a consensus that there must be adequate planning and implementation of strict policies to achieve it. This is to say that to achieve success in cities, the focus must be on influencing policy decisions, especially in the planning and design of transport systems. Sustainable transport systems encourage a shift towards more sustainable transport modes and acknowledge the interdependence between these modes. It aims to achieve a balance in the provision of traffic infrastructure and supports the integration and balanced development of all modes of transport. Sustainable, inclusive road design and mobility are very important for economic growth and providing access to productive opportunities (Murguía 2013). Meeting these targets would involve the adoption of a more integrated approach to sustainable mobility planning and policymaking to deal with the complexity of urban transport systems. Sustainable transport systems reduce the negative impact of transport on the environment, road safety, traffic congestion, air pollution and climate change and most importantly promotes healthy living. This aligns with the key benefits of NMT as discussed in the previous section.

The development of NMT including walking and cycling is a key element in effectively promoting inclusive, integrated and sustainable urban transport in Africa. It remains a viable option in meeting any sustainable approach to urban transport but is often ignored in the planning and design of transport systems, especially in Africa. They play important roles in sustainable living. Current policies focus on the construction of more roads which encourages the use of motorized vehicles. Additionally, the negative effects such as accidents and air pollution impact unequally on society, particularly the low-income groups (Attard 2020). For this group, NMT is the most common form of transport used for short trips (for example to and from bus stops/stations, short distances to the markets, mall etc.). It provides an essential connection between different modes of transport and serves as a feeder mode to public transport facilities. Thus, ensuring the successful use of NMT in an urban environment would require both motorized transport and NMT to work efficiently together so as to integrate the transport network (Simões 2014). Lack of integration among the different transport modes and lack of overall long-term, vision and planning are major obstacles in the provision of sustainable transport services in African cities. Modes with potential for integration must be identified. For example, this could be the integration of NMT modes with the public transport system as each provides their own unique services to commuters. By providing facilities and infrastructure for the NMT, the existing public transport can be effectively used and will lead to reduction in air pollution and other transport externalities. However, providing suitable infrastructure solutions for integrated transport will depend on the broader traffic, environmental and planning objectives and on funds available to undertake these ventures which are expected to benefit the wider community.

Therefore, going forward, the focus in African cities should be on improving existing NMT as one of the key transport modes, integrating it as an essential element of public transport, making it very safe by providing infrastructure and allocating adequate and sustainable funding for its development and promotion. NMT-friendly policies and infrastructure have the potential of contributing to achieving sustainability in transport—providing safe and clean mobility to all city dwellers (Tiwari 2015).

The four chapters in this section revealed that the major problem affecting NMT, and traffic safety is a general lack of investment in Infrastructure. Appreciating the enormous benefits of NMT modes, developing, and implementing friendly policies to sustain them and integrating them into the current transport system could go a long way as prospects and opportunities appear to exist for encouraging the use of NMT in Africa if the facilities needed for safe and convenient experiences are provided.

References

Attard M (2020) Mobility justice in urban transport-the case of Malta. Transp Res Proc 45:352–359

Global Mobility Report (2017) Tracking sector performance. http://www.sum4all.org

Kim P, de Jong R (2011) High time to change road investment patterns in. http://www.unep.org/Transport/sharetheroad/PDF/StR_article_UrbanWorldDec2010.pdf. Accessed 21 Dec 2020

Litman T (2012) Evaluating non-motorized transportation benefits and costs. Victoria Transport Policy Institute

Mokitimi MM, Vanderschuren M (2017) The significance of non-motorised transport interventions in South Africa—a rural and local municipality focus. Transp Res Proc 25C:4802–4825

Murguía RO (2013) Share the road: design guidelines for Non-motorised transport in Africa, Nairobi. UNEP Transport Unit

Simões V (2014) An evaluation of the inclusion of principles of corridor development, transit-oriented development and non-motorised transportation within Johannesburg's Corridors of Freedom: the case of Turffontein Corridor. https://wiredspace.wits.ac.za/handle/10539/20470. Accessed 10 March 2020

Tiwari G (2015) Role of non-motorized transport and sustainable transport in Indian cities. In: Dev S, Yedla S (eds) Cities and sustainability. Springer Proceedings in Business and Economics. Springer, New Delhi. https://doi.org/10.1007/978-81-322-2310-8_7

Transport Research Laboratory (2002) Scoping study: urban mobility in three Cities. SSATP Working Paper No. 70

Tseu DL (2006) Urban transport growth: the challenges ahead—the new realism and institutional changes. State Planning Unit, Chief Minister's Office, Sarawak

Uzondu CC, Jamson S, Lai F (2018) Exploratory study involving observation of traffic behaviour and conflicts in Nigeria using the traffic conflict techniques. Saf Sci 110(pt a):273–84

VTPI (2010) Online TDM encyclopedia. Victoria Transport Policy Institute www.vtpi.org/tdm

World Health Organization (2013) Projections of mortality and causes of death, 2015 and 2030. http://www.who.int/healthinfo/global_burden_disease/projections2015_2030/en/

World Health Organization (2018) Global status report on road safety 2018. World Health Organization, Luxembourg. https://www.who.int/violence_injury_prevention/road_safety_status/2018/en/

Yuri T (2002) Sustainable transportation system: a system dynamics approach. In: Proceedings of the 37th annual conference of the Canadian transportation research forum

Chinebuli Uzondu is a Lecturer and Researcher at the department of Transport Management Technology, Federal University of Technology Owerri Nigeria. She has previously worked as a Post-doctoral researcher with the Human Factors and Safety Group at the Institute for Transport Studies, University of Leeds, United Kingdom where she obtained her Ph.D. Her research seeks to understand road user behaviour through projects funded by various research organisations and the Nigeria and UK governments. This involves using both quantitative and qualitative research methods, to inform the design of transport infrastructure and/or technology interventions, and to evaluate the effectiveness of these interventions.

Chapter 16
Planning for Walkability in Johannesburg

Astrid Wood

Abstract Walkability is defined by the extent to which the built environment supports and sustains walking by providing pedestrian comfort, connectivity, and convenience. This chapter considers the steps taken in Johannesburg to develop pedestrian-friendly policies and projects. It focuses on three examples of walkability projects—the Grayston Pedestrian and Cycle Bridge, the Milpark Pedestrian Bridge, and the Westbury Pedestrian Bridge and Park. These examples provide valuable insights into the South African understanding of non-motorized transport, as well as how these experiences can provide lessons for cities around the world. This chapter, therefore, bridges the lacuna between empirical accounts of walking in Africa and theoretically rich discussions of walkability. In so doing, it problematizes both the concept of walkability as a planning instrument used to promote inclusivity, livability, and sustainability in the African context. Such a critical reading of the intertwined and overlapping practices of policymaking provides insight into the future of urban development and spatial transformation in (South/ern) Africa as well as across the global south.

Keywords Non-motorized transport (NMT) · Johannesburg · South African cities · Walkability

16.1 Introduction

For Walter Benjamin, the flâneur is an affluent resident of nineteenth century Paris who aimlessly strolls the city for entertainment (Lauster 2007). Such a romantic (and unrealistic) image stands in sharp contrast to the contemporary pedestrian scurrying across six lanes of fast-moving cars to get to a low-paid job, or the seller trudging along the unpaved, unimproved verge taking their wares to the market. And so it is not surprising that for the most part cities have been unashamedly auto-oriented and roadways designed to accommodate speed, efficiency, and circulation.

A. Wood (✉)
School of Geography, Politics and Sociology, Newcastle University, Newcastle upon Tyne NE1 7RU, UK
e-mail: astrid.wood@ncl.ac.uk

R. A. Acheampong et al. (eds.), *Transport and Mobility Futures in Urban Africa*, The Urban Book Series, https://doi.org/10.1007/978-3-031-17327-1_16

This chapter engages with concepts of non-motorized transport (NMT) and specifically walkability in the African context by considering its applicability in the South African city of Johannesburg. In so doing, it problematizes both the concept of walkability as a worldwide planning instrument used to promote inclusivity, livability, and sustainability.

This discussion of walkability is particularly relevant in African cities where pedestrian rates are higher than in their European and North American counterparts. On average, across African cities 50% of trips made on foot—with as many as 81% in Dakar or 70% of trips in Kinshasa and Addis Ababa and as few as 42% in Ouagadougou made on foot (Pendakur 2005). And in many African cities, rapid urbanization has been accompanied by a high incidence of poverty among urbanites. The simultaneous concentration of formal economic activity within the central business district(s) and dispersal of residential developments as well as the ambiguous role of government as both a transport policymaker and a policy regulator also exacerbates existing economic and social disparities (Parnell and Pieterse 2014). In South Africa however, pedestrianization and other forms of NMT are particularly noteworthy as they overlap with ongoing efforts to redress the economic and social inequalities leftover from decades of apartheid spatial planning.

These analyses add to ongoing conceptualizations of urban mobilities in the global south (Uteng and Lucas 2017) and specifically transport planning (Appelhans et al. 2021) and NMT in African cities (Mitullah et al. 2017). This is certainly not the first attempt to feature walking within discussions of NMT in African cities—studies of Praia, Cabo Verde (Anciaes et al. 2017) and Nairobi, Kenya (Nyamai 2021) serve as pertinent antecedents. Likewise a host of international (Montgomery and Roberts 2008; Pendakur 2005) and South African policy documents (Harber et al. 2018; Pieterse and Owens 2018) highlight the important role NMT and walking play in African development. By drawing on policies and projects in Johannesburg, this chapter aims to bridge the lacuna between empirical accounts of walking in Africa (Behrens 2005) and theoretically-rich discussions of walkability (Zavetoski and Agyeman 2015).

Empirically, this chapter considers the preparation of walkability infrastructure for pedestrians and pedestrian-oriented cities in accordance with Johannesburg's mobility plans. At the broadest level, transport policies aim to improve the quality of life for present and future generations; while in a more prescriptive sense, policies identify specific dimensions of paving width and amenities. This chapter calls for three overlapping tactics for designing and developing walkable infrastructure: first, high-profile, flagship projects can be used to publicize the principal role of walking in transport planning, e.g., the Grayston Pedestrian and Cycle Bridge; second, infrastructure needs to be integrated with city-wide principles and policies, e.g., the Milpark Pedestrian Bridge; and third, infrastructure should be developed in partnership with the local community and users, e.g., Westbury Pedestrian Bridge and Park. These three approaches illustrate how cities like Johannesburg should actually 'do' walkability.

Methodologically this chapter will analyze the practices by which city officials introduce walkability policies and planning. This chapter is the outcome of a recurring engagement with urban policy actors inside and outside government. Between 2012 and 2020, interviews were conducted with architects, consultants, politicians, and planners involved with NMT in South Africa as well as international transport experts. I also reviewed the relevant frameworks (e.g., City of Johannesburg 2009), legislation (e.g., National Department of Transport 2009), and international guides (e.g., Fang 2013); and the archives of the Johannesburg Development Agency, the implementation arm of the City of Johannesburg, has provided essential primary data for this chapter (e.g., JDA 2015a, 2018).

The next section of the chapter considers how walkability has been conceptualized across a range of geography and planning literature before outlining two major gaps in the literature: (i) research of the "downsides of walkability" and (ii) "walkability-with-people". From there the chapter turns to an empirical discussion of how to make cities more walkable by considering the pedestrian-friendly policies and pedestrian-friendly projects introduced in Johannesburg. Policies and frameworks as well as three infrastructure projects—the Grayston Pedestrian and Cycle Bridge; the Milpark Pedestrian Bridge and the Westbury Pedestrian Bridge and Park—provide valuable insight into the South African understanding of walkability as well as how these experiences can provide lessons for cities around the world. In conclusion, I will reflect on the main theoretical arguments that serve to problematize concepts of walkability.

16.2 Conceptualizing Walkability

Academics and planners have tried (unsuccessfully) to develop a clearer understanding of how to create and calculate the walkability of the city. In its broadest sense, walkability is understood as an alignment between "residents' desires and expectations for types of destinations, their willingness to walk a given distance and the quality of the required path" (Manaugh and El-Geneidy 2011); and walkability is often used as a planning instrument to promote principles of inclusivity, livability, and sustainability. In a more prescriptive sense however, walking refers to all forms of bipedal movement. Southworth's (2005, p. 248) definition of walkability as "the extent to which the built environment supports and encourages walking by providing for pedestrian comfort and safety, connecting people with varied destinations within a reasonable amount of time and effort, and offering visual interest in journeys throughout the network" is perhaps most applicable for urban research and practice.

Walkability can also be a planning tool used to measure the extent to which the urban realm can be characterized as connected (i.e., comprehensive network of pathways), convenient (i.e., direct routes without lengthy diversions), comfortable (i.e., no steep gradients, good weather, good lighting, feeling safe), convivial (i.e., the presence of landscaping and street furnishings) and conspicuous (i.e., legible routes with clear signage, integration with public transport) (Pharoah and Buchan 1997). Forsyth and Southworth (2008, p. 2) go on to suggest an association between walkable and "close" (i.e., a short distance between origin and destination), "barrier-free" (i.e., accessible for all bodies), "safe", (i.e., void of crime and cars), "full of pedestrian infrastructure and destinations" (i.e., sidewalks, raised pedestrian crossings, street furniture, trees), and "upscale, leafy, or cosmopolitan" (i.e., "pleasant for middle-class professionals who have other choices for getting around"). The quality of pedestrian facilities, roadway conditions, and land use patterns, community support, security, and comfort for walking are all often cited components of walkability (Ewing and Cervero 2010; Litman 2003).

Indeed walkability is a characteristic of all cities—although the scholarship generally focuses only on its historical development in European and North American cities (Forsyth and Southworth 2008). European cities of the middle ages were by necessity walkable with an average spread of just a half-mile. Cities remained walkable through the industrial era; Urbino, Italy for instance accommodated 30,000 people within an area of 300 acres warranting a density of dwellings and diversity of overlapping activities. Meanwhile, until the nineteenth century every point in Boston, USA was less than one mile apart (Southworth 2005). The introduction of transport services—the horse-drawn cart, the horse-drawn tram, the electric tram, and the automobile—fueled the dispersal of urban life and the accompanying decline in walkability. Logically then transport studies of the past century have favored the unimpeded circulation of the car and the proliferation of roadways (Alcaˇntara De Vasconcellos 2004). However, alarms emerged as early as the 1920s that walkability was being eroded by the car. Accordingly, early criticisms of auto-oriented living focused on the neighborhood level with the development of the "neighborhood unit" (Perry 1929), a concept used to delineate districts of the city and develop local amenities. There was concern among academics and practitioners that roadways were being used to segregate residential, business, and green spaces and that the pedestrian experience was being lost (Southworth 2005).In reaction to the continued proliferation of car-usage and low-density sprawl in the postwar period, scholarship continued to question the relationship between traffic and land use. In particular, authors advocate for a limiting of vehicle growth as well infrastructure to support walking, cycling, and public transport (Khayesi et al. 2010).

Increasingly global understandings of walkability can be found in a range of contemporary terminology: "universal design"—adding a dropped curb to make the urban realm accessible for disabled people developed by Selwyn Goldsmith (1963), "road diets"—converting a four-lane road to three lanes comprised of two through lanes and a center two-way left lane (Labuschagne and Ribbens 2014) and "complete streets"—coined in 2003 by Barbara McCann and the American advocacy organization America Bikes to consider the whole right-of-way, e.g., sidewalks, crossings, cycle lanes, road medians, and so forth—in the design, planning, operations, and maintenance of the road. A complete street includes cyclists, pedestrians, motorists, and "transit riders of all ages and abilities" in transport planning (McCann and Rynne 2017, p. 3). The concept was picked up in the United States by the American Planning Association and American Association of Retired Persons as well as by international advocates including the Bicycle Empowerment Network based in Cape Town and Embarq in Mexico City (Prytherch 2018). This more international and inclusive terminology according to Zavetoski and Agyeman (2015) is a form of "street-level spatial justice", a "democratization of the street" by decentering drivers and redistributing the right of the right-of-way to the public. And Litman (2015, p. 14) suggests that complete streets policies "help achieve equity objectives by giving non-drivers a fair share of road space".

It is important to note that while the scholarship on walkability remains interdisciplinary, encompassing discussions in urban planning, transport, urban design as well as geography, much of the existing literature is situated in Europe and North America even as the concept has increasing international relevance (Krambeck 2006; McCann and Rynne 2017). Indeed the African perspective of walkability is quite different because walking is usually out of necessity, not choice, and infrastructure is far less formalized. An African flâneur, as depicted in Ivan Vladislavić's novel *Portrait with Keys* (2006), is "anxiously alert, but nevertheless manages to immerse himself in the city" (Gaylard 2017, p. 57). For the most part, African cities deter flâneurism (Gaylard 2017, p. 60); indeed, "the people who walk Johannesburg daily are not flâneurs at all, but migrants, or workers, to whom the city still denies the right to public transport" (Gevisser 2014, p. 20). "It takes a person of some physical statute and courage to roam its streets" explains Gaylord (2017, p. 60), those "who pound the pavements on a daily basis do not do so with impunity and without trepidation". Rapid urbanization across African cities has exacerbated the difficulties of walking: in Addis Ababa for example, sprawling development in recent years has made a previously pedestrian-friendly city increasingly unwalkable. Likewise, rapid population growth in Nairobi, another historically walkable city, has led to increasingly high numbers of accidents between motorists and pedestrians.. This chapter thus aims to theorize the way in which the walkability concept has been applied in an African/global south city. This more global orientation allows for additional opportunities to critique the concept and consider its universal applicability.

An African/global south perspective on walkability warrants further reflection in two areas: the "downsides of walkability" and "walkability-with-people". First, for the most part existing research presumes walking as a universal good (Litman 2015; Southworth 2005). Walking is said to have many advantages: it is cost-effective, flexible to the changing economic and social opportunities, healthy for pedestrians and sustainable for the environment (Ewing and Cervero 2010). The theoretical thrust of this chapter, therefore, aims to depict walking and walkability as more than merely "active transport" which presupposes issues of health and environmentalism. Indeed, there are noteworthy downsides to walking—it is time-consuming, slow, uncomfortable and at times dangerous, and it is contingent on the physicality of the pedestrian as well as the urban conditions available. Some of these challenges (e.g., the urban conditions) can be overcome by policy and infrastructure but others (e.g., physicality of walking) cannot. Much of the discourse on walkability then should be aligned with concepts of transport-justice. A person is transport poor according to Lucas (2018) if any of the following conditions apply: the transport options do not reach either the origin and/or the destination, or the available transport modes are not suited to the person's physical capabilities, or the cost of transport pushes the household below the poverty line or the duration of the journey is excessive, or the conditions of the journey are unsafe or otherwise unhealthy. Under any of these circumstances, a person's quality of life, health or opportunities for social advancement will be restricted. Consequently walkability must be motivated by a pro-poor agenda that recognizes that walking is not always the most desirable option but it is often the only option.

A second motivation of the chapter is to advocate for planning processes that promote walkability-with-people. This would fulfill an egalitarian perspective on transport-justice, which reasons that transport has a social benefit for all users and therefore investment should be made to increase access for the maximum number (Van Wee and Geurs 2011). Some extant research considers the role of residents in walkability—indeed Appleyard's (1981) early work on livability emerged from his study of residential experiences in San Francisco and more recent research has tried to conceptualize the role of the built environment in promoting walkability (Leslie et al. 2005) or the effects of walkability on property values and investment returns (Pivo and Fisher 2011). Still most research presumes residents to be the users rather than the drivers of better urban planning. Instead, the literature preferences the role of urban planners in guiding walkability and while it is certainly important for cities to put forward frameworks and policies, the agenda should be built in partnership with local communities. Indeed there is opportunity as well as need to develop policy alternatives "that maximize the minimum level of primary goods to the people in the worst-off position" (Pereira et al. 2017, p. 175) so that "everybody should be well-off up to a certain minimum threshold" (2017, p. 477). This would fulfill a transport-justice mission as well as enable cities to develop transport solutions that actually work for the community utilizing them. The section that follows applies these concepts for further interrogation in a case study of Johannesburg.

16.3 Making the Case for Walkability in Johannesburg

Johannesburg is particularly unwalkable. Expansive low-density urban sprawl accompanied by unevenly allocated pavement, scarce street-level amenities, and inhospitable high-walls means that most pedestrians are poor and otherwise stranded. And yet, South African cities, like many cities of the global south, have a disproportionately high number of walkers comprised primarily of commuters marginalized from automobile usage and ownership.

Of the 3.5 million daily trips in Johannesburg 31% or about 800,000 trips are made by walking ("all the way") (City of Johannesburg 2013a). These "captive walkers" are pedestrians because they have no alternative (Montgomery and Roberts 2008). Data reveals that Johannesburg 66% of households do not own a car (City of Johannesburg 2009); and therefore 60% of households spend more than 10% of their income on motorized forms of public transport (including informal transit), up from 55% in 2014 (Gauteng Department of Roads and Transport 2020). These captive walkers are also "time poor" with the average pedestrian exerting significantly more time and energy on mobility (Montgomery and Robert, 2008). Over the past five years, walking time to the first public transport stop has increased from nine minutes to 14, while access to the final destination has increased from eight minutes to 14 (Gauteng Department of Roads and Transport 2020). Moreover pedestrians account for 40 percent of all road fatalities in South Africa (Road Traffic Management Corporation 2018). Such data not only triggers renewed attention to pedestrian behavior and supporting infrastructure but also the economic and social demands to end the legacy of spatial apartheid in South African cities.

Foot traffic has been a part of Johannesburg since the beginning. Morgan (2021) reveals that Johannesburg streets were originally multi-purpose spaces of business and leisure. In South Africa, in particular the public realm was a place of conflict in which racial segregation was highlighted by the banning of black people from walking on the pavement. And in post-war suburbanization—as in most cities around the world—cars, the symbol of modernity, came to dominate the streets. Apartheid-era town planners gave little thought to walking or walkability and for the next few decades planners were singularly focused on building asphalt highways to accommodate the white elite who could afford to drive (Wood 2020). Unsurprisingly, apartheid-era planners only gathered data for motorized trips and relatively little is known about non-motorized or off-peak travel in this era (Behrens 2005). Today, South African politics continues to draw heavily on the caricature of the wealthy white driver versus the poor black pedestrian as well as other associations between walking and poverty. A range of pedestrian-friendly policies and projects are required to remedy these perceptions, a topic to which we now turn.

16.3.1 Pedestrian-Friendly Policies

In the postapartheid era, transport has become a tool for social and spatial integration, and NMT increasingly recognized as a pro-poor approach. As part of these efforts, the City of Johannesburg has developed several walkability documents: the Framework for Non-motorized Transport (City of Johannesburg 2009) and the Complete Streets Design Guidelines 2013 (City of Johannesburg 2013b) as well as the city's Strategic Integrated Transport Framework (City of Johannesburg 2013a). These documents are backed by the 2040 Growth and Development Strategy (City of Johannesburg 2011), the primary document guiding Johannesburg's planning, which calls for greater social inclusivity and cohesion, reasoning that improvements to transport and mobilities will further these social goals.

These documents promote walking as both an access mode and an access sub-mode where appropriate and they call for new infrastructure to support walking as well as better integration with public transport. At the broadest level, the 2013 Strategic Integrated Transport Framework aims for transport infrastructure and systems to improve the quality of life for present and future generations, while in a more prescriptive sense, Johannesburg's Non-motorized transport Framework identifies walking as the preferred mode for short-distances with student-school trips not exceeding 30 min and convenience-related trips to shops and services of up to 15 min. Also importantly, it requires all new developments to incorporate sidewalks through the development site and immediately adjacent to the site boundary. No provisions are made for retrofitting walkability infrastructure into existing sites or the larger transport network.

In 2016, Johannesburg's African National Congress (ANC) Mayor Parks Tau (2011–2016) proposed an ambitious plan to "re-stitch" the city by integrating sustainable transport with high-density, mixed- residential, retail, and business developments as well as educational and leisure facilities. The Corridors of Freedom, as the plan was dubbed, aimed to integrate Johannesburg's economic and social divisions through a series of transport-led interventions (Wood 2021). "NMT is a crucial part of the transport system" explains the Strategic Area Framework for the Empire-Perth Development Corridor, one of the key planning documents for the Corridor of Freedom, and "everyone should have access to urban opportunities and mobility with the necessary safety precautions in place" (City of Johannesburg 2012, p. 26). In addition to upgrading parks, clinics, libraries, and sports facilities, the plans call for pedestrian-friendly interventions in particular pedestrian bridges across highways and railway lines (e.g., the Grayston Pedestrian Bridge and the Milpark pedestrian bridge, and the Westbury Bridge). This approach was selected for its potential to connect rich and poor, white and black, residential and commercial areas in adherence with the complete streets principles. It is to a discussion of this infrastructure that this chapter now turns.

16.3.2 Examples of Pedestrian-Friendly Projects

As a result of these pedestrian-friendly policies, a range of pedestrian-friendly projects have been introduced into new and upgraded developments in Johannesburg: improved sidewalks, crosswalks, and paths; pedestrian-oriented land use and building design; increased road connectivity with additional shortcuts for pedestrians and cyclists; traffic calming measures and vehicle restrictions; safety education and law enforcement; and vehicle parking policies and fuel taxes (JDA 2018). Pedestrian bridges are among the more visible outcomes of this substantial investment in walking infrastructure and policymaking. These projects align with ongoing investment in Johannesburg's Rea Vaya BRT (e.g., the Grayston Bridge is part of Phase 1C running along Louis Botha Avenue and the Milpark Bridge and the Westbury Pedestrian Bridge are part of Phase 1B operating between Soweto and the CBD) as well as policies promoting the Corridors of Freedom and Joburg 2040. At the other end of the spectrum and far less costly are trees and shrubs, lighting, benches, street sellers' stalls, and security cameras (JDA 2015b).

These walking amenities all subscribe to the complete streets concept which according to Christine Walters, then-Mayoral Committee for Transport, is "a modern global trend in urban design" in which "the road network and transport systems were regarded as social infrastructure that served the entire community, rather than, predominately, the private transport user" (JDA 2015a). Only by implementing both the visible and less-visible, expensive and less-expensive, material and non-material infrastructure can Johannesburg build pathways for mobility. The remainder of this section outlines an example of each of these approaches. Taken together they illustrate how Johannesburg has tried to actually do walkability albeit to varying degrees of success.

The first example, the Grayston Pedestrian and Cycle Bridge, which opened in March 2018 and forms part of the Great Walk, is a multimillion Rand high-profile investment to support walking and cycling linking Alexandra (a.k.a., Alex, a historically black area and now a very deprived neighborhood) and Sandton (Johannesburg's secondary CBD and Africa's wealthiest square-mile, home to international companies and the city's wealthy elite). The pedestrian bridge is far-reaching with a reported 10,000 domestic workers, gardeners, and shop workers walking across Johannesburg's busiest highway, the M1, every morning and evening. And the Great Walk is substantial—the 5.2-km journey was previously a dangerous and difficult trek with people scurrying along the concrete shoulder of a car-bridge and/or walking circuitous routes to avoid vehicular traffic (JDA 2015a).

Glitzy infrastructure like the Grayston Bridge is not only meant to provide a new pedestrian experience but also to transform the public transport system in both the wealthy and deprived parts of the city. The bridge according to Christine Walters, then-Mayoral Committee for Transport "literally bridges the divide between these two contrasting communities and creates a visual gateway into the economic heartbeat of our city". Sello Lemao, then-Member of the Mayoral Committee for Public Safety concluded that transport interventions such as the pedestrian bridge served to transform the city making it more "sustainable, livable and economically vibrant" (City of Johannesburg 2015). But it is important to realize that the pedestrian bridge is not the only intervention on the Great Walk. More mundane upgrades—the pavement on Katherine Street and Marlboro Drive was widended from 1.8 to 3-m and new streetlights and underground service infrastructure were installed (JDA 2015a)—are essential for the complete streets concept, however, these vernacular topographies are rarely promoted and valorized. This first Johannesburg-based example thus exhibits how flagship projects are used to broadcast the importance of walking but also include a host of more vernacular interventions.

The second case of Johannesburg's approach to walkability aligns infrastructure with international concepts (e.g., complete streets) and citywide policies (e.g., Corridors of Freedom) as well as between private and public sector organizations. Infrastructure such as the Milpark Pedestrian Bridge, completed in December 2019, was introduced as part of the Corridors of Freedom. The project is not only inscribed in the Empire-Perth Corridor of Freedom but it is also supported by the Milpark Precinct Urban Design Framework which was approved by Council in 2016, both which were conceptualized to attract density and diversity to the area (JDA 2020).

This particular intersection was selected because the pedestrian bridge, which crosses Barry Hertzog Avenue at the intersection of Empire Road, is located in an area that incorporates a mixture of land-uses including office, retail (e.g., 44 Stanley), residential, media (e.g., SABC and Media24), health facilities (e.g., Milpark Hospital), education (e.g., University of Witwatersrand and University of Johannesburg), hotels, and public space. High traffic volumes on both Empire Road and Barry Hertzog Avenue were seen as obstacles to pedestrian movement creating disconnections between institutions and landmarks. Consequently several of these businesses became partners in the project. This second Johannesburg-based case demonstrates the possibilities for introducing walkability when policies align and partners work in concert.

A third example of Johannesburg's implementation of walkability is the focus on local community. The Westbury Pedestrian Bridge and Park provides insight into the opportunity to build walkability through enhanced city-community partnerships. The pedestrian bridge includes an amphitheater and a park at a cost of R25 million and was completed in December 2016 (JDA 2016b). Situated between two deprived communities, Coronationville and Westbury, the infrastructure literally bridges a divide. Important for this third feature of Johannesburg-based walkability: the project came from the Westbury community itself who during a stakeholder consultation with the city requested a pedestrian bridge for people to move safely across Fuel Road.

The City commissioned a study that revealed that over 500 people per hour cross during the morning and evening peak. The community also expressed concern that the existing on-grade crossing was dangerous not only for pedestrians but also for school children. And they found the new Rea Vaya station difficult to access without an overheard connection. The pedestrian bridge thus also provides access to/from the Rea Vaya station but also includes a park and an amphitheater, amenities that the community requested. The bridge of course is not the only walkability project—complete streets-inspired infrastructure is planned for Kretzschmar Street, Dowling Avenue, and Steytler Road as well as additional pathways in neighboring Sophiatown and Wesdene (JDA 2016a). And note that the Westbury Bridge is also part of the Empire-Perth Corridor of Freedom and is supported by a precinct-level plan (Iyer Architects 2018).

What is more, these projects have yet to fulfill the broader promises of postaparthied social and spatial integration, commitments which were central to their support (McCool 2016).

With increased costs, long delays, and widely publicized accidents, critics increasingly note the under-utilization of these pedestrian bridges (Kekana and Allison 2019). In 2019, the City conducted a count of non-motorized transport across Grayston Drive. Of the more than five-thousand pedestrians crossing the M1 highway, only 26% used the pedestrian bridge. Instead people walk on the wider southern sidewalk (Loots 2019). The report goes on to suggest that widening existing sidewalks on either side of the vehicle bridge "might have been a more cost-effective way of solving the problem" (Loots 2019, p. 6). Moreover, such an approach would have better considered "the people who need to cross the bridge between Alexandra and Sandton", the report concludes (Loots 2019, p. 6). This is just one example in which walkability policies and projects failed to lead significantly shift the profile of the pedestrian, the pedestrian experience or the promises of post-apartheid social and spatial integration (McCool 2016). It is important to realize however that these are long-term investments whose realization is still many years into the future and patience and persistence is fundamental to making a change in any city.

16.4 Concluding Remarks

This chapter has argued that in Johannesburg at least, walkability is more than a planning instrument through which to build comfortable, connected, and convenient pathways but rather a far-reaching synonym for accessibility, inclusivity, public space, economic and social development, and livability—or as one planner put it, "all of those things that I would just consider to be good planning," (interview, 12/2019). Walkability also fit well within broader policy thinking on transit-oriented development and complete streets as well as the Corridors of Freedom policy. Indeed, no singular approach could cover all the possibilities or provide for all opportunities. Instead, walkable neighborhoods should be developed in partnership with existing infrastructure, policies, and the community. Together these three approaches provide an opportunity to reimagine walkability in the African context, not as a prescriptive mechanism for promoting walking but as a flexible concept that can and should be applied as appropriate with the local context.

For the future of transport research, this chapter further complicates the notion of walkability by challenging calculative studies of transport geography. Rather than associating walkability with a series of metrics, this chapter concludes that walkability should be about everyday people and everyday mobilities. Efforts to improve walkability in South Africa, in particular, overlap with efforts to redress the economic and social inequalities leftover from decades of apartheid spatial planning. Walkability is therefore part of a broader commitment to social and spatial integration.

For South African scholarship, as well as urban studies in general, this chapter confirms the importance of locally-made decisions. Indeed the (South) African experience of walkability provides valuable openings for understanding how to build and retrofit accessible, connected, and safe pathways on the continent. The examples in South African cities direct transport planners—and academics as well—to take not only the expensive flagship infrastructure but also the vernacular enhancements, the paved sidewalks, and pedestrian crossings, seriously for transport transformation. Indeed the empirical stories help us to generate a more complicated and richer understanding of walkability both within the continent and around the world.

References

Alca˘ntara De Vasconcellos E (2004) The use of streets: a reassessment and tribute to Donald Appleyard. J Urban Des 9(1):3–22

Anciaes PR, Nascimento J, Silva S (2017) The distribution of walkability in an African city: Praia, Cabo Verde. Cities 67:9–20

Appelhans N, Scholz W, Baumgart S (eds) (2021) Transport planning and mobility in Urban East Africa. Routledge, Abingdon

Appleyard D (1981) Liveable streets. University of California Press, Berkeley

Behrens R (2005) Accommodating walking as a travel mode in South African cities: Towards improved neighborhood movement. Plan Pract Res 20(2):163–182

City of Johannesburg (2009) Framework for non-motorized transport. Department of Planning and Urban Management, Johannesburg

City of Johannesburg (2011) Johannesburg 2040: growth and development strategy. Office of the City Manager, Johannesburg

City of Johannesburg (2012) Strategic area framework: empire Perth development corridor. Johannesburg, South Africa

City of Johannesburg (2013a) Strategic integrated transport framework. Department of Transport, Johannesburg

City of Johannesburg (2013b) Complete streets design guidelines. City of Johannesburg, Johannesburg

City of Johannesburg (2015) Construction of the 'Great Walk' Bridge commences. Johannesburg Development Agency, Johannesburg

Ewing R, Cervero R (2010) Travel and the built environment: a meta-analysis. J Am Plann Assoc 76(3):265–294

Fang K (2013) Urban design manual for non-motorized transport-friendly neighborhoods. World Bank, Washington, D.C.

Forsyth A, Southworth M (2008) Cities afoot—pedestrians, walkability and urban design. J Urban Des 13(1):1–3

Gauteng Department of Roads and Transport (2020) Gauteng province household travel survey report. Gauteng Department of Roads and Transport, Tshwane, South Africa

Gaylard G (2017) An African Flaneur? walking in Johannesburg in portrait with keys. In: Moreillon O, Muller A, Stiebel L (eds) Cities in flux: metropolitan spaces in South African literary and visual texts. Lit Verlag, Zurich, pp 57–72

Gevisser M (2014) Lost and found in Johannesburg. Jonathan Ball, Johannesburg

Goldsmith S (1963) Designing for the disabled: the new paradigm. Routledge, London

Harber J, Parker A, Joseph K, Maree G (2018) Taking streets seriously. Gauteng City Region Observatory, Johannesburg, South Africa

Iyer Architects (2018) Project five: corridors of freedom. In: Pro landscaper Africa, pp 45–46

JDA (2015a) Great walk will link Alex and Sandton over the M1. Johannesburg Development Agency, Johannesburg

JDA (2015b) Construction of R130m Alex-Sandton bridge gets under way. Johannesburg Development Agency, Johannesburg, South Africa

JDA (2016a) Westbury's new Corridor of freedom landmark starts taking shape. Johannesburg Development Agency, Johannesburg

JDA (2016b) Westbury Park & Pedestrian Bridge open for use by the community. Johannesburg Development Agency, Johannesburg

JDA (2018) Non-Motorised Transport (NMT) Project under construction-Westbury development renewal precinct–NMT phase 4. Johannesburg Development Agency, Johannesburg

JDA (2020) Milpark Pedestrian Bridge reaches practical completion. Johannesburg Development Agency, Johannesburg, South Africa

Kekana M, Allison S (2019) Bridges only work when both sides cross them. YES! Magazine

Khayesi M, Monheim H, Nebe JM (2010) Negotiating "Streets for All" in urban transport planning: the case for pedestrians, cyclists and street vendors in Nairobi, Kenya. Antipode 42(1):103–126
Krambeck HV (2006) The global walkability index
Labuschagne K, Ribbens H (2014) Walk the talk on the mainstreaming of non-motorised transport in South Africa. Tshwane, South Africa, pp 191–206
Lauster M (2007) Walter Benjamin's myth of the "Flâneur." Mod Lang Rev 102(1):139–156
Leslie E, Saelens B, Frank L, Owen N, Bauman A, Coffee N et al (2005) Residents' perceptions of walkability attributes in objectively different neighbourhoods: a pilot study. Health Place 11(3):227–236
Litman T (2003) Economic value of walkability. Transp Res Rec 1828(1):3–11
Litman T (2015) Evaluating complete streets: the value of designing roads for diverse modes, users and activities. Victoria Transport Policy Institute, Vancouver, Canada
Loots A (2019) Great walk: assessment of NMT interventions by the JDA. Johannesburg Development Agency, Johannesburg
Lucas K (2018) Editorial for special issue of European transport research review: transport poverty and inequalities. Eur Transp Res Rev 10(17):1–3
Manaugh K, El-Geneidy A (2011) Validating walkability indices: how do different households respond to the walkability of their neighborhood? Transp Res Part D 16:309–315
McCann B, Rynne S (2017) Complete streets: best policy and implementation practices. American Planning Association, Washington, D.C.
McCool A (2016) A walk to freedom: can Joburg's bridges heal the urban scars of apartheid? The Guardian
Mitullah WV, Vanderschuren M, Khayesi M (eds) (2017) Non-motorized transport integration into urban transport planning in Africa. Routledge, London
Montgomery B, Roberts P (2008) Walk urban: demand, constraints and measurement of the urban pedestrian environment. World Bank, Washington, D.C.
Morgan N (2021) Could Covid-19 change how cities view their streets? New Frame
National Department of Transport (2009) National land transport act
Nyamai DN (2021) Urban mobility and spatial justice: prospects of non-motorized transport in Nairobi. In: Appelhans N, Scholz W, Baumgart S (eds) Transport planning and mobility in Urban East Africa. Routledge, London, pp 31–42
Parnell S, Pieterse E (eds) (2014) Africa's urban revolution. Zed Books Ltd., Johannesburg
Pendakur VS (2005) Non-motorized transport in African cities. World Bank, Washington, D.C.
Pereira RHM, Schwanen T, Banister D (2017) Distributive justice and equity in transportation. Transp Rev 37(2):170–191
Perry C (1929) The Neighborhood unit-a scheme of arrangement for the family-life community, regional survey of New York and its environments. Committee of Regional Planning of New York, New York
Pharoah T, Buchan K (1997) Putting London back on its feet. London Planning Advisory Committee, London
Pieterse E, Owens K (2018) Johannesburg: confronting spatial inequality. World Resources Institute, Washington, D.C.
Pivo G, Fisher J (2011) The walkability premium in commercial real estate investments. Real Estate Econ 39(2)
Prytherch D (2018) Law, engineering and the American right-of-way. Palgrave Macmillan
Road Traffic Management Corporation (2018) State of the road report. National Department of Transport, Tshwane, South Africa
Southworth M (2005) Designing the walkable city. J Urban Plan Dev 131(4):246–257
Uteng TP, Lucas K (eds) (2017) Urban mobilities in the global south. Routledge, London
Van Wee B, Geurs K (2011) Discussing equity and social exclusion in accessibility evaluations. EJTIR 11(4):350–367
Vladislavic I (2006) Portrait with keys: the city of Johannesburg unlocked. Portobello Books, London

Wood A (2020) Tracing the absence of bike-share in Johannesburg: a case of policy mobilities and non-adoption. J Transp Geogr 83:1–7

Wood A (2021). Problematizing concepts of transit-oriented development in South African cities. Urban Studies

Zavetoski S, Agyeman J (eds) (2015) Incomplete streets: processes, practices and possibilities. Routledge, London

Astrid Wood is an Associate Professor in the School of Geography, Politics and Sociology at Newcastle University in the United Kingdom. Dr Wood is an urban geographer specializing in infrastructure, governance and transport. She is the author of *How Cities Learn: Tracing Bus Rapid Transit in South Africa* (2022) as well as over 30 peer-reviewed book chapters and journal articles on transport and urban studies.

Chapter 17
Exploring Barriers and Prospects of Bicycle Transportation: A Case Study of a Ghanaian University Campus

Daniel Atuah Obeng and Emmanuel Komla Junior Dzisi

Abstract Cycling as a form of sustainable transportation continues to experience only marginal growth in Ghana owing to the perception of it being a mode of commute for the poor. Besides this factor are the lack of supportive infrastructure and the negative cultural stereotypes that limit individuals' adoption of cycling. University campuses, however, present a distinct opportunity to provide a different perspective on bicycling. This is partly because university campuses are relatively smaller communities in their own right, having spatially separated activities that require members to travel on a daily basis. The pro-active educational environments of campuses, however, make them ideal settings for championing sustainable transportation practices. Additionally, universities provide a critical mass of young people, many of whom are willing to experiment with new ideas, including, perhaps, culturally limited modes of transport, such as bicycling. This chapter shares insights from a recent study that sought to evaluate the factors underlying bicycle ridership on the campus of the Kwame Nkrumah University of Science and Technology, Ghana, and the prospects for bicycle use among students. The study considered current psychosocial perceptions and some specific infrastructural challenges limiting the use of bicycles. The results showed that bicycle ownership and use on the University campus was very low (7.4%) although, the majority of the respondents (85%) also expressed willingness to use bicycles on campus if the required infrastructure was provided in support of bicycling, to guarantee rider safety. It is recommended that the university implements some minor infrastructural improvements (such as the covering of open roadside drains) that help to increase the roadway width available to cyclist, and improve feelings of safety. Further, an on-campus bicycle share scheme could be piloted to make bicycles available to interested cyclists and, through this, help determine the real adoption intentions of commuters on campus.

Keywords Bicycle · Transportation · University · Ghana · Sub-Saharan Africa

D. A. Obeng (✉) · E. K. J. Dzisi
Department of Civil Engineering, Kwame Nkrumah University of Science and Technology, Kumasi, Ghana
e-mail: obengatuah@yahoo.co.uk

R. A. Acheampong et al. (eds.), *Transport and Mobility Futures in Urban Africa*,
The Urban Book Series, https://doi.org/10.1007/978-3-031-17327-1_17

17.1 Introduction

Due to the adverse impact of greenhouse gases on the environment and human health, sustainable transportation practices, of which bicycle transportation is a part, are being encouraged within the transportation sector as part of the global effort to ensure environmental sustainability. Bicycle transportation is ideal for short trip distances and offers society an emission-free low noise alterative to motorized transport. Compared to motorized vehicles, bicycles are an affordable mode of transport and can be used by most people who cannot afford personal motorized vehicles.

On university campuses, in particular Balsas (2003) is of the view that there are compelling reasons for greater bicycle use as the pro-active educational environments of such places provide ideal settings for championing sustainable transportation practices. Besides, short trip distances favor bicycle use (Tiwari and Jain 2013), and the majority of trips to and between activity centers on university campuses fall into this category. On University campuses, however, (Balsas 2003) argues that the neo-traditional town setting of these places makes them unique in their land use and function. Additionally, the sense of community that can be created around bicycles could foster their use. Campuses are also places where people of different backgrounds, incomes, lifestyles, and attitudes come together to live, study, work, and recreate. In this sense, university campuses build societies that are at once transitory and lasting and have an ideal human scale (Ojeda et al. 1997). On university campuses, the levels of prejudice against bicycling for instance could also be lower, since people from different backgrounds (including those from less bicycle-friendly places and those from more bicycle-friendly places) can be found. The traditional campus also concentrates a variety of functions within reach of pedestrians (Dulken 1992; Turner 1995). Campuses are often self-contained neighborhoods where classrooms, offices, apartments, student centers, child care facilities, performance halls, art galleries, gymnasiums, swimming pools, sports arenas, and shopping places are all in close proximity. This makes the use of non-motorized modes of transport such as walking and cycling more feasible. Additionally, campuses have their own streets, squares, and open spaces, where people can stroll and get together. Although most campuses do not totally exclude the automobile, walking is the expected way to get around even though other ways of transportation may also be possible. This makes them a good example of a 'people's place' (Balsas 2003). According to Huang et al. (2012), if universities implemented policies that encouraged non-motorized transport on campuses, they would be contributing to the sustainable development of society. However, the promotion of on-campus bicycling must be done against the backdrop of understanding the psychology and culture of bicycling. Goetzke and Rave (2011) describe bicycle culture as a social phenomenon where a higher bicycle mode share motivates others to also ride a bicycle. This is an inducing effect apparently encouraged by the general feeling of being safe in a group due to improved visibility and group dynamics (Pucher et al. 2010). Aside this, Goetzke and Rave (2011) also believe that more students may take bicycling if it comes into vogue or is seen as the practical means of travel on campuses. Therefore, bicycling must be encouraged for its many health benefits which are similar to those of walking, for a given trip length.

Mobility and modal choices are influenced by the socio-economic status of the trip maker, but in the case of the bicycle as a transport mode, it generally breaks this bound. Nevertheless, the perception and response of the masses toward its use also play a key part in its mode share (Dill and Voros 2007; Fernández-Heredia et al. 2014). In addition, socio-cultural considerations may discourage the use of bicycles despite the fact that economic constraints may make bicycles essential in making daily travel easier and cheaper for the urban population. According to Dias Batista (2010) and Nkurunziza (2013), in some cultures, the bicycle is considered a vehicle for the poor. However, in others, it may not be cultural biases or some externalities but the lack of the appropriate transportation infrastructure that may be militating against greater bicycle use.

The extent to which some of these factors would play out in a university setting in Ghana is not clear. Nonetheless, bicycle transportation requires bicycle lanes or a complete network of bicycle tracks that are completely physically segregated from motorized traffic for safety reasons (Tiwari and Jain 2013). The potential for bicycle use cannot be fully harnessed when the infrastructure for that mode of transport does not exist or is poor. However, when cities invest in road infrastructure that is biased in favor of motorized transport, the environment is rendered more hostile for pedestrians and bicyclists (Tiwari 2001) who may genuinely feel intimidated and unsafe using the system. The need for the safety of the cyclist cannot be truer in Ghana where cyclists form a minuscule proportion (<3% in urban areas) of road users (Acheampong and Siiba 2018). In spite of this, they account for 3.7% of all road traffic fatalities (Afukaar et al. 2003) due to a general lack of appropriate supportive infrastructure for bicycles in urban areas where motorized traffic poses greater risks to cyclists. The limited bicycle use on most university campuses in Ghana is no doubt a reflection of the non-even and generally low bicycle ownership and use in the country as a whole. Except for the northern part of Ghana (Northern, Upper East, and Upper West Regions) where bicycles are a part of the culture to the extent that about 63% (Acheampong and Siiba 2018) of households in the rural areas of these regions own bicycles, bicycle use within the southern and mostly populous parts of the country is very limited as there generally exist, among other reasons, strong anti-cycle prejudices among some ethnic groups (Grieco et al. 1992) and parents (Amoako-Sakyi and Owusu 2011) brought up mainly in a road use regime that does not favor safe bicycle use. There as such seems to be justification for such prejudices as according to Quarshie (2007) most motorists contacted in a survey tended to see cyclists on the road as a nuisance.

17.1.1 Conceptual Framework for the Study

As a guide for the study, a conceptual framework was developed following a review of potential factors that could impact bicycle use on the campus. The following

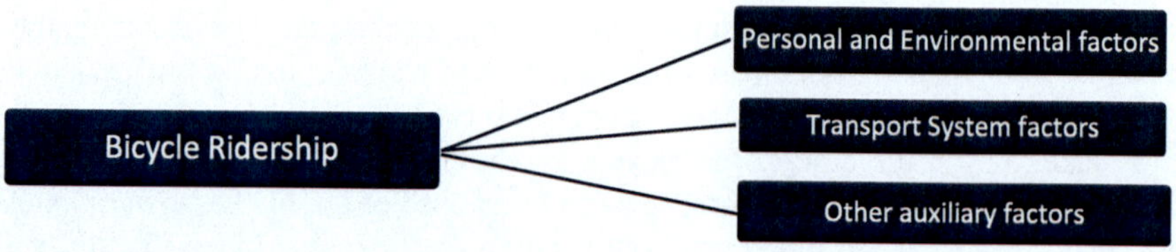

Fig. 17.1 Conceptual framework for the study based on factors that influence cycling

factors: health, weather conditions, terrain, bicycle ownership, inadequate cycling infrastructure, and speeds of motor vehicles on campus, identified based on the review of literature, as some of the underlying causes of low bicycle ridership on university campuses, were used to design questionnaires in this study in order to establish the extent to which the factors influenced the decisions of respondents to bicycles on campus. Questions were categorized under three main headings as follows using the previous work of de Sousa et al. (2014), Fernández-Heredia et al. (2014); and Dill and Voros (2007) as a guide. The eventual conceptual framework developed shows how a combination of these factors can influence cycling choice (Fig. 17.1).

Following the development of the framework, the authors sought to evaluate the prospects of on-campus cycling at KNUST, by assessing the bicycle use intentions of students on the campus. In Sect. 17.2, the materials and methods used in the collection and analysis of relevant data are identified. In Sect. 17.3, the results deduced following the analysis are outlined, while further discussion of these results and their implications are covered in Sect. 17.4. Conclusions and some recommendations are made in Sect. 17.5, and a list of relevant references is provided in References.

17.2 Materials and Methods

17.2.1 The Case Study Area and its Transportation System

The Kwame Nkrumah University of Science and Technology (KNUST) which was the site for this study was founded in 1951 and is one of the largest in Africa. The University campus is located about eight kilometers from the Central Business District of Kumasi and covers an area of about eighteen square kilometers (Niboi 2013). The main Campus has also been sub-divided into six broad land use categories. These are as follows:

- Academic/Teaching Area;
- Commercial Area;
- Senior Members Housing Area;
- Senior Staff and Junior Staff Housing Areas;
- Halls and Hostels Areas;
- Administration/Library/Great Hall Area (KNUST Planning Department 2014).

The map in Fig. 17.2 shows the key activity zones on the campus and their locations. The University campus has about 70 km of road infrastructure comprising mainly of asphalt concrete and bituminous surfaced pavements. Despite having a wide expanse of the road network, key activity zones that are highly patronized by staff and students are concentrated within a smaller area of the university campus. The majority of the other roads on the campus also lead to staff residences, on the eastern and western ends of the campus. Since most activity zones are concentrated within a relatively smaller area, travel distances between these zones are usually within 5 km, making most places easily accessible by walking and bicycling. Separated pedestrian walkways are available along most roads linking halls of residences to the main teaching and learning area/zone. However, pedestrians may occasionally share sections of the carriageway which do not have pedestrian walkways with motorists. Cyclists on the other hand have no designated cycling facilities and are required to share the carriageway with motorists. Although uncommon, cyclists can also be occasionally spotted using some of the pedestrian walkways particularly when the volumes of pedestrian traffic are low. In such instances, it is presumed that cyclists do this to lower their risk of collision with motorists, even if it requires them to be alert enough not to run into pedestrians as well. The university also has a shuttle bus service comprising of small 16-seater minibuses. These minibuses traverse specific corridors of the campus and charge users a fixed fee of about 13 US cents per trip*(2016 exchange rate). As a result of this, most students prefer walking, as it minimizes their overall expenditure on campus. Some students however use the shuttle service regularly, while others resort to it for relatively longer trips, or when absolutely necessary.

From a transport planner's standpoint, the introduction of bicycling on a university campus such as KNUST's may seem plausible and perhaps worth pursuing, but as to whether there is a latent demand for bicycle use to the extent that any investment in the provision of the necessary bicycle infrastructure could spur bicycle use and be justified is a matter that needs extensive interrogation, considering existing prejudices and the generally low use of that mode in the country as a whole. In light of these considerations, this study, therefore, explored the prospects that exist for the introduction of on-campus bicycle transportation at the Kwame Nkrumah University of Science and Technology in Ghana and the constraints that may hinder bicycle use on the campus.

17.2.2 Survey: Questionnaire Design and Data Collection

The questionnaire used in the study asked 26 questions in total. Some of these questions were Likert scale items, which respondents were expected to rate on a scale of 1 to 5. The preamble for questions (4a to 4i) was 'On a scale of 1 (very little) to 5 (very much), how much do the following factors influence your decision to cycle?', and the preamble for the rest (4j to 4n) was 'On a scale of 1 (very little) to 5 (very much), how much do the following factors encourage your decision to

Fig. 17.2 Map of key activity zones on the KNUST campus

cycle?'. The Google Forms platform was used to design and distribute questionnaires online to reach practically all students. Questions centered on mode use on campus, frequented locations, bicycle ownership, ability and willingness to ride, access to bicycle if the necessary infrastructure was put in place, etc. The questionnaires were kept active online for a period of 30 days. Text messages that included the web link of the online questionnaire were sent to all students (about 35,000) through the University Information Technology Service (UITS) with permission from the Dean of Students Office for ethical reasons. These were also supported with WhatsApp messages posted to various KNUST student platforms to encourage higher student participation in the survey. Students were particularly targeted because they represent the largest group on campus that walk, and, therefore, it was assumed that if their needs were fully understood and addressed perhaps that could give a boost to bicycle use on campus. Staff of the university were excluded from the survey mainly because most live off-campus and are not likely to shift mode and commute to work by bicycle in a city where the road infrastructure does not support that mode. In spite of these efforts to encourage student participation in the survey, only 330 responded to the questionnaires. Table 17.1 shows the 15 Likert scale items that formed part of the questionnaire.

Table 17.1 Items that formed the questionnaire

Personal and environmental factors	Physical well-being (health)
	Weather conditions
	Terrain
	Bicycle ownership
	Are bicycles fashionable?
Transport system factors	Lack of routes
	Lack of lanes
	Speeds of vehicles
	Narrow roadways
	Secure bike storage
Other auxiliary factors	Exclusive lanes
	Education on bicycling would encourage its use
	Safer bicycle infrastructure
	Friend factor

17.2.3 Approach for Analysis

After sorting and data cleaning, 300 responses (8.6% of the student population) were considered valid for analysis. Based on the nature of the observations, some results were represented with descriptive statistics. SPSS was therefore employed in conducting this analysis. Additionally, some exploratory factor analysis was performed to help identify clusters of related variables and thus help in the grouping of study variables into fewer ones that could be more easily understood. The factor analysis in particular was performed to observe which factors were the most influential in the low bicycle ridership among students and was done as well using SPSS.

17.3 Results

17.3.1 Respondents' On-Campus Travel Behavior

Four major zones on campus were identified as most frequented by students: the Academic/Teaching Area (43%), the Commercial Area (28%), Halls and Hostels Areas (16%), and the Catholic Church/Protestant Chapels (10%). It should be noted that the GUSSS Hostel is a prime on-campus student residential facility where students reside, but the greater majority of students in the University live in private off-campus housing scattered within the immediate communities abutting the University. Distances between these locations range from 0.5 to 2 km which would not deter cycling. Identification of the most-frequented zones also provides an indication of how intervention in terms of developing bicycle routes on campus might need to be planned in order to ensure that those zones formed part of the nodal points within the route network.

Table 17.2 provides a summary of the primary modes used by students for commuting on campus. As was expected, the majority of the respondents indicated walking (71%) and the Campus Shuttle (21%) as their primary modes of travel on campus, with very little use of other modes; bicycles accounted for only 4% of the modal share.

Bicycle ownership among students was established to be very low at only 7.4%. In this study, it was interesting to realize that bicycle ownership was higher among male respondents as compared to female respondents, perhaps reflecting the influence that some of the negative stereotypes about cycling had on the use of this mode by females. The ability to ride bicycles was also higher among male respondents, further giving some credence to this hypothesis. As was indicated early on, the low bicycle ownership is an apparent reflection of the low bicycling culture characterizing the country as a whole with the exception of the three northern regions of the country where bicycle use among the population is over 30%. In addition, there is very little incentive in owning a bicycle on campus when the environment does not support its

Table 17.2 Descriptive Statistics of survey respondents.

Gender	*Percentage*
Male	70.9%
Female	29.1%
Mode of commuting on campus	*Percentage*
Walking	71%
Bicycles	4%
Shuttle	21%
Motorcycles	1%
Private vehicle	3%
Walking	71%
Most-frequented campus locations	*Percentage*
CCB	43%
Commercial area	28%
Catholic church	10%
Swimming pool	3%
GUSSS Hostel area	16%
Willingness to cycle if infrastructure improved	
Yes	80%
No	20%
Bicycle use in case of improved parking	
Yes	84%
No	16%

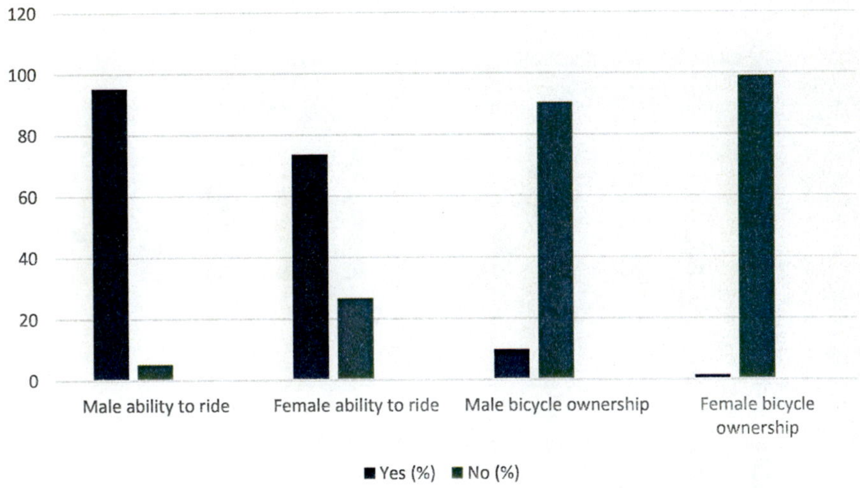

Fig. 17.3 Ability to cycle and Bicycle Ownership across genders

use. It may only be speculated that ownership might increase (even among both sexes) if measures were put in place to support bicycle transportation and people began to see their utility for commuting on campus. In terms of ability to ride a bicycle, as high as 88% of the respondents gave positive responses but there was no means by which their stated ability could be verified. Granting that the responses were genuine, then there appeared to be some degree of optimism that bicycle transportation on campus if promoted could have appeal to a substantial proportion of the student populace provided other factors that favored and influenced the use of the mode existed. Figure 17.3 shows the ability to cycle, and current bicycle ownership levels across genders.

17.3.2 Factors Underlying Low Bicycle Use

The following charts (Fig. 17.4a to n) show the percentage of respondents that selected various points on the Likert scale, in response to the items on the questionnaire.

From the Likert scale data, the following factors were identified as limiting bicycle transport on campus; lack of access to bicycle, lack of secure bicycle storage/parking, lack of lanes, etc. Factor analysis was subsequently conducted to ascertain the impacts of these specific variables on bicycling intention. Based on the Kaiser–Meyer–Olkin measure of sampling adequacy of 0.802 and Bartlett's test of sphericity value of 928.169, the sample was considered adequate for the factor analysis. A scree plot (Figure 17.5) was used as the basis for the selection of the factors, and the use of the

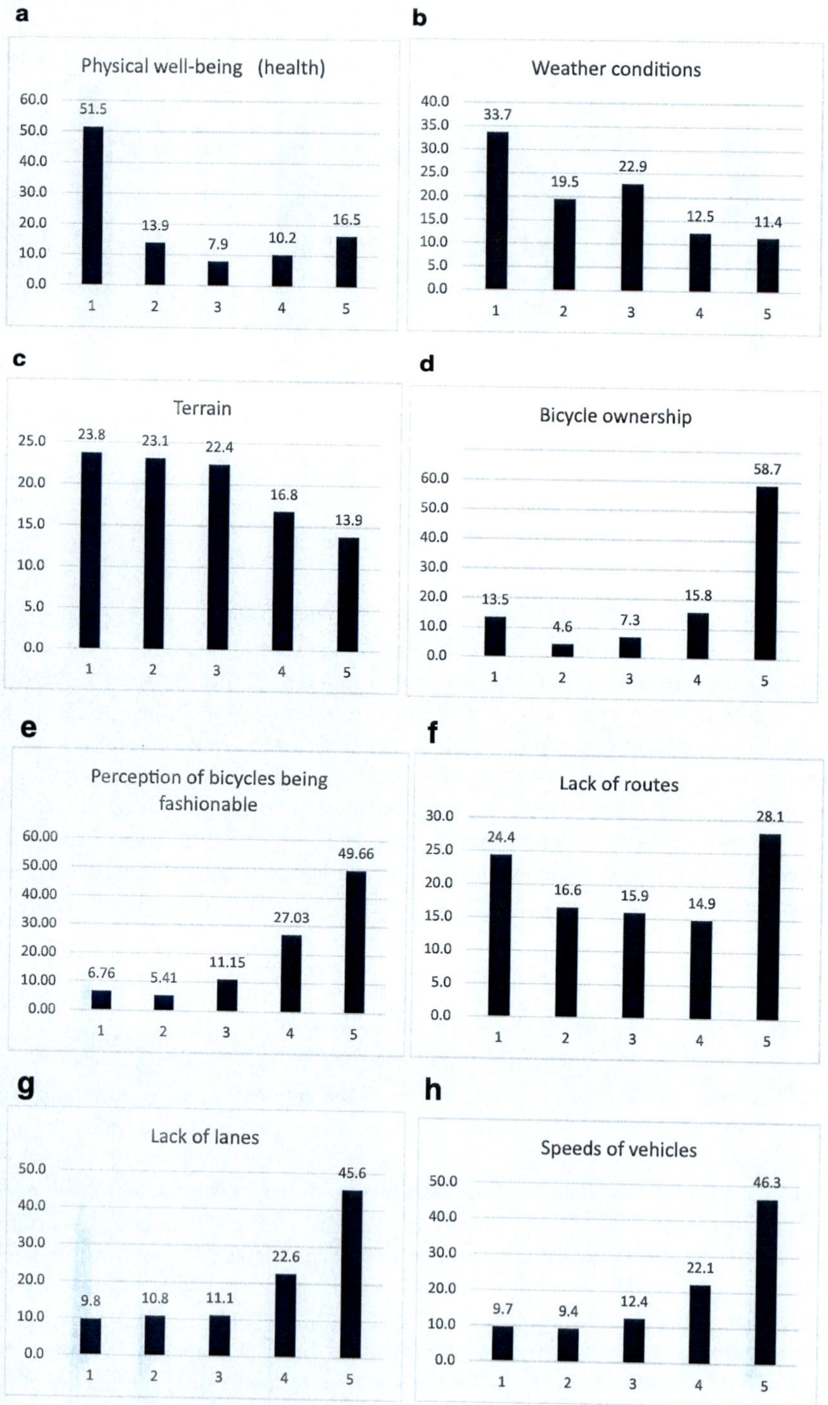
a
Physical well-being (health)
60.0
50.0
40.0
30.0
20.0
10.0
0.0
51.5
13.9
7.9
10.2
16.5
1 2 3 4 5
b
Weather conditions
40.0
35.0
30.0
25.0
20.0
15.0
10.0
5.0
0.0
33.7
19.5
22.9
12.5
11.4
1 2 3 4 5
c
Terrain
25.0
20.0
15.0
10.0
5.0
0.0
23.8
23.1
22.4
16.8
13.9
1 2 3 4 5
d
Bicycle ownership
60.0
50.0
40.0
30.0
20.0
10.0
0.0
13.5
4.6
7.3
15.8
58.7
1 2 3 4 5
e
Perception of bicycles being fashionable
60.00
50.00
40.00
30.00
20.00
10.00
0.00
6.76
5.41
11.15
27.03
49.66
1 2 3 4 5
f
Lack of routes
30.0
25.0
20.0
15.0
10.0
5.0
0.0
24.4
16.6
15.9
14.9
28.1
1 2 3 4 5
g
Lack of lanes
50.0
40.0
30.0
20.0
10.0
0.0
9.8
10.8
11.1
22.6
45.6
1 2 3 4 5
h
Speeds of vehicles
50.0
40.0
30.0
20.0
10.0
0.0
9.7
9.4
12.4
22.1
46.3
1 2 3 4 5

◂**Fig. 17.4** **a** Physical well-being (health). **b** Weather conditions. **c** Terrain on campus. **d** Bicycle ownership. **e** Bicycles are fashionable. **f** Lack of routes. **g** Lack of lanes. **h** Speeds of vehicles. **i** Narrow roadways. **j** Secure bike storage. **k** Exclusive bicycle lanes. **l** Education. **m** Safer bicycle infrastructure. **n** Friends' use of bicycles

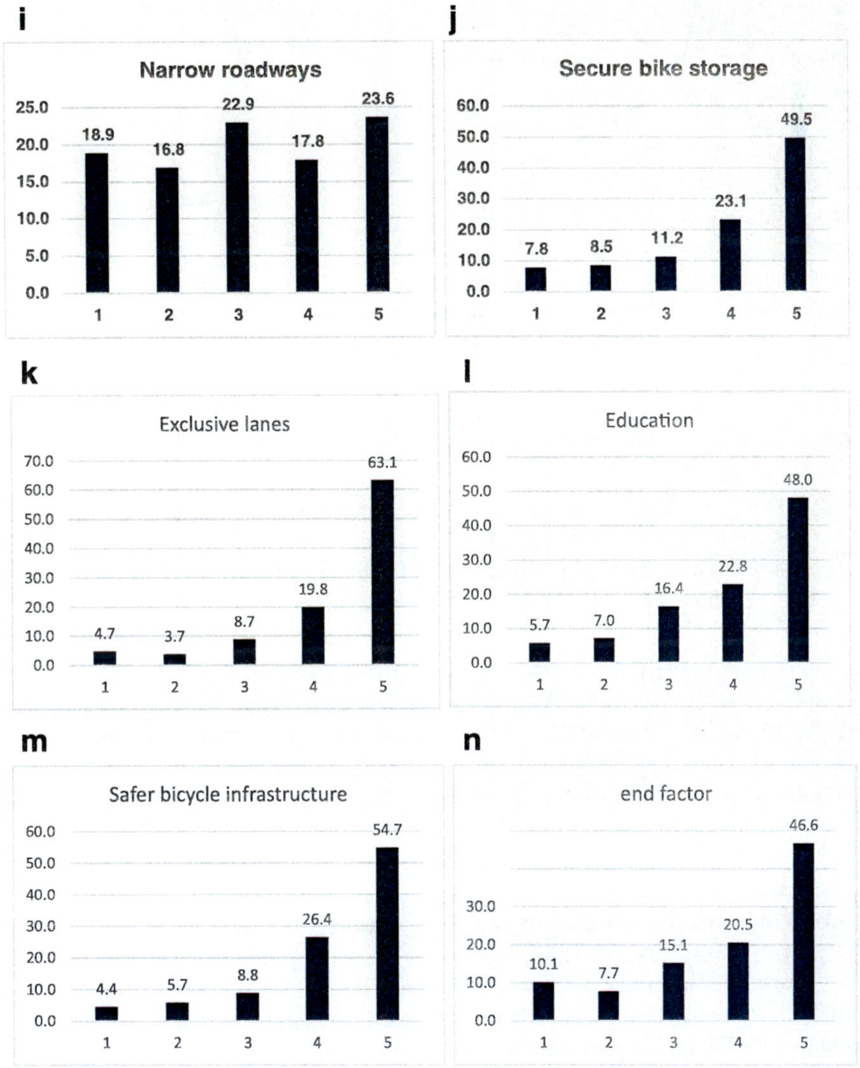

Fig. 17.4 (continued)

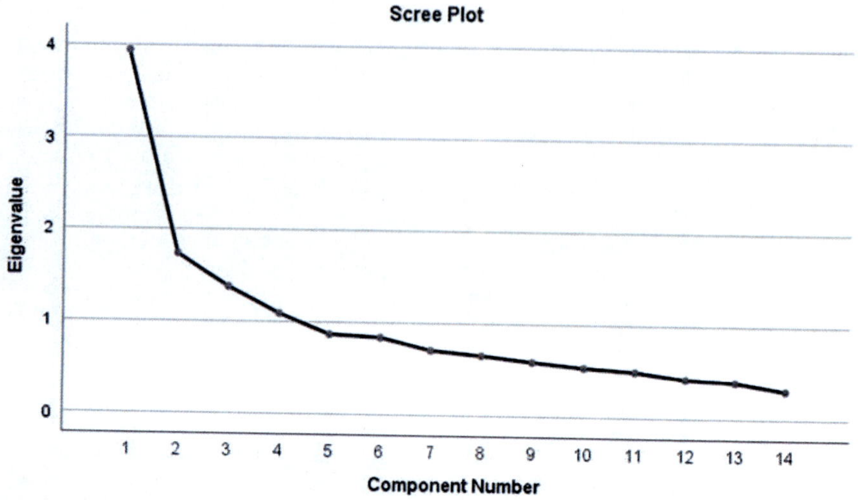

Fig. 17.5 Scree plot

Guttmann-Kaiser criterion suggested that only factors with an Eigenvalue greater than unity needed to be retained.

Following the Eigenvalue analysis of the correlation matrix, four components were considered significant in explaining the total variance of the factors underlying low bicycle use on campus, namely dangerous riding conditions (28.2%), unexplored promoting factors (12.4%), health and environmental factors (9.8%), and supportive factors (7.8%).

Table 17.3 shows the total variance explained, and Table 17.4 shows the rotated factor loadings and communalities for these factors. The components extracted cumulatively explained a little over 58.2% of the variation in the data set and passed the cumulative proportion of variance criterion which requires that the extracted components altogether should explain at least 50% of the variation. These factors are further discussed in the ensuing sections (Tables 17.3 and 17.4).

17.3.2.1 Factor #1: Dangerous Riding Conditions

Component 1 from Table 17.4 was named as ‘Dangerous riding conditions’. The variables ‘lack of routes’, ‘lack of lanes’, ‘speeds of vehicles’, and ‘narrow roadways’ from Table 17.4 loaded unto this factor. The component together explained 28.2% of the total variance of the factors underlying low bicycle use on the KNUST campus. Since safety is one of the most important considerations people would make before choosing cycling, it can be said that the perception of risk while using bicycles could be determined as a possible deterrent for its use on campus. Where there is a lack of these infrastructure, cyclists could regard the risk of using bicycles as high and consequently be less inclined toward using bicycles on campus. With regard to

Table 17.3 Total variance explained

Component	Initial eigenvalues			Extraction sums of squared loadings			Rotation sums of squared loadings		
	Total	% of Variance	Cumulative %	Total	% of Variance	Cumulative %	Total	% of Variance	Cumulative %
Physical well-being	3.946	28.183	28.183	3.946	28.183	28.183	2.513	17.953	17.953
Weather conditions	1.734	12.385	40.569	1.734	12.385	40.569	2.428	17.343	35.296
Terrain	1.377	9.837	50.405	1.377	9.837	50.405	1.871	13.364	48.661
Bicycle ownership	1.092	7.803	58.208	1.092	7.803	58.208	1.337	9.548	**58.208**
Are bicycles fashionable?	0.871	6.219	64.428						
Lack of routes	0.842	6.012	70.439						
Lack of lanes	0.709	5.061	75.501						
Speeds of vehicles	0.657	4.695	80.196						
Narrow roadways	0.594	4.241	84.437						
Secure bike storage	0.539	3.849	88.286						
Exclusive lanes	0.503	3.593	91.879						
Education on bicycling would encourage its use	0.427	3.052	94.931						
Safer bicycle infrastructure	0.400	2.857	97.789						
Friend factor	0.310	2.211	100.000						

Extraction Method: Principal Component Analysis

Table 17.4 Rotated component matrix

	Component			
	1	2	3	4
Physical well-being (health)			**0.765**	
Weather conditions			**0.802**	
Terrain			**0.694**	
Bicycle ownership				**0.708**
Are bicycles fashionable?		**0.674**		
Lack of routes	**0.731**			
Lack of lanes	**0.682**	0.434		
Speeds of vehicles	**0.654**			
Narrow roadways	**0.733**			
Secure bike storage				**0.502**
Exclusive lanes		**0.745**		
Education on bicycling would encourage its use		**0.594**		
Safer bicycle infrastructure		**0.702**		
Friend factor		0.415		**0.544**

Extraction Method: Principal Component Analysis. Rotation Method: Varimax with Kaiser Normalization

a. Rotation converged in 6 iterations. Factor loadings with values less than 0. 4 have been removed

the lack of bicycle-only paths/lanes, perhaps respondents were seeking separation of bicycle traffic from motorized traffic to guarantee rider safety. In addition, the speeds of motor vehicles and narrow roadways on campus both emphasize concern for rider safety in mixed traffic.

17.3.2.2 Factor #2: Unexplored Cycling Promoting Factors

As earlier identified, the safety of riding a bicycle is a key determinant of the willingness of students to cycle on campus. Under the second factor, designated bicycle lanes and safer bicycle infrastructure, as well as the need for education on bicycles, and the perceptions of bicycles being fashionable formed part of the variables that comprised the factor. The safety that could be enjoyed by cyclists and the potential of having on road facilities that attend to cycling exclusively is important to promoting bicycling with respondents. Since these promoting factors are non-existent, it possibly has an effect on the low ridership that exists, as cyclists could consider themselves limited in their route choices using that mode. Perceptions of bicycles being fashionable, as well as the need for education on bicycles, were also considered important avenues that were unexplored in promoting its use. Together, these contributed to the factor variance of 12.4%. Interestingly, the results from the survey indicated a 76.9% affirmation that bicycle use was perceived as fashionable, which potentially could have a positive effect in causing some students on campus to be attracted to this mode if it became available, and equally suggest that this positive perception has not been adequately taken advantage of in promoting on campus cycling.

17.3.2.3 Factor #3: Health and Environmental Factors

The third principal component, which is made up of three main factors; physical well-being (health), weather conditions, and terrain, could explain 9.8% of the total variance and recorded Eigenvalues of 0.765, 0.802, and 0.694, respectively. Physical well-being (health) was a factor examined in this study, with the aim of understanding whether the health or the physicality of cycling was reason enough for its non-use by respondents. From the responses, about 26% of respondents indicated they were held back by their health or the physicality involved in cycling. The weather condition was also identified as a factor that could impact respondents' choice to cycle. The results showed that close to 53% of respondents thought the weather was not a significant deterrent, but merited consideration as one that has a direct effect on bicycle use. The rolling terrain on campus produced an Eigenvalue of 0.694 indicating that it was of concern as a factor that could limit the use of bicycles.

17.3.2.4 Factor #4 Shared Facilities and Experiences

The fourth most significant component was a lack of shared facilities and experiences. The variables that reflected under this factor were lack of bicycle ownership, secure bicycle storage, and the impact of peer influence on respondents' decision to cycle. Dill and Voros (2007) consider access to bicycles as a sociodemographic factor that influences the decision of an individual to ride a bicycle. This seems to suggest that many students may probably consider cycling if they have access to bicycles. Access to bicycles may be improved using cycle-hire programs even if students do not necessarily own bicycles (Shaheen et al. 2012). Therefore, to promote cycling on KNUST Campus, perhaps a way that could help is for the University to franchise a dealership in bicycles to set up shop on campus to, among other things, hire out bicycles and offer related services such as shared parking facilities to students and other interested persons. The factor also highlights the role of peer influence and shared experiences of cycling in improving ridership on campus.

17.3.3 Willingness to Cycle

On willingness to cycle if bicycle transportation was introduced, 85% of the respondents gave positive responses although a slightly lower percentage (72%) indicated they had considered the prospects of using a bicycle on campus. These responses were necessary to provide an indication of interest in on-campus bicycle transportation among students and also to establish the level of publicity and advocacy that might be needed to whip up interest in and enthusiasm for this type of transportation if it were to be considered in the future. While the percentage of respondents favorably disposed toward potential bicycle use appeared encouraging, it did not necessarily mean that the same would actually take to cycling if it were introduced, as willingness is only an expression of intent. The major driving force would probably be how favorably potential users perceived that mode vis-à-vis other competing modes such as walking and the Campus Shuttle or to improve physical fitness.

17.4 Discussion and Implications

The aim of this study was to evaluate the factors underlying bicycle ridership on the campus of the Kwame Nkrumah University of Science and Technology, Ghana, and the prospects for bicycle use among students. The study considered current psychosocial perceptions and some specific infrastructural challenges limiting the use of bicycles. The study attributes the underlying causes of low bicycle ridership on the KNUST Campus to four major components, namely dangerous riding conditions, unexplored promoting factors, health and environmental factors, and shared facilities and peer influence. In a university setting, where activity centers are in close proximity to each other, bicycle transport comes as a cheaper alternative to other transport modes. Unfortunately, in Ghana, patronage of bicycles is generally low and represents a meager 8.6% of all modes for commuting (Quarshie 2007). It is, however, more accepted by persons within the age cohorts of 30 years and below and, therefore, should be an attraction for KNUST students, as close to 90% of them fall within this age group. As has been made clear by the study, the lack of good bicycle routes, lack of bicycle-only paths, speeds of motor vehicles, and narrow campus roads create dangerous situations for riding bicycles and contribute to the low bicycle ridership. It is equally evident that good bicycle infrastructure on campus that guarantees rider safety will encourage bicycle use as major activity centers are within cycling distance from each other. The development of bicycle transportation across Africa, and Ghana in particular, has encountered numerous constraints. Challenges include the unavailability of funds to develop the much-needed transportation infrastructure, the overall safety of cyclists, and the perceptions and general apathy of people toward cycling as a whole (Dias Batista 2010; Nkurunziza 2013). In this study, over 80% of respondents were inclined to cycle in the event of improved infrastructure and the creation of safer conditions for bicycling, stressing the importance of such interventions.

The physical health of respondents was generally not a major reason why respondents did not cycle, as about 65% of respondents opined that they were fit enough to cycle. About 27% of respondents thought that the physicality of cycling, and other health concerns, could be ample reasons for them not to cycle. Additionally, according to Ahmed et al. (2012), bad weather for instance can be a disincentive and could reduce the incidence of cycling drastically. In this study, weather conditions (and possible sweating resulting from cycling in warm weather conditions) were also identified as a factor that could impact respondents' choice to cycle.

The results showed that 53% of respondents indicated that the weather was not a significant deterrent to their use of bicycles, however, quite a significant number of respondents (24%) perceived it as meriting consideration in their use of the mode. Practically, such concerns could be resolved, for example, through the planting of trees to provide shade along planned cycling corridors. In terms of topography, respondents indicated that the terrain on campus played a role in their willingness to choose bicycles. Since flat terrains are usually preferred for cycling as compared to riding on rolling terrains such as those on campus, the terrain could have made the idea of cycling on campus appear more challenging. About 30% of respondents thought the terrain could pose a challenge to them using bicycles, while 47% thought the terrain was not much of an impediment. Generally, there is a positive perception about the use of bicycles as an efficient transport mode among students. There are also subjective indicators and concepts which although may not be measurable are indicative of the intention to use bicycles. In this study, the perception of bicycle use as a fashionable form of commute is seen as an influential factor that defines students' affinity for peer group bicycling. The survey results show that about 77% of respondents affirmed bicycling as fashionable. It is believed therefore that this perception of bicycles could be used in influencing students to embrace bicycle transportation. The cost of purchasing a used bicycle may not be a deterrent as the average price of a bicycle in Kumasi is about GHC250.00 (US$ 43.00) and the possibility also exists of arranging for the hiring of bicycle services on campus. On the University Management side, commitment must be shown toward the provision of the needed infrastructure to enhance the safety, speed, and convenience of this green transport mode. A deliberate policy that encourages bicycling on campus must be put in place to incentivize students to embrace this mode and change the existing situation.

From an attitudinal point of view, on-campus cycling seems feasible. However, from an infrastructural point of view, the environment is not supportive. Generally, the study provides evidence on the willingness of commuters to use the mode and, as well, shows that other factors such as the physical health of respondents did not serve as limiting factors in their possible use of bicycles. Rather, the lack of supportive physical infrastructure such as bicycle-only paths and storage for bicycles, among a host of others, could serve as barriers to the use of bicycles. This study nonetheless provides evidence of the feasibility of bicycling on a university campus in Ghana despite the negative stereotypes that currently exist around this form of transport in the country. Findings of the study when viewed in the context of the efforts to increase the use of sustainable modes of transport suggest that such modes, particularly bicycling, could be nurtured in places such as universities. Efforts to create a different narrative about bicycles as an example could ideally begin at university campuses, where supportive infrastructure, policies, and bicycle sharing schemes could make the mode attractive to younger demographics. As university campuses become safe havens for cycling, more young people may interact with the mode during their university years, creating a new sense of appreciation for this form of transport. Bicycles and bicycling could gain a new role on campuses as a viable mode by which students can commute. The most significant impact of the change in narrative and perception of bicycles could come well beyond most students' campus years, where their interactions with this form of transport in university could make former university students more patient with cyclists in traffic, viewing them less as a nuisance, and more as a 'road vehicle' with equal rights. The rippling effect of this could allow the larger society to be more tolerant of bicycle users, making cycling safer even for more cautious individuals. The push for more infrastructure for cycling could also follow, where people who have seen the use of this mode of transport in society eventually advocate for more infrastructure, creating slowly, but certainly, a more cyclist-friendly society.

Although these discussions present an idealistic perspective of the introduction of cycling, they can be viewed as low-risk approaches to improving the conditions surrounding this mode. The unfavorability with which bicycles are viewed, however, presents a significant hurdle for the changes that can be made.

Improvements could also be made on the infrastructure front to improve cycling. Policies that ensure that all open drains along most roads are closed could make the thought of cycling less risky. Since most roads in parts of Ghana have narrow widths, the sealing off of open drains to flush with the roadway could increase the perceived roadway available for traffic. This additional section created by these efforts could be cordoned off and dedicated to cyclist activity. Such policies could ensure that a safe space is created for cyclists and that a reduction in real and perceived risks is created. The introduction of policies that focus on educating drivers to show more discretion and respect toward cyclists could also go a long way in improving the interactions between these road users. Where cyclists can be treated as part of a hierarchy of road users who have equal access to the roadway as vehicles, significant drops in the heckling and disrespect shown toward its users could impact the desire of others to use it.

Future transport planning regimes in Ghana and across Urban Africa as a whole could also take bicyclists into consideration and provide designated cyclists infrastructure and more mixed-use facilities that can allow these vehicles to be used in traffic. The combination of these factors could well improve cycling in Ghana and make it much more attractive for even non-cyclists. These findings and policy perspectives also have implications for other developing countries, particularly those within Africa, where similar problems exist.

Where long-term policies can be instituted to support cycling and other forms of non-motorized transport, the development of cities will bear these modes of transport in mind and accommodate them for a more diverse and multimodal transport system. Where policies for these modes of transport lack, however, this lack of supportive policies and infrastructure will continue to stifle the possible growth of non-motorized modes of transport.

17.5 Conclusion

In conclusion, this study confirmed that bicycle use on the KNUST campus, whether for recreation or commuting, is very limited, which to a large extent reflects the fact that only a small percentage of the people in Ghana as a whole are favorably disposed toward that mode of transport for commuting. It was established that even though most respondents indicated that they were capable of cycling and had previously considered using bicycles on campus prior to the survey, the current environment did not support nor guarantee the safety of that mode. On the positive side, greater than 85% expressed their willingness to ride bicycles if a supportive environment for bicycling is created on the campus. In the view of the majority of respondents, the creation of off-street bicycle lanes, education on bicycle safety, and peer influence were necessary to spur bicycle use on campus. Potential riders also wanted to see parking/storage facilities on campus that guaranteed the safety of their bicycles. While the percentage of respondents favorably disposed toward potential bicycle use appeared encouraging, it did not necessarily mean the same would embrace and shift to bicycle transportation if it was introduced, as willingness is only an expression of intent. That notwithstanding and despite the relatively low bicycle ownership on campus, prospects appeared to exist for bicycle transport to have appeal to students if it was introduced and more students got into the riding culture.

References

Acheampong RA, Siiba A (2018) Examining the determinants of utility bicycling using a socio-ecological framework: an exploratory study of the Tamale Metropolis in Northern Ghana. J Transp Geogr 69:1–10

Afukaar FK, Antwi P, Ofosu-Amaah S (2003) Pattern of road traffic injuries in Ghana: implications for control. Inj Control Saf Promot 10(1–2):69–76

Ahmed F, Rose G, Figliozzi M, Jakob C (2012) Commuter cyclist's sensitivity to changes in weather: insight from two cities with different climatic conditions. In: Transportation Research Board Annual Meeting, Washington, DC, USA

Amoako-Sakyi RO, Owusu SA (2011) Moving on two wheels: a comprehensive study on bicycle behaviour among Ghanaian children. SBHA 76(1):107–125

Balsas CJ (2003) Sustainable transportation planning on college campuses. Transp Policy 10(1):35–49

Dias Batista E (2010) Bicycle sharing in developing countries: a proposal towards sustainable transportation in Brazilian median cities

de Sousa AA, Sanches SP, Ferreira MA (2014) Perception of barriers for the use of bicycles. Procedia Soc Behav Sci 160:304–313

Dill J, Voros K (2007) Factors affecting bicycling demand: initial survey findings from the Portland, Oregon, region. Transp Res Rec 2031(1):9–17

Dulken D (1992) The very model of a suburban community, what we can learn from the university campus. Planning 58(8):24–25

Fernández-Heredia Á, Monzón A, Jara-Díaz S (2014) Understanding cyclists' perceptions, keys for a successful bicycle promotion. Transport Res Part a Policy Pract 63:1–11

Goetzke F, Rave T (2011) Bicycle use in Germany: explaining differences between municipalities with social network effects. Urban Stud 48(2):427–437

Grieco M, Turner J, Kwakye EA (1994) A tale of two cultures: ethnicity and cycling behavior in urban Ghana. Transp Res Rec 1441:101

Huang Z, Xia Y, Zhang M, Li X (2012) Integration of multi-modal travel in historical campus-The case of Wuhan University. In 2012 6th International association for China planning conference (IACP) (pp 1–6). IEEE

Tiwari G, Jain D (2013) Promoting low carbon transport in India: NMT infrastructure in India: investment, policy and design

KNUST Planning Department (2014) knust land use plan proposals (2014–2024)

Niboi BE (2013) Investment opportunity in Kumasi, Ghana-KNUST Student Accommodation/Real Estate

Nkurunziza A (2013) Sustainable transport in Dar-es-Salaam: the potential for BRT and cycling from a user perspective

Ojeda OR, Kohn W, Moore CW (1997) Campus & community: moore ruble yudell: architecture & planning. Rockport

Pucher J, Dill J, Handy S (2010) Infrastructure, programs, and policies to increase bicycling: an international review. Prev Med 50:S106–S125

Quarshie M (2007) Integrating cycling in bus rapid transit system in Accra. In: Highway and urban environment (pp 103–116). Springer, Dordrecht

Shaheen S, Guzman S, Zhang H (2012) Bikesharing across the globe. City Cycling 183

Tiwari G (2001) Pedestrian infrastructure in the city transport system: a case study of Delhi Geetam Tiwari. Editorial Board 7(4):13–18

Turner PV (1995) Campus. An American Planning Tradition, The Mıt Pres Cambridge, Massachusets and London

Daniel Atuah Obeng is a Senior Lecturer and MSc Programmes Coordinator of TRECK-ACE Impact Project, KNUST, Ghana. He has an undergraduate degree in Civil Engineering from KNUST, an MSc degree in Transportation Planning and Policy from the University of Newcastle upon Tyne, UK, and a PhD degree in Civil Engineering from the KNUST, Ghana. His research area spans transport and highway engineering with particular interest in road traffic safety, public transport and sustainability of transport systems. Dr. D. A. Obeng also has considerable consultancy experience in the planning, development, management, monitoring and evaluation of transport infrastructure projects.

Emmanuel Dzisi is a PhD candidate of the Kwame Nkrumah University of Science and Technology, Ghana where he is researching the feasibility of introducing mobility-as-a-service (MaaS) for public transport options in Ghana. Emmanuel is also interested in finding sustainable solutions for improving urban mobility, particularly for the transport vulnerable. His research interests span the fields of bicycling, walking and MaaS in the context of African cities and their development.

Chapter 18
The Provision of Non-Motorised Transport Infrastructure in the City of Lusaka: An Analysis of Policy, Practice

Raymond Lukomona and Wilma S. Nchito

Abstract This chapter presents an analysis of the provision and use of Non-Motorised Transport (NMT) in the city of Lusaka. NMT includes all forms of travel that are human powered such as walking and cycling. The chapter presents results from an academic study that examined NMT infrastructure along selected roads in Lusaka in relation to their state, challenges, and integration into city policies. Using a case study approach, the study covered four roads within the city of Lusaka which were Alick Nkhata, Burma, Dedan Kimathi Roads, and Independence Avenue. Data was collected through direct observations and in-depth interviews with key informants and NMT users. The sample was 40 people comprising 9 key informants and 31 NMT users. Data was analysed through narrative and content analyses. Descriptive statistics are used where appropriate. The chapter reveals that NMT infrastructure on the selected roads is inadequate. Walkways translate into 44.3% of carriageways on the selected roads. NMT users are thus subjected to the dusty, disjointed, and uneven foot tracks which are barely passable after heavy rains. The chapter will further show that NMT users face challenges such as difficulties in crossing roads due to the inadequacy of crossing facilities on the selected roads. An analysis of City Plans, Road Project documents, and related legislation revealed that NMT is not adequately provided for.

Keywords Non-motorised · Transport · Planning · Policy · Users

This chapter is based on a dissertation entitled 'Non-motorised Transport (NMT) Infrastructure along selected roads in Lusaka: Status, Experiences and Policy' submitted to the University of Zambia by Raymond Lukomona in partial fulfilment for the award of Master of Science in Spatial Planning.

R. Lukomona
Department of Physical Planning, P.O Box 110082, Solwezi, Zambia

W. S. Nchito (✉)
Department of Geography and Environmental Studies, University of Zambia, Lusaka, Zambia
e-mail: wnchito@unza.zm

R. A. Acheampong et al. (eds.), *Transport and Mobility Futures in Urban Africa*, The Urban Book Series, https://doi.org/10.1007/978-3-031-17327-1_18

18.1 Introduction

It is estimated that by 2052, cities in Africa will be home to 60% of the global urban population (Pieterse 2011). This rapid rise in urban populations will need to be planned for and managed in a sustainable manner; otherwise, African cities will be prone to congestion, diseases, and zones of increased human hazards. The majority of people in these rapidly expanding cities will continue to rely on NMT but there remains limited attention paid to the planning of NMT. Walking, cycling, and the use of handcarts are often the only available or affordable means of transport for residents in cities of the global south. NMT also presents benefits to urban transport systems of cities all over the globe, and some of these benefits include minimal space requirements, negligible pollution, minimal energy consumption, and improved social interaction making it the most sustainable transport option (United Nations Human Settlements Programme (UN-Habitat) 2010). Another benefit is the reduction in emissions in light of climate change and increased respiratory disease burdens. Despite these benefits, governments in developing countries continue making large investments in major road projects which increase the capacity and accessibility of vehicles. These large projects often neglect NMT infrastructure and sometimes reduce accessibility for NMT users. Due to the lag in infrastructure development, cities in these countries have the potential to realign the future trajectory of people's mobility through the equitable provision of transport systems (Lucas and Porter 2016).

Lusaka, the capital city of Zambia, is home to approximately 2 million inhabitants who are mostly dependent on NMT. Despite NMT being a major means of transport, both national and local authorities have concentrated investments on road infrastructure that favours motorised transport (MT) users who are in the minority. The city has in the past ten years embarked on various projects aimed at widening and improving the quality of pre-existing roads around the city. City authorities, especially in the global North, have taken some commendable strides to reverse car dependency by promoting and formulating policies, plans, and legislation aimed at making NMT a preferred transport means (Hogan 2015). On the contrary, global south cities such as Lusaka seem to be moving in the opposite direction by neglecting NMT through the continued expansion of city roads to accommodate more vehicular traffic. This is not a sustainable solution for the future. Studies show that the demand for motor vehicles in global south cities especially Sub-Sahara African (SSA) will continue to increase in the next few decades (Freund and Martin 1999; Zambia Institute for Policy Analysis and research (ZIPAR) 2014). This is while climate change is demanding a reduction in the use of motorised transport to reduce carbon emissions (UN Environment 2019). The bulk of scholarly work on the subject matter in Sub-Saharan Africa is skewed towards MT options. Going forward more studies will need to be carried out to understand how global south cities such as Lusaka have incorporated NMT in their plans, policies, and road development projects. This will provide a basis for future developments which will adequately meet the needs of the people. The lack of detailed localised information creates a situation where solutions are transplanted from the developed world and do not meet local needs.

Using empirical evidence, the chapter shows that the NMT infrastructure in the city is mostly in a bad state and does not improve people's mobility in and around Lusaka. The chapter highlights the problems faced by pedestrians and other NMT users like cyclists and cart pushers because generally, the few available routes are not interlinked, forcing users to intermittently use the main road carriageway. Transport policies have completely left out provisions for hand carts (these are wheelbarrows with metal extensions welded on them to increase their carrying capacity) which were found to be a common feature on some routes and play an invaluable role in the movement of goods around the city.

Considering that the modal split in Lusaka is 65% of walking, 23% use of public transport, 10% use of private cars, and only 2% of cycling, the city authorities need to assign more resources to the provision of NMT infrastructure (Japan International Cooperation Agency (JICA) 2009). Ultimately, the study demonstrates that all the selected roads can accommodate NMT infrastructure owing to good terrain (Lusaka is generally flat), availability of space, and favourable climate.

This chapter further looks at how NMT issues are integrated into the plans and policies of the City of Lusaka. It provides a contextualised understanding of the socio-economic impacts of different mobility solutions and considers what can be done to improve the future of NMT in the city.

18.2 Research Methods, Data Collection, and Analysis

The study used a case study approach in order to obtain detailed context-specific information on the subject matter. There was limited documentation on the location of NMT infrastructure and this had to be collected from in-depth interviews and direct observations. The selection of the roads included in the study was done after NMT infrastructure was identified to avoid the selection of a road that did not have NMT infrastructure. Semi-structured interviews were conducted to collect data from key informants in the following organisations: Lusaka City Council (LCC); Road Development Agency (RDA); Ministry of Local Government and Housing (MLGH); Ministry of Communications, Transport, Works and Supply (MCTWS); Road Traffic and Safety Agency (RTSA); and the Zambia Police (ZP). Data was collected from key informants related to aspects such as planning, works, designs, and construction of roads in the City of Lusaka. Semi-structured interviews were conducted with NMT users to capture their experiences in the use of the selected roads. Interviews with NMT users were mostly recorded (with consent) as this enabled the interviews to take place while the user continued to their destination. Where a research participant was uncomfortable, interview data was captured through note-taking. A checklist was used to take note of the availability and state of walkways, cycle tracks, foot tracks, pedestrian crossings, footbridges, traffic lights, humps, signage, and shoulders on the selected roads. Photographs of salient features on the selected roads were captured using a digital camera. Considering the nature of the targeted interviewees who were NMT users and key informants, this study utilised non-probability sampling techniques of convenience and purposive sampling. The study produced an updated map of the selected area showing the extent and state of NMT infrastructure.

18.2.1 Study Setting

The study was carried out in Lusaka, Zambia's capital city. The original layout of the city which was planned in the 1930s incorporated the 'garden city' concept and the plan created by Adshead a British planning consultant conspicuously left out recommendations for transport (Collins 1986). The city from its inception as a sprawling capital has not catered for NMT. With time it has grown into a sprawling metropolis covering 418 Kilometres^2. Lusaka City has a population of approximately 2,900,000 inhabitants (Macro Trends 2021). Due to the haphazard nature of the more recent developments, poorer residents cover large distances to get to work or school. The car dependency of the city is evidenced by the concentration on road expansion by the road agency. The study focused on four roads within the City of Lusaka as shown in Fig. 18.1. These are Alick Nkhata, Burma, Dedan Kimathi Roads, and Independence Avenue.

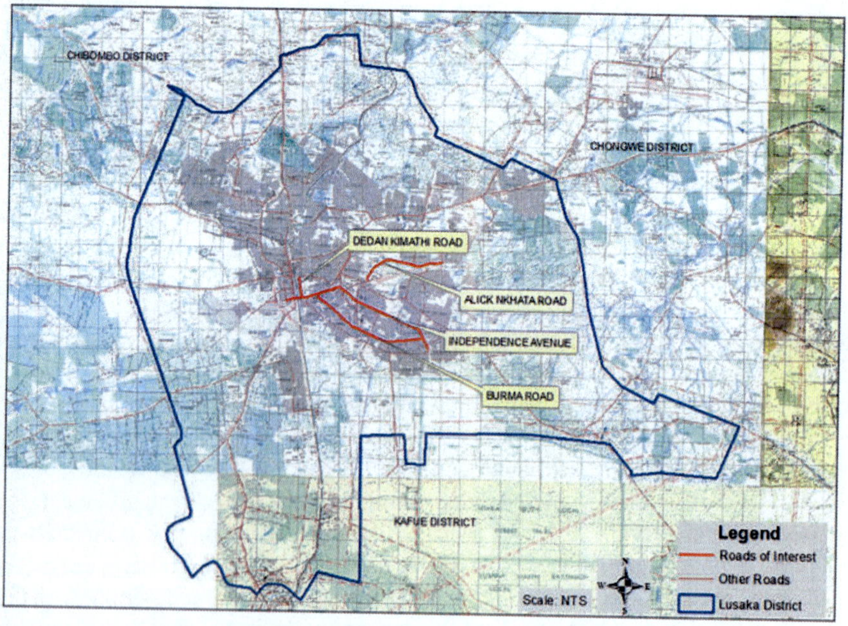

Fig. 18.1 Locations of the Selected Roads in Lusaka City. (*Source* Field Survey, 2015)

Table 18.1 Selected roads for the study within the City of Lusaka

No	Road name	Road class	Length (km)	Surface type	Condition
1	Independence	Secondary	6.893	Bitumen	Good
2	Burma road	Secondary	6.047	Bitumen	Good
3	Alick Nkhata	Tertiary	4.26	Bitumen	Good
4	Dedan Kimathi	Tertiary	0.98	Bitumen	Good

Source JICA (2009)

The roads studied were selected for the following reasons:

- *Alick Nkhata Road* connects four densely populated informal settlements namely Kalingalinga, Mtendere, Helen Kaunda, and Kalikiliki to areas of employment such as the low-density residential areas of Ibex Hill and Kabulonga and to retail areas.
- *Independence Avenue* was selected due to the existence of a walkway and being the road with the highest private vehicle transport in the City of Lusaka (JICA 2009). Independence Avenue also has a multiplicity of land uses along it such as residential, commercial, and institutional making it a very busy road attracting all types of road users.
- *Burma Road* connects the medium-cost residential areas of Chilenje and Kabwata to the Lusaka City Central Business District (CBD). The road was also selected because anecdotal evidence suggested that the road originally had a cycle lane and pedestrian walkway. The Road was also one of the roads targeted for expansion under the L 400 National Road Project (Road Development Agency (RDA) 2014).
- *Dedan Kimathi Road* connects Independence Avenue to Church Road and other major roads. Dedan Kimathi road presents a unique opportunity of linking a major commercial area of Kamwala and the City's only Inter-city Bus Station/terminus. The road presented a unique opportunity to understand the experiences of wheelbarrow users who provide informal courier services to and from Inter-City Bus Terminus. Other relevant details about the selected roads are summarised in Table 18.1.

18.3 Research Findings

18.3.1 State of NMT Infrastructure on Selected Roads

Table 18.2 shows that none of the selected roads has all the identified NMT infrastructure, and in some cases, infrastructure such as cycle track and pedestrian stairway is not present on any of the selected roads. Walkways are available on all but one of the selected roads (Table 18.2), however, the available walkways do not cover the full length of the roads. Pedestrians and other NMT users are expected to use the road

Table 18.2 NMT infrastructure on the selected roads

Road	Infrastructure											
	Road length (kilometres)	Foot trail (length)	Walkway (length in metres)	Cycle track	Crossing facilities (None)	Kerb Ramps/ guard rail	Pedestrian stairways	Street lighting	Traffic lights (No.)	Humps \dips (No.)	Signage	Shoulder Width (metres)
Alick Nkhata	4.26	√	–	–	1	#	–	–	3	3	–	1[a]
Burma	6.047	–	6047	–	8	√	–	√	0	0	√	1
Dedan Kimathi	0.98	–	30	–	0	–	–	–	0	0	–	0
Independence	6.893	√	1970[r]	–	0	^	–	√+	3	0	*	0
Total	18.18		8.047									

Source Lukomona (2015)

[a] Infrastructure is not available the whole length and is not uniform where it is available

[r] Discontinued and dilapidated walkways are also available as part of the underpass, infrastructure is only available for a stretch of 50 m at UN Complex building

^ Infrastructure is only available as part of underpass but is damaged by vehicles

* Available but inappropriate

√+ Infrastructure available only from the avenue's junction with Nationalist Road to Kabulonga Road

[#] Kerb ramps provided in front of UN offices for a distance of 50 m shoulders at certain points. These road shoulders are often narrow, irregular, and/or absent in some cases. This was confirmed by 29 of the 31 NMT users interviewed on the selected roads. A 50 m stretch of a well-protected and paved walkway was initially created along Alick Nkhata Road by the United Nations (UN) in front of their office complex. Soon after the study was completed expansion works started on Alick Nkhata Road and the road has been transformed into a dual carriageway. The result has been a reduction in the space for NMT since a drainage was created on the side of the road taking up the space previously allocated to the walkway.

Due to the absence of walkways, in some cases, such as portions of Alick Nkhata Road and Independence Avenue, NMT users end up using undesignated dusty foot trails. This is more problematic for cyclists, wheelbarrows, and wheelchair users who are left with no choice but to risk their lives by using the main carriageway which is meant for vehicular traffic. Tiwari (2002) refers to this group of road users as 'captive pedestrians' because they have no option but to make do with the adverse road conditions since no other transport option is not feasible to them. The increase of wheelbarrows on the roads is noticeable, and since they take up a considerable amount of space, they can cause accidents as motorists try to avoid them. It is not uncommon to find goods tipped from a wheelbarrow scattered across a road.

18.3.1.1 Crossing Points

Other key NMT infrastructure is crossing facilities such as pedestrian crossing and/or traffic lights and pedestrian bridges and underpasses. NMT users identified traffic lights as the safest place to cross the selected roads. Conversely, traffic lights on the roads under study are often not the nearest and preferable crossing point for NMT users. As a result, it is not surprising that several NMT users were seen crossing at undesignated but shortest points which are not safe. However, it should be noted that the placement of traffic lights is largely based on engineering standards that aim at making the movement of MT easier as opposed to NMT. Besides being inappropriately situated, some pedestrian crossings are faded and barely visible to motorists and NMT users alike. In addition to being faded, most pedestrian crossings on the roads under study have no signage to warn motorists, making them unsafe. Road signage is an integral part of any good road network and necessary for its optimal operation. In commenting on the issue of road signage, one key informant from RTSA stated the following;

> I have always said the difference between life and death could be a single road sign. The presence or absence of just a single road sign can make that important difference.

All the key informants acknowledged that road signage was inadequate on the selected roads or was misplaced and inappropriate, in cases where it was provided. Misplaced and inappropriate signage was observed on Independence Avenue, near its junction with Dedan Kimathi Road. At this particular intersection, a pedestrian crossing sign has been placed right next to a speed limit sign of 65 km/h while the pedestrian crossing itself is less than 100 m away and after the intersection. Such inappropriate placement of signage and NMT infrastructure poses a danger to NMT users and makes the spot susceptible to accidents. Although none of the roads in the study has a footbridge, users tend to shun them where they are available. This is usually because they are not appropriately located or not built to make it easier for differently abled road users.

Members of the general public were of the view that vandalism was the cause of missing signage on roads in the city. This is also noted in official reports and publications of the Road Development Agency (RDA) where the agency warned members of the public against vandalising road traffic signage and other road installations (RDA 2013, 2014, 2015). However, 15 NMT users disputed this notion and claimed that some road signs were made of poor-quality materials and thus could not last. The rest agreed with the official position on the matter. To curb vandalism of road signage, road markings can be marked on the pavement or road surface as is a normal practice globally. However, the use of long-lasting high quality long lasting paint should be recommended.

Within a three-hour period of observations along Dedan Kimathi Road, 57 wheelbarrows were seen ferrying cargo between the Kamwala trading area and Inter-city Bus Terminus. However, Dedan Kimathi Road has no designated space for use by wheelbarrows or parking space forcing them onto the carriageway. All wheelbarrow users confirmed that they had no designated parking space within or outside Inter-city Bus Terminus but parked illegally near bus loading and offloading bays. They complained of occasional harassment from local authority officials who demanded parking fees and/or chased them from the bus terminus. A senior officer from RTSA acknowledged the prevalence of wheelbarrows on Dedan Kimathi Road and added that there was an urgent need to regulate them as they were a ‘menace’ and cause of accidents and congestions. This is indicative of the absence of bottom-up user and non-user consultation in transport planning as noted by Lucas and Porter (2016). None of the commercial entities on the selected roads has made parking provisions for bicycles or wheelbarrows. Both RTSA and LCC officials admitted that there was no management plan for wheelbarrows on Dedan Kimathi Road despite the role they played in facilitating the movement of goods for small- and medium-scale enterprises. Neither the LCC nor RTSA has statistics on the number of wheelbarrows in the city, a situation that complicates any feasible management of this NMT equipment. An attempt by the local authority to register the wheelbarrows in 2020 was quickly halted to allow for ‘further consultation’. The lack of statistics on the number of wheelbarrows or bicycles that use the selected roads consolidates the notion by Gemzøe (2006) that NMT users remain invisible to city authorities.

The summary presented above shows that NMT infrastructure on the selected roads is inadequate. The paved walkways for instance only add up to a mere 8,047 m while the whole combined length of the carriageway, covering 18,180 m, is tarred. Therefore, the length of the paved walkway is only 44.3% of the total length. This implies that the selected roads are mainly conducive for motorists at the expense of other road users. This is against the idea of ‘complete streets’ which seeks to make streets work for existing and future users, of all ages and abilities, regardless of how they travel not just those using a motor vehicle (National Complete Streets Coalition 2011). As a result of this bias, other road users on the selected roads face a number of challenges which are discussed in the following section.

18.3.2 Challenges of NMT Use in the City

NMT users on the selected roads face various challenges which remain unknown to City Authorities. Challenges faced by NMT users emanate from four broad areas which relate to the state and inadequacy of NMT infrastructure as discussed in Sect. 18.3.1. These areas are the safety of roads for NMT use, the appeal of the NMT environment, space availability, and road user awareness.

18.3.2.1 Road Safety

All research participants agreed that the selected roads are not safe for NMT users. Although police records for road traffic accidents (RTAs) are not disaggregated by district or road, the views of research participants resonate with the general trend of RTAs in Lusaka Province which confirm that NMT users comprise the majority of all accident victims. Close to 57% of RTA victims in Lusaka Province are NMT users (Zambia Police [ZP] 2015). This shows how risky it is for NMT users to move from one point to another in Lusaka City. The implication of these statistics is that city roads are generally unsafe for NMT users. All the 30 NMT users interviewed agreed with this sentiment as did the key informants.

According to the Zambia Police Report on RTAs for 2014, over 1,135 RTAs recorded in Lusaka Province involved pedestrians who were hit by vehicles while trying to cross the roads (ZP 2015). The situation could even be worse on roads such as Dedan Kimathi which do not have traffic lights or pedestrian crossings to regulate traffic. A simple action such as crossing the road can be life-threatening to NMT users on City of Lusaka roads including those in the study due to the inadequacy of NMT infrastructure. On the other hand, 29 NMT users cited excessive speeding as a major challenge to their safety. This claim was supported by official reports from the Zambia Police which attribute over 1,323 RTAs recorded in the year 2014 in Lusaka Province to excessive speeding (ZP 2015). Despite excessive speeding being an issue of concern, it was observed that some roads, such as Alick Nkhata and Dedan Kimathi, do not have signs indicating the speed limit making it difficult for one to know the allowable speed. When asked about this observation, an RTSA official stated that 40 km/h was the maximum speed limit for all urban roads in Lusaka. This claim is in contrast to the observations on Independence Avenue and Burma Road where the speed limits are above 40 km/h. Speed limit setting is an important aspect of traffic management that should not be left to the discretion of motorists as it has a significant bearing on the lives of road users especially NMT users who bear the highest brunt of RTAs. It may be useful to set the speed limit of 30 km/h in areas where there are high volumes of NMT as this is the maximum speed at which humans have a high survival rate when it is human and motor vehicle accidents (Goyal 2014).

18.3.2.2 The Attractiveness of the Road Environment to NMT Users

The state of the surfaces on the selected roads also presented different challenges to NMT users during the different seasons of the year. During the rainy season, road edges are flooded and muddy forcing NMT users onto carriageways. During the dry windy season, dust and dirt were identified as the main challenges faced by a majority of NMT users interviewed. These situations make NMT use unattractive to potential users and are a disincentive to private motorists who would ideally be urged to shift to either NMT or public transport. Road reserves of the selected roads have no tree canopy or cover to protect users from direct sunlight and/or rains apart from the 800 m walkway on Independence Avenue stretching from the Woodlands roundabout to the road's junction with Nationalist Road which is partially covered by a tree canopy. Ordinarily transport and road authorities prescribe and ensure details for the protection of NMT users from adverse weather elements included in design guidelines for NMT infrastructure.

The attractiveness of the environment is also negatively affected by the increased presence of vehicles on the selected roads making NMT users feel unsafe since the available infrastructure does not adequately meet their needs. All NMT users interviewed had noticed an increase in the number of vehicles making the roads more dangerous for them. The increase in the number of vehicles was also confirmed by an official from RTSA who stated that

> On average, we register over 16,000 vehicles per quarter nationwide……we register over 330 vehicles per day and over 200 of that number is in the City of Lusaka alone. This means that slightly over 50% of the total vehicle population is in Lusaka City.

A study by Chikuba (2014) which confirmed the increase in the number of vehicles highlighted the fact that the majority are imported second-hand motor vehicle. The increase in private car ownership is encouraged by the neglect of transport options such as NMT and public transport which manifests in affluent households purchasing more private cars to meet their transport needs. Public transport in Lusaka City is characterised by failure and inefficiency resulting in it not being a preferred option especially for affluent households (JICA 2009; Chikuba 2014). NMT infrastructure is inadequate in places like bus stops and markets which are more accessible to pedestrians and other NMT users due to congestion.

18.3.2.3 Encroachments on Road Reserves

It was observed that road reserves and shoulders on the selected roads are highly contested spaces that are usually encroached on by billboards, mobile money booths, and other informal activities. This observation was validated by 27 NMT users, whereas only 2 officials acknowledged this encroachment on road reserves as a challenge to NMT users. The presence of such activities on spaces meant for NMT points to the failure of LCC to adequately manage trading places in the city given the availability of designated space for such merchandise (Nchito 2011). Given that the city authorities expressed ignorance of the activities that encroach road reserves on the selected roads exposes a critical gap and perhaps justifies arguments for the involvement of NMT users in the design of city roads (Gemzøe 2006; Dewar 2011). There is a need to find a middle ground where the needs of NMT and of vendors will be attended to. The informal sector is the main provider of livelihoods in Zambia, and therefore, the needs of vendors should also be incorporated in the design of road infrastructure.

18.3.2.4 Inadequate Road User Education

Of the NMT users interviewed, only 3 are aware of road user regulations such as the Highway Code and other legislation. Most NMT users interviewed assumed that road traffic regulations and laws were only meant for motorists and not them. This finding was echoed by an RTSA official who noted that

> Most pedestrians assume that the Highway Code is just meant for motorists. it is for all road users. You see it in the manner that pedestrians behave on the road....you will see some Evelyn Hone College students crossing that Zebra cross adjacent to their school on Church Road with headphones in their ears or busy talking on the phone assuming that a motorist has seen them and will automatically stop for them.

In the study, several NMT users were observed using the wrong side of the road and talking on the phone while crossing the road. The most prominent challenge is safety concerns as a result of difficulties in crossing the selected roads, excessive speeding by motorists, and the unattractiveness of the road environment especially during the rainy season forcing NMT users onto the carriageway competing for space with speeding vehicles.

18.4 Integration of NMT Infrastructure in Policies of the City of Lusaka

Like any city, road development in Lusaka is guided and influenced by policies, legislation, and plans, which are translated into projects. The main legislation dealing with transport and NMT are the Transport Policy of 2016, the Road Traffic and Public

Roads Act, and The Urban and Regional Planning Act of 2015, and the most recent and relevant is the NMT Strategy of 2019.

18.4.1 Transport Policy, 2016

The City of Lusaka does not have a specific localised transport policy. However, there exists a Zambia Transport Policy of 2016. The Policy provides broad guidelines on road development for the whole Country. It only makes broad references to sustainable transport but does not specifically mention NMT and its related infrastructure (Government of the Republic of Zambia (GRZ) 2016). The policy focuses on road, rail, maritime and inland waterways, and air transport. It also has a section on inter-modal transport and only mentions NMT in relation to rural non-motorised and intermediate transport. The absence of NMT in relation to urban areas speaks volumes and is indicative of a vehicle-focused development paradigm which is probably why the government has focused on massive road projects.

18.4.2 Road Traffic and Public Roads Acts

Despite the lack of explicit reference to NMT or its infrastructure in the transport policy, the Road Traffic Act No. 11 and Public Roads Act No. 12 of 2002 propose some explicit provisions for NMT and its infrastructure. The focus of the Road Traffic Act is to manage and control all road traffic which includes NMT users and thus stipulates how road users ought to conduct themselves on public roads (GRZ 2002a). The Public Roads Act, on the other hand, makes provisions for road development and administration. With regard to specific NMT infrastructure, Sect. 52 of the Public Roads Act, 2002 provides that "…a road authority shall provide where it shall consider it necessary or desirable, for the safety of accommodation of pedestrians and pedal cyclists, proper and sufficient footpaths by the side of roads under its control …" (GRZ 2002b). Unlike other provisions of the Public Roads Act relating to MT, Sect. 52 is not prescriptive. The section suggests that the provision of NMT infrastructure is at the discretion of a road authority who may even decide not to erect it. The section seems to suggest that NMT infrastructure should be provided only to ensure the safety of NMT users and not merely on its own merit as a viable transport option.

The absence of prescribed specifications for infrastructure such as walkways and cycle tracks in the Public Roads Act reinforces the notion that NMT infrastructure is not important. Apart from giving the standard widths of public roads, the Act should have prescribed minimum widths for NMT infrastructure. Such omissions negatively affect the provision of NMT infrastructure. It should be emphasised that NMT infrastructure should be provided as a preferred and readily available transport option for the majority of road users as opposed to it being seen merely as a means of avoiding accidents on roads meant for vehicles as the Public Roads Act implies.

18.4.3 *NMT Strategy, 2019*

Despite the Transport Policy of 2016 being silent on NMT, the government produced an ambitious NMT Strategy in 2019. Although the Strategy is more aligned to the Road Traffic and Public Roads Act, it highlights the benefits of NMT and specifies the various elements of NMT infrastructure required in Zambian cities. The Strategy states that

> The Government will create urban street design guidelines, known as the Zambia Urban Street Design Manual (ZUSDM). The ZUSDM will include detailed standards and design guidelines for footpaths, cycle tracks, carriageway, Bus Rapid Transit (BRT), and other street elements (GRZ 2019: 39).

The Strategy is clearly based on best practices, using examples from around the world but it is yet to be seen if it will be implemented. The urban street design manual is yet to be completed and road projects continue without the guidelines. Some newly expanded roads which were not part of the study have walkways, but their usability is diminished by the placement of streetlight or billboard poles in the middle of the walkways. This makes it impossible for users like cyclists and wheelbarrow pushers to use the already narrow walkways. The presence of deep drainages lined with concrete next to many new walkways also poses a hazard to pedestrians if they were to fall into them. Similarly, some recent roads have walkways, but these tend to be narrow and discontinuous making it difficult for an NMT user to use them from the beginning to the end of their journeys. Where NMT infrastructure is available, there is no segregation of modes (Mitullah and Opiyo 2017a).

18.4.3.1 Planning Legislation and Plans for the City of Lusaka

The City of Lusaka and the rest of the Country use the Urban and Regional Planning (URP) Act of 2015 to prepare plans which guide all land developments including road transport (GRZ 2015). The URP Act repealed the Town and Country Planning (TCP) Act of 1962. The URP Act like its predecessor has made general reference to sustainable infrastructure while explicit prescriptions have been made for MT infrastructures such as parking and loading bays. Essentially, planning legislation has not adequately provided for NMT infrastructure.

In terms of City plans, there are basically 3 plans that are responsible for guiding the development of Lusaka City and these are the Doxiadis Plan, 1972, **the** Lusaka Comprehensive Master Development Plan (LCMDP) of 2009, and the Strategic Plan

2010–2015. These plans were all developed based on the TCP Act, 1962, which was focused on motorists. It is thus not surprising that the Doxiadis Plan made recommendations for road infrastructure that would only favour motorists. Commenting on the Doxiadis Plan, Nachengwa (1990: 79) remarks that "…one of the most important shortfalls of the plan was the failure, despite having emphasized the urgency of expanding public transport, to take into account the needs of the majority of the city's population-cyclists and pedestrians. Instead, private car ownership was assumed, leading to the proposal for a tremendous network of freeways and expressways".

The LCMDP which is slightly more recent was prepared by Japan International Cooperation Agency (JICA). This plan attempted to redress the shortfalls of the Doxiadis Plan. Provisions for NMT infrastructure in the JICA-sponsored plans were made through two objectives namely *i) provision of safe and comfortable pedestrian routes* and *ii) promotion of NMT* (JICA 2009). The construction of pedestrian facilities at black spots, sidewalk, and bicycle road networks was proposed in the plan, and based on those proposals, 3 projects related to NMT infrastructure were on the list of 34 projects earmarked for implementation in the short term (2010–2015). However, it is surprising that all 3 NMT projects were deemed to be of 'medium and low relevance' and thus of low priority (JICA 2009; 53).

The most recent is the Strategic Plan (2010–2015) which seems to fully acknowledge the importance of NMT infrastructure. Through this plan, LCC makes an audacious but unrealistic commitment that "by 2015, the city of Lusaka will have closed the gap in infrastructure renewal of roads, sidewalks, traffic control signals, traffic signs and streetlights" (LCC 2011;18). This goal seemed ambitious given the Council's limited financial capacity. The Council does not have adequate financial and human resources to achieve that goal. The gap with regard to NMT infrastructure remains unfilled. In the absence of a large citywide transport study, it is difficult to know the extent of the 'gap' that needs to be filled.

As earlier noted there have been a number of road development projects in the city. Apart from the Ring Road projects, a majority of these road infrastructure development projects are as a result of political pronouncements and are not in any of the City Plans discussed earlier. An example is the L400 Road Project in which 408 kms of Lusaka Township Roads (Burma Road included) have been earmarked for tarring, expansion, and/or rehabilitation (RDA 2013). Distances and specific details of NMT infrastructure to be built under this project and other major road projects are not available in all the project documents reviewed. However, observations on Burma Road reveal that NMT infrastructure has been retrofitted covering the whole road length. Another mega project is the Lusaka Decongestion Project (LDP) (2018–2021) which includes the widening of 120.7 kms of city roads, improvement of nine junctions, and the construction of four flyover bridges all geared towards improving the motorist's experience and not the NMT users. The external drivers of these projects are not interested in NMT as is usually the case (Mitullah and Opiyo 2017b).

18.4.3.2 Towards the Future: Potential of the Selected Roads to Accommodate NMT Infrastructure

All the respondents were of the view that all the selected roads have sufficient space to accommodate NMT infrastructure. Analysis of the widths of the selected roads using GIS confirmed that the space is adequate to accommodate both NMT and MT infrastructure. The widths of the roads under study are not uniform due to encroachments. Prevalent activities observed on the selected roads are car washes, restaurants, kiosks, and hoardings which the LCC regards as temporary and whose land lease renewal is at the local authority's discretion. Even at their narrowest points, the selected roads are wide enough to accommodate NMT infrastructure. All the selected roads except Alick Nkhata Road are more than 30 m wide at their narrowest points. This is positive in terms of possibilities for future NMT infrastructure.

The inadequacy of NMT infrastructure on the selected roads despite the availability of space is evidence of the continued focus on MT. It is important for the city to have specific guidelines for road and NMT infrastructure design and not leave the design process to road engineers who may omit critical issues which are important to NMT users. The LCC and RDA currently rely on regional and international standards when designing NMT infrastructure for the city. A holistic and realistic approach should be used during the planning and design phase of road infrastructure development. In a city like Lusaka, one would have to take into consideration the needs of NMT users as well as vendors such as the ubiquitous mobile money booths.

18.5 Conclusion

The chapter has shown that the NMT infrastructure on the selected roads is inadequate and where it is available is in a bad state. The fact that 65% of Lusaka residents walk as a mode of transport but are not provided for in terms of infrastructure speaks volumes to the perception of development. There is a need to change people's perceptions of NMT by making it more attractive and user-friendly. The roads in this study did not have adequate NMT in terms of the quality and length of the available walkways. The crossing points for NMT users are poorly labelled while speed limits are not often clearly indicated for motorists. It is clear that NMT infrastructure has not received the attention it deserves from the government. In a world where climate change has taken centre stage, the goal should be to keep people out of cars. This will only be feasible if going forward, NMT infrastructure receives the same level of funding as motor transport.

NMT users should not be left behind in terms of road user education. The study has disclosed that pedestrians and other NMT users think that road safety rules do not apply to them. Ignorance of road safety is hazardous for both NMT users and motorists alike. NMT users should be educated on road safety while they are still young; hence, schools should be targeted. Much needs to be done to increase awareness of the benefits of NMT use to users, society, and the environment. A lot can and should be done to sensitize NMT users on their responsibilities and obligations when using roads.

It is clear that the selected roads are not safe for NMT users. Contributing factors to this situation are inadequate crossing facilities, inadequate signage, and excessive speeding. The scenario put forward in this chapter also results in reducing transport options for most vulnerable road users (NMT users) and should thus convince policy and decision-makers to prioritize infrastructure that supports walking and other NMT modes of transport instead of widening carriageways. The aim should be to maintain the use of private cars at 10% but improve the state of NMT infrastructure and carefully plan NMT routes. The pleasant tropical climate experienced in Zambia allows the use of NMT for most parts of the year.

Being the first planned new capital in Africa (Williams 1986), there are many lessons from Lusaka's past which can be used to create a better transport future for the city dwellers. As this chapter has shown, the initial neglect of NMT in preference for motor vehicles has been perpetuated and the distances covered by the poor are getting longer. Planning of NMT should enable users to use the infrastructure from their origins to their destinations. Extensive research in the transport needs of citizens needs to be carried out as does the training of transport planners. The NMT strategy which has been developed should be implemented in order for NMT in Lusaka to become an accessible option for many.

References

Central Statistical Office (CSO) (2012) 2010 Census of population and housing. National descriptive tables. Central Statistical Office, Lusaka

Chikuba Z (2014) Lusaka needs public transport reforms. The ZIPAR Quaterly, pp 4–5

Collins J (1986) Lusaka: The historical development of a planned capital, 1931–1970. In: Williams JG (ed) Lusaka and its environs. The Zambia Geographical Association, Lusaka, pp 95–137

Dewar D (2011) The relationship between spatial planning and transportation planning in Southern Africa and its consequences for human settlement. World Academy of Science, Engineering and Technology, pp 5–20

Freund P, Martin G (1999) Driving south: the globalization of auto consumption and its social organization of space. Montclair State University, New Jersey

Gemzøe L (2006) Are pedestrians invisible in the planning process? Copenhagen as a case study. Royal Academy of Fine Arts, Copenhagen

Government of the Republic of Zambia (GRZ) (2019) Non-motorised transport strategy. Ministry of Transport and Communication, United Nations Environment Programme and Institution for Transportation and Development Policy. Government Printers, Lusaka

GRZ (2016) The zambia transport policy (Revised). Government Printers, Lusaka

GRZ (2015) The urban and regional planning act, No. 3 of 2015. Government Printers, Lusaka

GRZ (2002a) Road traffic act No. 11 of 2002a. Government Printers, Lusaka

GRZ (2002b) Public Roads Act No. 12 of 2002b. Government Printers, Lusaka

Goyal VS (2014) Planning for non-motorised transport in urban areas. In: National capacity building workshop for sustainable and inclusive development, August 12–13, 2014. GIZ, Dhaka, pp 1–59

Hogan MK (2015) London mayor approves Europes first bike superhighway. http://inhabitat.com/london-mayor-approves-europes-first-bike-superhighway/. Accessed 19 Feb 2015

Japan International Cooperation Agency (JICA) (2009) The study on comprehensive urban development plan for the City of Lusaka in the Republic of Zambia: Final Report Volume II-Chapter 1 Urban transportation. Ministry of Local Government and Housing, Lusaka

Lucas K, Porter G (2016) Mobilities and livelihoods in urban development contexts: introduction. J Transp Geogr 55:129–131

Lukomona R (2015) Non-motorised Transport (NMT) infrastructure along selected roads in Lusaka: status, experiences and policy. University of Zambia, MSc. Dissertation

Lusaka City Council (LCC) (2011) Lusaka city council corporate strategic and operational plan [2010–2015]. Lusaka City Council, Lusaka

Macro Trends (2021) Lusaka population. https://www.macrotrends.net/cities/23277/lusaka/population. Accessed 30 March 2021

Mitullah MV, Opiyo R (2017a) Non-motorised transport infrastructure provision on selected roads in Nairobi. In: Mitullah WV, Vanderschuren M, Khayesi M (eds) Non-motorised transport integration into urban transport planning in Africa. Routeledge, London, pp 90–111

Mitullah MV, Opiyo R (2017b) Institutional framework for walking and cycling provision in Cape Towm, Dar es Salaam and Nairobi. In: Mitullah WV, Vanderschuren M, Khayesi M (eds) 2019. Non-motorised transport integration into urban transport planning in Africa. Routeledge, London, pp 189–205

Nachengwa L (1990) A history of public road transport in Lusaka, 1889–1987. Masters Dissertation, University of Zambia

National Complete Streets Coalition (2011) Complete streets. www.completestreets.org. Accessed 5 July 2014

Nchito WS (2011) Formalizing trading spaces and places: an analysis of the informal sector in Lusaka, Zambia. In: Banik D (ed) The legal empowerment agenda: poverty, labour and the informal economy in Africa. Abingdon, Ashgate, pp 87–106

Pieterse E (2011) Rethinking African urbanism from the slum. https://lsecities.net/wp-content/uploads/2011/11/2011_chw_5060_Pieterse.pdf. Accessed 30 March 2021

Road Development Agency (RDA) (2015) Vandalism of road furniture worries RDA. ROADS, 12th ed. RDA, Lusaka

RDA (2014) Vandalism and encroachment worries RDA. ROADS, 10th ed. RDA, Lusaka

RDA (2013) 400 kilometres of Lusaka City roads to be rehabilitated. ROADS, 6th ed. RDA, Lusaka

Tiwari G (2002) Planning for bicycles and other non motorised modes: the critical element in city transport system. Indian Institute of Technology, Delhi

UN-Habitat (2010) Planning sustainable cities UN-Habitat practices and perspectives. UN-Habitat, Nairobi

UN Environment (2019) Calculating the potential climate value of Non-Motorised Transport projects in African Cities. UN Environment, Nairobi. http://www.unep.org/Transport/ShareTheRoad

Williams JG (1986) Introduction. In: Williams JG (ed) Lusaka and its environs. The Zambia Geographical Association, Lusaka 95–137:1–4

Zambia Institute of Policy Analysis and Research (ZIPAR) (2014) Used motor vehicle imports and the impact on transportation in Zambia:Working Paper No. 21. ZIPAR, Lusaka

Zambia Police (2015) Zambia police service annual police report, Lusaka Zambia

Raymond Lukomona is a Planner with over 15 years of experience in the field. Raymond holds a Master's degree in Spatial Planning while his Bachelor's degree is in Urban and Regional Planning. He also has various certificates in GIS and Project Management. Raymond has championed the integration of non-motorized transport in city plans. He has presented on this subject to the National Planning Conference of the Zambia Institute of Planners and is a Fellow of the Institute which is the highest membership category. His research interests are in sustainable transportation, environment, planning law and history and smart cities.

Wilma S. Nchito is an Urban Geographer who has studied the contemporary patterns and material aspects of Zambia's urbanisation which is in transition. She has worked closely with the local authority and a Slum Dwellers International (SDI) affiliate on several research projects to co-produce knowledge for inclusive upgrading. She has published relevant research outputs on the informal sector in Zambia, climate change and adaptation in the city, settlement upgrading. She has extensive knowledge of the city of Lusaka and has worked on several issues including local area planning, gentrification, access to water, climate mitigation, street vending and urban transport development. She is currently the Director of Research and Postgraduate Studies at the University of Zambia.

Chapter 19
The Future of Non-motorised Transport in Urban Africa

Chinebuli Uzondu and Anderson Aja Etika

Abstract Non-motorised transport (NMT) modes are an important and integral component of urban transport across the world. Besides the provision of basic mobility, affordable transport and physical fitness, they serve to reduce negative environmental impacts of transportation. However, despite the NMT being a dominant mode of transport in most African rural and urban areas, it continues to be largely neglected in terms of relevant policies, planning and provision of infrastructures. Governments in African nations remain unsustainably focused on expanding the road networks and increasing motorization, with NMT as modes borne out of necessity for the poor. Non prioritization of NMT carries a huge road safety burden of fatalities and injuries to these vulnerable road users and an increasing car-culture. Thus, it is imperative that a future narrative is drawn for this key transport mode. This research seeks to examine the current challenges faced by NMT users in Sub-Saharan Africa and provide insightful policy ideas and infrastructure development strategies to make walking, cycling and other NMT transport modes more convenient, safe, pleasant and convincing in urban Africa. Through consultations with relevant stakeholders and experts, the book chapter recommends a pathway for the integration of NMT to urban mobility plans in African cities and towns.

Keywords Non-motorised transport · Infrastructures · Nairobi · Accra · Lagos · Kampala

C. Uzondu (✉)
Department of Transport Management Technology, Federal University of Technology, Owerri, Imo State, Nigeria
e-mail: Chinebuli.uzondu@futo.edu.ng

A. A. Etika
Department of Civil Engineering, Cross River University of Technology, Calabar, Nigeria
e-mail: andersonetika@crutech.edu.ng

R. A. Acheampong et al. (eds.), *Transport and Mobility Futures in Urban Africa*,
The Urban Book Series, https://doi.org/10.1007/978-3-031-17327-1_19

19.1 Introduction

Globally, it is now estimated that over 50% of the world's population live in towns and cities (World Bank 2020). This has resulted in increased transport demands to meet urban mobility needs. This increase in transport demand could be regarded as a reflection of economic growth. However, managing this enormous demand for mobility has become a major challenge for most countries as such increase comes with other socioeconomic problems such as traffic congestion, environmental pollution, traffic crashes and fatalities, noise pollution, among others.

According to the World Health Organisation, the African region has the highest rates of road traffic deaths in the world (26.6/100,000 people). It is even more troubling that Africa has the highest proportion of pedestrian and cyclist mortalities with 44 deaths (WHO 2018). The heavy burden of deaths borne by these road users reflects the prioritization of infrastructure for cars and other motorised transport (MT) and overlooking of Non-Motorised Transport (NMT) modes. According to VTPI (2010), NMT includes walking, bicycling, and variants such as small-wheeled transport (cycle rickshaws, skateboards, scooters and hand carts), however, the focus of this study will be on walking and cycling.

These modes are gradually becoming an integral part of urban transport systems in most developed countries of the world. Besides the provision of basic mobility, affordable transport, and physical fitness, they serve to reduce the negative environmental impacts of transportation. However, despite NMT and particularly walking, being a dominant means of transport in most African urban and rural areas, it continues to be largely neglected in terms of relevant policies, planning and provision of infrastructures. The road infrastructures within most African cities, towns and even villages are majorly designed for MT with little attention to NMT. The lack of dedicated space for cyclists, pedestrians and other NMT users in most African Cities and Towns means they share the roadway with MT users, thus exposing them to road traffic crashes and fatalities.

The high cycling culture and safety in countries such as the Netherlands, Germany and the Scandinavian countries of Denmark and Sweden have been attributed to their transport and land use policies which prioritise non-motorised and public transport facilities. However, cycling and walking in most low-income countries is borne out of necessity for the "poor". In these countries, NMT infrastructure is either non-existent or provision is poor. For example, there is a lack of pedestrian crossings, lack of cycle lanes, streets with open drains exposing pedestrians to hazards, and poor lighting of streets. According to Vanderschuren et al. (2017), transport planners mostly concentrate on improving urban automobility and rarely make adequate provisions for pedestrians. This is consistent with Mitullah et al. (2016) where they concluded that NMT integration into urban transport planning in Africa clearly shows that transport and urban planning remain situated in an automobile-dependent transport planning and global city development which pays no adequate attention to walking and cycling. Additionally, transportation planning in these countries tends to overlook and undervalue many active transportation benefits (see Table 19.1),

Table 19.1 Cost and benefit of non-motorised transport

	Improved active transport conditions	More active transport travel	Reduced automobile travel	More compact communities
Benefits	• Improved user convenience, comfort and safety • Improved accessibility for non-drivers, which supports equity objectives • Option value • Higher property values • Improved public realm (more attractive Streets)	• User enjoyment • Improved public fitness and health More local economic activity • Increased community cohesion (positive interactions among neighbours) • More neighbourhood security	• Reduced traffic congestion • Road and parking facility cost savings • Consumer savings • Reduced chauffeuring burdens • Increased traffic safety • Energy conservation • Pollution reductions • Economic development	• Improved accessibility, particularly for non-drivers • Transport cost savings • Reduced sprawl costs • Open space preservation • More liveable communities • Higher property values • Increased security
Cost	• Facility costs • Lower traffic speeds	• Equipment costs (shoes, bikes, etc.) • Increased crash risk	• Slower Travel	• Increases in some development costs

Source Litman (2021)

resulting in underinvestment in these modes, which reduces overall transport system diversity and efficiency (Litman 2021).

The primary purpose of this study is to highlight the current existing conditions of NMT interventions, constraints and opportunities in selected African cities. The research has underlined the importance, role, benefits and influencing factors of NMT in order to have a clear understanding of the significance of NMT and its required adequate facilitating infrastructure in African cities.

19.2 Role of NMT in Improving Urban Mobility of African Cities

NMT remains the principal mode of transportation in most African cities, particularly for local travels. While this is largely not by choice but rather by non-availability of affordable and accessible options, its importance in aiding mobility cannot be over-emphasised.

(i) Emissions are growing rapidly in most developing countries as the use of motor vehicles increases. However, NMT produces no air pollution, no greenhouse gases, and little noise pollution.
(ii) While a well functional public transport provides more efficient use of road space, cyclists use less than a third of road space while pedestrians use less than a sixth of road space used by private vehicles thus combating congestion.
(iii) Besides serving as a transport mode, NMT also provides aerobic exercise which could help address the health challenges associated with lack of exercise such as obesity and mental diseases. Walking and cycling can also serve as an alternative to the traditional gym.
(iv) According to the World Health Organisation (2018), Africa currently has the highest road fatality rates with pedestrians and cyclists being overrepresented. Thus, promoting NMT will help reduce the annual road deaths from crashes in African cities.
(v) Increasing walking and cycling in African cities will greatly reduce their dependence on fossil fuels and the cost of policies such as subsidies. For example, Nigeria, Africa's largest oil exporter, imports all its fuel spending over $300 Million as subsidies (Olurounbi and Clowes 2021).
(vi) It is important that those who already walk very long distances in poor conditions because they cannot afford other transport modes are better protected. Therefore, encouraging safe cycling and walking is key in improving accessibility for the lower class and promoting social cohesion. Getting to workplaces in most African cities remains a difficult task for most low-income earners. This group spend a large part of their income on fares for trips which are most times <3 kms. Improving NMT infrastructure will serve to create a more cohesive society as persons of all incomes will be able to have access to bikes or walk in a pedestrian area.

19.3 Trends and Conditions of NMT

NMT modes are very common worldwide, but their needs are rarely recognised and planned for in most transport infrastructural projects. Additionally, NMT users form a significant percentage of road traffic crash victims worldwide as there is a general lack of infrastructure in most cities. However, some cities in Europe and England have been redesigned to accommodate these modes. According to Singh (2018), the number of non-motorised trips varies greatly in developed countries, with walking and cycling making up less than an eighth of daily trips in countries such as the US and Canada, and over 20% in most European countries. Some places such as the Netherlands and Denmark in Western Europe have very high bicycle ownership. This could be because of transport policies formulated and being implemented in these countries which are in favour of NMT and public transport rather than MT. These countries also have very low death rates from road traffic accidents.

Recently, governments have begun showing interest in and promoting NMT in developing countries. In some parts of Latin America like Mexico, NMT constitutes about 39% of the trips in cities. Similarly, NMT is the principal mode of transportation in most cities in Africa. It is known that most trips begin and end with NMT and therefore, the modal share of walking and cycling could be higher than published figures. Figure 19.1 shows urban modal shares of selected African countries. For example, the average share of walking in Accra is 40% and Lagos is about 38%. In Nairobi walking and cycling account for about 43% of all trips and in Kampala, it is 44%. Most of the inhabitants in these cities walk because they cannot afford MT. Therefore, this mostly affects the low-income population groups who invariably cannot afford public transport. As seen in Fig. 19.1, bicycle utilization rate is very low. This could be because of the unsafe road environment created by MT, lack of NMT infrastructures and high cost of acquiring a bicycle. Therefore, where more than 38–40% (UN Environment 2016) of the population already walk, these existing trips could be embedded into the sustainable development goals (SDGs) targeted for African cities.

According to ICE (2000), experience from some developed countries such as the Netherlands, Denmark and Germany has shown that consistent and substantial funding sustained over extended planning and implementation horizon is required in the development and promotion of a comprehensive and operational NMT network. It has also been about integrated land use and transport planning that recognises the need for compact activities and mixed-use development with housing. Additionally, these countries made walking and cycling part of transport and infrastructures planning function with a dedicated strategy and budget—which is mostly not obtainable in African cities. Therefore, for improved mobility, cities must develop NMT infrastructures and facilities that are safe, secure, accessible, convenient and attractive to the users.

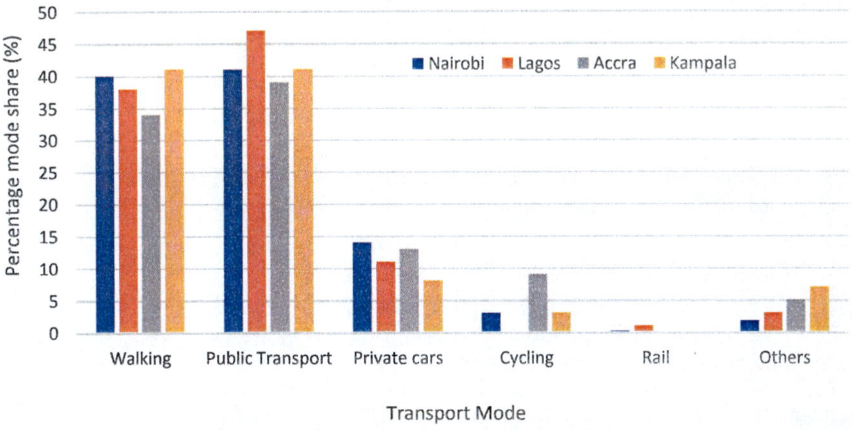

Fig. 19.1 City level statistics of modal share (*Source* Compiled by authors from city authorities, published and local unpublished documents and unofficial documents)

19.4 Method

The methodology involves a review of primary and secondary sources of information. This focused on examining the literature on trends and challenges of NMT globally. A qualitative approach was then used to identify and understand the current situation of NMT in selected Sub-Saharan African cities (Lagos, Nairobi, Kampala and Accra). Semi-structured interview templates were designed to elicit views and trends on NMT across the four cities from transport consultants, government officials and academics. A total of twenty-two (22) responses were received. Using Deductive Content Analysis (DCA), data were analysed, and discussions were made on the various topics of interest.

19.5 Case Studies: NMT in Selected African Countries

A qualitative approach was taken to understand the situation in selected African cities. The study was carried out in 4 different cities in Africa (Kampala in Uganda, Lagos in Nigeria, Accra in Ghana and Nairobi in Kenya) and is based on available data, discussions/interviews held with key government, non-government, and private sector stakeholders focusing on NMT. Information relating to the current condition of NMT, challenges and opportunities was sought from the respondents and formed the basis of the study. Below, major key findings are presented and discussed which gives some ideas of the condition of NMT and how it could be improved in Africa.

19.5.1 Brief Background of the Cities

These four cities are different from each other; however, they share some common features such as:

(i) low-income countries with low car ownership (WHO 2018)
(ii) a rapidly growing urban population
(iii) inadequate and deteriorating road infrastructure
(iv) poor facilities for NMT especially for walking and bicycling
(v) poor and inadequate formal public transport system resulting in the use of informal transport
(vi) growth in the use of minibuses, taxis, motorcycle services etc.
(vii) over dependence on informal or semi-formal transport services for most motorised trips

Nairobi

Nairobi is a hub for commerce, transport, and economic development in Africa. It has an estimated population of over 4 million and an annual growth rate of 3.95% (World

Bank 2016). Transportation is mainly road-based. However, the city has experienced rapid urbanisation which has overstretched the present road infrastructure. Traffic congestion has increased in recent years because of the increasing use of private cars. The NMT infrastructure is not developed, and this is a major challenge, especially to the low-income earners who mostly rely on them. The actual contribution of NMT to transport in the city is not well documented as there are widely varying data, however, walking and public transport are the dominant means of transport.

Lagos

Lagos is one of the fastest growing cities in the world with an estimated population of 18 million people and an annual growth rate of 6% (NBS 2016). It is the economic and commercial centre of Nigeria, and the transport system is predominantly road-based. Rising urbanisation has led to an increased demand for transport services and infrastructure and imposes challenges on the urban transport systems. Official data on walking and cycling are not available except those compiled from local unpublished sources and some official sites. Available records show that non-motorised transport modes account for an estimated 40% of all trips in the city. Almost all public transport trips involve walking, as a result most commuters walk as a part of their daily trips.

Accra

Accra is the most populated region in Ghana with more than 4 million people, and an annual growth rate of 4% (WHO, 2021). Though Accra has the least land size of 1.4% of the total land area of Ghana, similar to Nairobi and Lagos, Accra has not been able to absorb the great pressure exerted from rapid urbanisation. This has resulted in traffic being often gridlocked on poorly maintained, potholed roads. Congestions are made worse by street hawkers, who gather at busy junctions to sell their wares and the dominance of old buildings, which inhibit traffic flow, particularly around busy areas. While most of the population depends on the use of informal transport, reports show that NMT accounts for about 43% of all trips (see Fig. 19.1).

Kampala

Kampala is the second most populated Higher Local Government in the Greater Kampala Metropolitan Area with a projected population of 2.9 million people (Twinokwesiga 2020) and an annual growth rate of 5.4%. It is the administrative and economic capital of Uganda. The city is characterized by road-based public and private transport systems which is unable to meet the demand of the constantly growing urban population and the number of people who travel to the city for work and business every day. This has resulted in huge traffic congestions in the city. Even though most intra-urban trips are made by foot, the NMT infrastructure is not very well developed.

19.5.2 Current Condition

Increasing motorisation, combined with unplanned and inadequately maintained infrastructure has made NMT unsafe in most African cities. The needs of NMT users are rarely considered in road design, construction and improvements as there is a general lack of infrastructure. Road accidents adversely affect NMT users, especially pedestrians and cyclists because they continuously share the road space with high-speed vehicles. According to TRL (2002), the little NMT space available (footpaths and cycle lanes) is obstructed or occupied illegally by parked cars, and makeshift shops. NMT in Africa is currently in a very poor state exacerbated by lack of support from policymakers. The focus has always been and is still on expanding the road network which promotes car use and increases motorisation. Motorised transport is greatly prioritised over NMT, and most transport policies favour motor vehicles compared to NMTs. Even though some countries boast of some NMT infrastructures, reports from our study which are similar to the cities of interest show that most of them are in a very poor state (see Fig. 19.2).

For example, in Lagos, the roads are bad and lack adequate parking for motor vehicles resulting to the available footpaths frequently being occupied by motor

Fig. 19.2 State of some NMT infrastructures in some African cities

vehicles. Cities such as Lagos, Owerri and Calabar have wide streets that lack pedestrian crossings, making it difficult and risky for pedestrians to cross the road. A study by Uzondu et al. (2018) shows that the most common type of conflict observed among pedestrians in Owerri is the crossing conflict which is due to lack of pedestrian crossing facilities. There are little or no footpaths and cycle lanes, which also means that pedestrians and cyclists share the road space with other vehicles which impairs their safety and make movement very slow and difficult. Because of the very bad drainage system, the streets are often flooded, and this contributes to the difficulty in using NMT.

The situation is similar in Kampala where respondents reported limited NMT facilities and infrastructure to support different NMT modes. Reports show that most roads are narrow and poorly demarcated, and walkways are in a very poor state and not safe. These put the NMT users at risk of being knocked down by speeding cars. The interesting thing is that there are vast ideal road reserves that can be utilised efficiently but are not being considered.

In Accra, reports show that there is a disregard of law and order as the available walkways have been converted to illegal car parks. Most transport Infrastructure is not built to accommodate sufficiently the safety of NMT as pedestrians and cyclists use a part of the road as a walkway and for cycling.

Nairobi reports very low investment in NMT facilities and infrastructure. There are footpaths, footbridges and cycle lanes but they are not adequate and not up to the stipulated national standard. Like the situation in other cities, the available infrastructure is always occupied by parked cars and motorcycles and there are no penalties for these. There are no adequate policy and legislation on the design, use and maintenance of NMT infrastructure. This affects NMT users a great deal thereby exposing them to danger.

19.5.3 Opportunities for Improving NMT Infrastructure in Selected Cities

Studies have shown that there are huge opportunities not only for providing NMTs but also integrating them into the existing infrastructure in many African cities.

The study in Kampala shows that there are vast ideal road reserves that can be utilised efficiently to provide adequate and state of the art NMT facilities. There are great opportunities to design and promote appropriate infrastructure. Additionally, the pandemic has highlighted the great need for the provision of NMT facilities and thus should be a priority.

Additionally, Accra respondents reported that cycling may be the biggest opportunity for NMT in Africa. As there is more awareness of cycling today than it was 10 years ago, notwithstanding the present lack of adequate facilities that could make cycling safe and comfortable. With proper advocacy and sensitization there is an opportunity to get policymakers to integrate ideas that boost cycling and other NMT programs into road designs and planning. Also, construction of new roads should make provision for safe walkways and cycle paths. Apart from that the government and policymakers can integrate policies and programs on safe walking spaces into existing transport policies, incorporate these into road design and encourage the use of NMT based on its environmental, social and health benefits (see Table 19.1).

Similar to Kampala and Accra, Lagos has land spaces that can be utilised efficiently to provide adequate infrastructure for NMT. Cities can incorporate walkways and cycling paths into every road design to encourage the use of NMT.

In Nairobi, respondents emphasized the prioritization of NMT by the authorities and policymakers as lands are available to achieve these. NMT spaces can be created from the existing road networks and developed to link poor neighbourhoods to employment centres, schools and social amenities. Some roads and recreation areas could be made pedestrian and cyclists friendly by closing them up to motorists and restricting the use of cars in those areas.

19.5.4 Challenges in NMT Development

Even though there are huge opportunities for investment and provision of NMT infrastructure in African cities, there are also challenges which respondents believe could impede or delay the provision of these facilities.

Respondents in Kampala stated that a considerable number of people have moved to the urban areas, and this has contributed to traffic congestion in the cities as people can only use their private cars or public transport. Walking and cycling are not very common because of poor public perception, risk associated with sharing the road space with high-speed vehicles and lack of NMT facilities. This could be because little or no resources are allocated to them. Additionally, stakeholders' recommendations are rarely considered in policymaking, planning and implementation of NMT infrastructures as the political class doesn't appreciate the need to allocate resources to NMT. Road users are not aware of existing NMT policies because there are no sensitisations about the policies. Provision of NMT facilities have been shown to be more cost-effective compared to motor vehicle infrastructures yet they are rarely provided.

Similarly, in Accra, respondents attributed the non-provision of these infrastructures to lack of political will. They reiterated that policymakers do not prioritize the provision of safe walkways and are not willing to implement most NMT policies, pedestrian safety is not prioritised in the infrastructure development process. People are not aware of the NMT policies as participants noted that they are largely lacking, even where they are available, the infrastructure built is evidence that they are not being adequately used.

Reports from Lagos highlight lack of funding and adequate road space hinder the provision of NMT and NMT facilities. However, according to Uzondu et al. (2018), roads are currently being designed, built, expanded and improved without considering the needs of NMT users. The notion that motor vehicles come first has also affected road transport policy development as road infrastructures are largely designed to accommodate motor vehicles.

In Nairobi, stakeholder recommendations are rarely considered in transport planning. To improve the safety condition of NMT users, resident associations and NGOs try hard to lobby those in authority to see what could be done. There is massive corruption which has hindered the development of these projects.

These are in line with ICE (2000) on barriers to implementing a successful NMT policy which shows that the transport system in most developing countries is usually private-vehicle-oriented as the population perceives walking, cycling (and public transport) as the transportation mode for the poor and those who cannot afford private vehicles. Also, those in the upper-class of the population often have a disproportionate decision power, which makes NMT-focused policy risky. Where there are no NMT facilities, NMT users have a higher risk of being involved in accidents.

19.6 Design of NMT Environment

Traditionally, in most countries, road infrastructure planning is focused on the uninterrupted flow of motorised transport. The lack of usable, dedicated pedestrian and cyclist spaces in most African cities means that pedestrians and cyclists are forced to share the carriageway with fast-moving vehicles compromising NMT user safety. Thus, For NMT modes to be viable, and convenient, cities need to rebalance the distribution towards pedestrians and cyclists. Addressing this imbalance requires cities to redesign neighbourhoods and traffic systems to work in favour of cyclists and pedestrians, and to discourage car use. This will involve the provision of NMT infrastructure which encompasses shared spaces, slow-speed, footpaths, cycle tracks, and greenways on which to travel. There is a strong association between NMT use and high-quality NMT infrastructure that is separated from fast and heavy motor vehicle traffic (Pucher and Buehler 2008). Therefore, accommodating NMT would mostly involve the provision of separate travel spaces and measures to reduce vehicle speed. It is important to provide dedicated space for NMT users particularly on streets with moderately fast motorised vehicle traffic. This can be achieved by the introduction of barriers such as kerbs, landscaping or vehicle parking between pedestrian footpaths and cycle lanes along streets and motorised traffic (see Fig. 19.3) and pedestrianisation of streets. This is likely to be more effective than using paint to demarcate footpaths and cycle lanes, particularly in the African context with limited enforcement of traffic rules. There is also a need for road intersections to provide dedicated facilities for NMT users, and for traffic light signalling to prioritise pedestrians and cyclists.

Traffic calming measures such as vertical deflections (speed humps, speed tables, and raised intersections), horizontal shifts, and roadway narrowing could serve to reduce vehicle speed and enhance the street environment for NMT users. However, a by-pass area for cyclists should be included.

Fig. 19.3 Separated cycle lanes and pedestrian walkways

19.7 Policy Development to Encourage NMT Use in African Cities

Most road spaces in African cities are often developed for motor vehicles rather than people. Pedestrians and cyclists are not considered in the design and construction of road infrastructure despite that every journey either begins or ends with walking or cycling. Respondents believe that there are policies, but these are not being implemented. There are existing laws that could be leveraged upon to meet these needs, but they are often neglected. These laws could be modified to meet current demands instead of making new ones. Where there are walking and cycling facilities, motor vehicle drivers misuse them. Additionally, some available NMT infrastructure have been converted to illegal parks as guidelines developed for their use are not strictly enforced. Motor vehicle speeds are not regulated and therefore pose a great risk for NMT users who share the road space with them. Furthermore, there is a general lack of interest by stakeholders which have hindered policy development. According to VPTI (2010), past studies have shown that about 5–10% of car trips can be replaced by NMT if good policies are in place. There is an urgent need to develop appropriate policies to encourage NMT use in Africa. It must be made attractive with policy packages consisting of investments in infrastructure, awareness campaigns, strict enforcement etc. and must be sustainable.

Most countries in Africa have developed NMT policies in the past but most of them are not being fully implemented. Two of these polices from Nairobi and Lagos are presented below.

In 2015, the Nairobi City County Government developed a Non-Motorised Transport policy (Nairobi NMT 2015) that strives to facilitate a transport environment where all transport modes are of equal importance. It aims to develop and maintain a transport system that fully integrates NMT as part of the city's transport system. it will put in place laws and regulations to ensure that NMT facilities are not encroached by the MT modes and other street users.

The objectives of this policy are to:

(i) Increase mobility and accessibility;
(ii) Increase transport safety;
(iii) Improve infrastructure for NMT;
(iv) Increase recognition and image of NMT in Nairobi County; and
(v) Ensure that adequate funding/investment is set aside for NMT infrastructure (Nairobi City County Government NMT 2015).

The Lagos state government developed a Non-motorised Transport policy that aims to support increased accessibility by prioritising walking, cycling, and public transport. This would be achieved by creating a safe and pleasant network of footpaths, cycle tracks, greenways, and other facilities to serve the general population. According to the policy (Lagos State NMT 2018), this will achieve the following:

(i) **Enable equitable access for all** by improving access and mobility for all residents; promoting social and economic empowerment through the provision of improved low-cost mobility; facilitating safe access for children; enabling gender equity through the provision of non-motorised transport (NMT) and public transport facilities that are safe for women to use; enabling inclusion of persons with disabilities by creating NMT facilities that follow principles of universal design; and by creating a changed culture that accepts the use of walking, cycling, and public transport as acceptable and aspirational means to move around in the city.
(ii) **Optimise the use of resources such as space, funds, time, and energy** by investing in NMT and public transport modes that consume fewer resources per person-trip compared to personal motor vehicles (PMV) and by encouraging dense, compact, and mixed-use development that contributes to shorter trips and allows more people live and work close to PT facilities.
(iii) **Improve road safety and personal security** by improving management of traffic conflicts; reducing road crashes, and deaths; and creating public spaces that are safe at all times of the day for all users.
(iv) **Reduce local and global environmental impacts** of Lagos's transport system by expanding the use of zero-pollution NMT modes and low-pollution motorised modes, helping to improve the city's air quality.
(v) **Enable community participation by involving residents, businesses, and other stakeholders** in the preparation of designs to foster the community's active use and sense of ownership of these spaces.

These are very comprehensive policies that if fully adopted and implemented would improve NMT in these cities and contribute to safer urban mobility in Africa.

19.8 Conclusion and Policy Recommendation

This book chapter provides an overview of the importance of NMT and its contribution to sustainable mobility. From the research, it can be concluded that well-planned implementation and integration of NMT infrastructure could offer a safe and viable transport system. NMT currently accounts for a large percentage of trips in most developing countries, ironically NMT facilities are nearly non-existent in these countries. While there have been some improvements in most African cities, more can be done in terms of improving NMT facility implementations in these cities. The successful implementation of NMT projects will involve cooperation among multiple stakeholders. There must first be committed interest on the part of the policymakers. This means that they must be willing to make policies that would favour NMT in different African countries. Policies that are sustainable and would be implemented and enforced, in addition to raising awareness and sensitizing people of the needs and benefits of using NMT modes. These could be achieved by developing an integrated transport policy that includes NMT development, policy-makers should prioritize implementation of NMT policies and programmes considering that they are vital aspects of the transport system, and proper urban planning with some focus on pedestrian-oriented development which would include NMT facilities in road design and provision of policies to strengthen them. For sustainability, it is necessary to strongly recommend measures that will ensure that these policies are implemented. Transport development projects funded by development partners can, for example, come with requirements for NMT designs/assessments prior to approval. Public sensitization and awareness-raising could be done using influencers and through social and print media. However, integrating NMT modes in current and future road improvements could increase awareness.

Specifically, to improve the future of NMT in Africa, the following are recommended:

(i) More Investment and funding are needed to provide state of the art facilities. Therefore, governments need to address this and provide innovative ways of accessing funds that will aid the planning and implementation of NMT infrastructure.
(ii) NMT infrastructure should be included in all road infrastructure plans, designs and construction including when existing facilities are being renovated.
(iii) Developing ideas that boost NMT into road transport planning and designs.
(iv) Designing, building and improving sidewalks, cycle paths and all networks required to keep users of these modes safe.
(v) NMT could be integrated with conventional motorised transport through shared space.
(vi) Providing different traffic calming measures to reduce vehicle speed.
(vii) Proper sensitisation and strict enforcement are needed to make people understand and adhere to NMT guidelines.
(viii) More research is needed, especially, to develop evidence-based strategies needed to plan, develop and sustain these modes.

(ix) All NMT infrastructure should be developed and designed based on well-researched and validated NMT procedures.

This study has addressed NMT use in African cities by examining the role, current state, challenges and opportunities to improve a transport mode which majority of the population use, yet no adequate provision has or is being made to improve the infrastructure needed to support it. It has clearly shown that there are opportunities that could lead to the successful implementation and integration of NMT in the studied areas, but this will depend largely on adequate development of associated NMT facilities and infrastructure among other things. It is therefore very crucial that stakeholders and transport planners work together to develop innovative strategies tailored to address these challenges which will have a greater likelihood of success.

References

Interface for Cycling Expertise (2000) The significance of non-motorised transport for developing countries: strategies for policy development. https://www.gtkp.com/assets/uploads/20091124-204932-9999-non_motor_i-ce.pdf

Lagos Non-Motorised Transport Policy (2018) Empowering pedestrians and cyclists for a better city. https://wedocs.unep.org/bitstream/handle/20.500.11822/25415/LagosNMTPolicy.pdf?sequence=3

Litman T (2021) Evaluating active transport benefits and costs: guide to valuing walking and cycling improvements and encouragement programs. Victoria Transport Policy Institute

Mitullah WV, Vanderschuren M, Khayesi M (eds) (2016) Non-motorized transport integration into Urban Transport Planning in Africa, 1st edn. Routledge. https://doi.org/10.4324/9781315598451

Nairobi City County Government (2015) Non-motorized transport policy "towards NMT as the mode of choice". https://wedocs.unep.org/bitstream/handle/20.500.11822/25413/NairobiCity_percent20NonMotorizedTransportPolicy.pdf?sequence=1&isAllowed=y. Accessed 27 June 2021

National Bureau of Statistics (2016) National Population Estimates. Available from. http://Nigerianstat.Gov.Ng/Report/0

Oluroubi R, Clowes W (2021) Nigeria fuel subsidies near $300 million a month, NNPC says. Bloomberg. https://www.bloomberg.com/news/articles/2021-03-25/nigeria-fuel-subsidy-hits-nearly-300-million-a-month-nnpc-says

Pucher J, Buehler R (2008) Making cycling irresistible: lessons from the Netherlands, Denmark and Germany. Transp Rev 28(4):495–528

Singh K (2018) Mobility and NMT in sustainable urban transport development: role of city developers. Draft Policy Paper. Eleventh Regional EST Forum in Asia Mongolia

TRL (2002) Scoping Study: Urban Mobility in Three Cities, SSATP Working Paper N0. 70

Twinokwesiga A (2020) Kampala: the first port of call during the COVID-19 pandemic in Uganda. Innovative Governance of Large Urban Systems (IGLUS). https://iglus.org/kampala-the-first-port-of-call-during-the-covid-19-pandemic-in-uganda/

Vanderschuren M, Jennings G, Khayesi M, Mitullah WV (2017) Introduction: challenges and opportunities for non-motorised transport in urban Africa. In: Mitullah W, Vanderschuren M, Khayesi M (eds) Non-motorised transport integration into urban transport planning in Africa. Routledge, Abingdon, pp 1–10

UN Environment (2016) Global Outlook on Walking and Cycling 2016. UN Environment, Nairobi

Uzondu C, Jamson S, Lai F (2018) Exploratory study involving observation of traffic behaviour and conflicts in Nigeria using the Traffic Conflict Technique. Saf Sci 110 (part A):273–284

Victoria Transport Policy Institute (2010) Nonmotorized transportation planning: identifying ways to improve pedestrian and bicycle transport. TDM Encyclopaedia
World Bank (2020) Urban development. https://www.worldbank.org/en/topic/urbandevelopment/overview1
World Bank (2016) Republic of Kenya Urbanization Review. http://documents1.worldbank.org/curated/en/639231468043512906/pdf/AUS8099-WP-P148360-PUBLIC-KE-Urbanization-ACS.pdf
World Health Organisation (2018) Global Status Report on Road Safety 2018. Geneva. https://www.who.int/publications/i/item/9789241565684
World Health Organisation (2021) Urban Health Initiative. Geneva. Available from https://www.who.int/initiatives/urban-health-initiative/pilot-projects/accra

Chinebuli Uzondu is a Lecturer and Researcher at the department of Transport Management Technology, Federal University of Technology Owerri Nigeria. She has previously worked as a Post-doctoral researcher with the Human Factors and Safety Group at the Institute for Transport Studies, University of Leeds, United Kingdom where she obtained her Ph.D. Her research seeks to understand road user behaviour through projects funded by various research organisations and the Nigeria and UK governments. This involves using both quantitative and qualitative research methods, to inform the design of transport infrastructure and/or technology interventions, and to evaluate the effectiveness of these interventions.

Anderson Aja Etika is a Senior Lecturer, Researcher and currently the Head of Department, Civil Engineering at the Cross River University of Technology (CRUTECH), Calabar, Nigeria. He is a Chartered Civil Engineer and has worked in private practice as a Project Engineer for several projects. Anderson's research interests primarily lie in Road Safety, Speed Management, Highway Engineering and Transport Planning and Policy related areas. He currently leads the Highway and Transportation Engineering theme, at CRUTECH.

Part V
ICT, Platforms and New Technologies

Chapter 20
ICTs, Digital Platform Mobility Services, and Transport Decarbonisation in African Cities: An Introduction

Ranford A. Acheampong

Globally, the profound impacts of innovation and new technologies on cities are becoming evident. Urban transportation systems and the ways in which individuals meet their everyday mobility needs are undergoing radical changes due to technological advances and the attendant proliferation of digital platforms and ICT-mediated mobility solutions. In this final Part of the book, we explore the emergence, diffusion and implications of technological innovation for transportation and mobility in urban Africa. The three chapters in this section cover important issues related to the rise and impacts of digital platform/app-based mobility services; the role of ICTs in ensuring safety and security of public transport users; and the energy needs and implications of the transition toward eco-friendly transportation futures through electrification. Chapter 21 focuses on the important issue of personal security on public transport. In this Chapter, Ryseck and Behrens, based on research conducted in South Africa explore the opportunities and challenges that ICTs present in ensuring passenger safety and security on public transport. They argue that in the context of high levels of crime in South Africa's cities and fragmented authority inherent in traditional methods such as policing, ICTs could play a crucial role in gathering a more nuanced understanding of passengers' security concerns; providing passengers with valuable information; and implementing security-enhancing measures that target all stages of public transport journeys. Ultimately, they recognise as crucial, the need to overcome barriers of technology use and institutional distrust to harness the opportunities ICTs present in ensuring passenger safety and security on public transport. In Chap. 22, Acheampong explores the geography of the diffusion of app-based/digital platform mobility services in Africa and their emerging socio-environmental impacts. The chapter maps known digital platform/app-based mobility solutions across the continent to identify the main players/operators, cities of operation and the factors driving

R. A. Acheampong (✉)
Department of Planning and Environmental Management, University of Manchester, Manchester, England
e-mail: ransfordantwi.acheampong@manchester.ac.uk

R. A. Acheampong et al. (eds.), *Transport and Mobility Futures in Urban Africa*, The Urban Book Series, https://doi.org/10.1007/978-3-031-17327-1_20

the diffusion of these new forms of mobility. The chapter then utilises evidence from surveys conducted in Ghana to provide empirical insights into who the users (and non-users) of internet-based ride-hailing are, the associated travel behaviours and the emerging safety and security implications. In the final chapter (Chap. 23) of this section, Booysen and colleagues address the important topic of eco-friendly transport transitions through electrification. Recognising the need for renewable energy to decarbonise the paratransit system in African cities, the authors explore the implications of this transition for the already fragile electrical grids of the continent's cities. They model and evaluate the energy requirements of decarbonising paratransit through electrification, based on research in South Africa. Ultimately, the researchers provide scenarios of energy needs and electric charging opportunities that could inform planning of electric transport infrastructure systems in South Africa and other African countries with ubiquitous paratransit systems.

In the sections that follow, we set the wider context for the chapter contributions in this section by painting a portrait of the radical changes being triggered by recent technological advances and the presence of Transportation Network Companies (TNCs) and their digital platform/app-based mobility solutions globally. We also offer a brief discussion on decarbonisation of the transportation systems of African cities and the implications for energy supply systems and planning for electric charging infrastructure systems.

20.1 ICT, Digital Platforms and New Mobility Solutions

Recent advances in Information and Communication Technology (ICT), artificial intelligence and related digital technologies have enabled the proliferation of digital platforms that are transforming many facets of society with profound disruptive consequences. Various terminologies have been coined to capture and describe this recent phenomenon. One of the commonly used terminologies is 'digital platform economy'. In the broad sense, the platform economy refers to the creation and increased utilisation of sophisticated online architectures, to enable a wide range of human activities, as well as the profound changes these ICT-mediated structures are bringing about in how we access services, work, socialise and create value in the economy (Kenney and Zysman 2016). The 'sharing economy' a closely related term also has emerged to describe ICT-mediated consumption practices whereby consumers replace personal ownership of an asset, such as a car, with access-based alternatives, such as on-demand shared-mobility services (Barnes and Mattson 2016; Bardhi and Eckhardt 2012). More recently, 'platform urbanism' has emerged as an umbrella terminology to refer to the increasing presence of digital platform enterprises in urban areas, and the ways in which they are changing how people live in and experience the city (Leszczynski 2020). Lee et al. (2020) extend this conception of platform urbanism further as the intersection of urban areas and digital technologies, resulting in (re)-configuration of urban space around platform architectures that leverage increased forms of data capture, programmability, automation, and third party value generation.

The transportation sector is one area where Transportation Network Companies (TNCs) are deploying service-providing platforms to offer on-demand mobility solutions. Disruptive changes are unravelling globally due to the growing presence of digital platform/app-based mobility solutions, especially in cities. Mobility service-providing platforms have not only challenged but exposed lapses in existing transport governance regimes. Their emergence has posed major challenges to regulatory agencies globally and disrupted the transport industry with major ramifications for employment and livelihoods, especially in the traditional taxi sector (see e.g. Gomez-Morantes et al. 2019; Papadopoulos and van Eck 2021). As we will later see in this section, app-based mobility solutions are having an increasing presence in African cities. Yet, in most African countries where major TNCs such as Uber and Bolt operate, regulatory regimes are either so outdated that they have no relevance to govern this new phenomenon or non-existent. In a few countries, public transport governance regimes are now starting to institute the necessary regulatory governance responses to regulate TNCs and their app-based mobility solutions effectively.

In addition to the challenges that digital mobility platforms pose to transport governance, there is growing evidence of their impact on urban transportation systems, individual travel behaviours and wider sustainability impacts. App-based mobility solutions, such as ride-hailing, ride-sharing and bike-sharing are enabling individuals to meet their everyday mobility needs in much more flexible and convenient ways than conventional alternatives. In the process, there is some evidence that certain trips that individuals would have undertaken using private cars are being replaced with on-demand, app-based alternatives such as ride-hailing. By replacing private car use, on-demand alternatives could be helping to reduce overall parking demand (Henao and Marshall 2019), minimize the overall amount of driving per person (Rayle et al. 2016) and even complement public transport by offering quicker first-and-last-mile connections (Circella and Alemi 2018). On the flip side, we also know that app-based solutions, such as ride-hailing compete with rather than complement existing public transport services, replace walking and induce new trips (see e.g. Acheampong et al. 2020; Dias et al. 2019; Alemi et al. 2018; Rayle et al. 2016).

While issues of safety and security in transportation are long-standing, it is also clear that the presence of ICT-mediated mobility solutions new risks and concerns have come to the fore with the advent of ICT-mediated services. As the chapter by Ryseck and Behrens in this section will show, ICT presents opportunities to address safety and security risks associated with everyday use of public transport. That notwithstanding, there is also growing concern about safety and security risks associated with mobility platforms that match passengers to services, such as ride-hailing. Like all infrastructure systems, these platforms have been shown to be vulnerable to breaches, with consequences that range from individual privacy breaches and exposure to malicious and criminal activities (see e.g. Young and Farber 2019; Acheampong 2021).

Given that globally, digital platforms in general and app-based mobility solutions, in particular, are relatively new phenomena, our understanding of their wider societal impacts is still emerging. In Africa in particular, research on the app-based mobility phenomenon is rather limited. This implies that there is more that we do not yet know about the unfolding impacts. The first two chapter contributions in this section are therefore intended to contribute to the on-going global discourse on the future of cities in general and their transportation systems in particular, as ICT, artificial intelligence and the digital platforms they enable pervade various facets of society.

20.2 Beyond ICT and Digital Platforms: Transitioning to Eco-Friendly Transportation in African Cities

Yet another inevitable transition that African cities must face, in addition to the digital transformation currently underway, is decarbonising their transportation systems. It is estimated that transport currently contributes up to 10% of total greenhouse gas emissions in Africa, and this is expected to increase in line with the continent's fast-rising motorization levels (Conzade et al. 2022). Between 2010 and 2016 alone, it is estimated that transport emissions in Africa grew by 84% (SLoCaT 2018).

The expected increase in transport-related greenhouse gas emissions in Africa will result from a confluence of long-standing and emerging factors. While most African cities are still in the early stages of motorization, the demand for road transport is expected to increase significantly in the coming decades, because of continued urban population growth and an expanding middle class (Collett and Hirmer 2021). Moreover, similar to global historical trends, urban planning and infrastructure development in Africa today privilege car-centric development that induces more car use (Odhiambo et al. 2021). While on-demand, ICT-mediated mobility solutions are offering alternatives to private car use, as we will see in the first chapter of this section, not only are the fleet of vehicles on offer fossil-fuelled but the overall impact of on-demand mobility options on car use appears to be marginal.

Yet another contributing factor to current and future transport-related emissions is the fact that in most Sub-Saharan African countries, an estimated 85% of all four wheel vehicle sales are used vehicles (United Nations Environment Programme 2020). As developed countries rid of unwanted used Internal Combustion Engine (ICE) vehicles as part of decarbonisation measures, the continent further risks becoming the dumping ground for them (Conzade et al. 2022). With bans on the import of used vehicles instituted in only a few countries such as Egypt, South Africa, Seychelles and Sudan (United Nations Environment Programme 2020), polluting ICE vehicles will in the short to medium run continue to contribute to transport emissions in urban Africa.

Against the backdrop of the foregoing on the one hand and the need to address the associated environmental, public health and economic impacts, on the other hand, African cities must as matter of urgency, develop and implement strategies to transition to cleaner source fuel and eco-friendly transportation systems. Like the rest of the world, the emerging consensus is that electrification of the transportation system constitutes one of the effective pathways toward decarbonisation. Electrification of the transportation system is seen as a better option to biofuels because the latter produces substantial tailpipe emissions and competes with land for food and agriculture (Collett and Hirmer 2021). Electrification also does not require completely new infrastructure being developed as electric vehicle charging infrastructure can easily be connected to existing electricity systems (ibid).

While the potential benefits of electrification of transportation systems are obvious, what is less obvious is how African cities can design and implement context sensitive, fit-for-purpose transition policies, plans and strategies. Firstly, for widespread adoption and use of electric vehicles (EVs) in African cities, African states must resolve the underlying problems of unreliable electricity supply, by increasing generation. A second fundamental problem stems from the very nature of public transport systems in Africa. Dominated by paratransit systems characterised by second-hand minibuses and two-wheelers, the public transport systems of African cities are fundamentally different from those of Global North cities. This means that models and practices that have proved effective elsewhere might not necessarily work in African cities. Instead, African cities must experiment with and evolve transition models and pathways bespoke to their unique transportation situations, taking into account the wider implications for crosscutting issues including employment, livelihoods and energy security. Moreover, while some evidence suggests a growing awareness of the public regarding EVs and their potential benefits, the up-front costs associated with EV ownership are currently prohibitive in Africa (Conzade et al. 2022).

Thus, for African cities to leverage and realise the benefits of electric transport a combination of supply- and demand-side policy strategies are critical. On the demand side, government support through subsidies could help individuals and transport operators to overcome the prohibitive up-front costs of switching from ICEs to EVs. On the supply side, increasing electricity generation and developing the required electric charging infrastructure will be vital to electrification. As Collett et al. (2020) note, institutional, policy and systems integration in the currently siloed sectors of electricity and transportation are critical pre-conditions for the transition to eco-friendly transport futures in Africa. To this end, the final chapter in this section makes valuable contributions by offering data-driven estimation and analysis of what the energy demands and charging requirements would be for a typical paratransit system.

In conclusion, the central message of the chapters in this section is that technological advancement presents challenges and opportunities for meeting the everyday mobility needs of the growing urban population in Africa and for making transportation systems socially and environmentally sustainable. Together, they provide valuable insights about the emerging impacts, positive and negative, of new ICT-mediated mobility solutions, the possibilities that ICT and digital technologies present for improving passenger security, and the prospects, challenges and pathways toward decarbonising transport in African cities.

References

Acheampong RA (2021) Societal impacts of smart, digital platform mobility services—an empirical study and policy implications of passenger safety and security in ride-hailing. Case Stud Transp Policy 9(1):302–314

Acheampong RA, Siiba A, Okyere DK, Tuffour JP (2020) Mobility-on-demand: an empirical study of internet-based ride-hailing adoption factors, travel characteristics and mode substitution effects. Transp Res Part C: Emerg Technol 115:102638

Alemi F, Circella G, Handy S, Mokhtarian P (2018) What influences travelers to use Uber? Exploring the factors affecting the adoption of on-demand ride services in California. Travel Behav Soc 13:88–104

Bardhi F, Eckhardt GM (2012) Access-based consumption: the case of car sharing. J Consum Res 39:881–898. https://doi.org/10.1086/666376

Barnes SJ, Mattsson J (2016) Understanding current and future issues in collaborative consumption: a four-stage Delphi study. Technol Forecast Soc Chang 104:200–211

Circella G, Alemi F (2018) Transport policy in the era of ridehailing and other disruptive transportation technologies. In: Advances in transport policy and planning, vol 1. Academic Press, pp 119–144

Collett KA, Hirmer SA (2021) Data needed to decarbonize paratransit in sub-Saharan Africa. Nat Sustain 4(7):562–564

Collett KA, Byamukama M, Crozier C, McCulloch M (2020) Energy and transport in Africa and South Asia. Energy Econ Growth

Conzade E, Engel H, Kendall A, Pais G (2022) Power to move: accelerating the electric transport transition in sub-Saharan Africa. McKinsey & Company. https://www.mckinsey.com/industries/automotive-and-assembly/our-insights/power-to-move-accelerating-the-electric-transport-transition-in-sub-saharan-africa

Dias FF, Lavieri PS, Kim T, Bhat CR, Pendyala RM (2019) Fusing multiple sources of data to understand ride-hailing use. Transp Res Rec 2673(6):214–224

Gomez-Morantes JE, Heeks R, Duncombe R (2019) A multi-level perspective on digital platform implementation and impact: the case of easytaxi in Colombia. In: International conference on social implications of computers in developing countries. Springer, Cham, pp 195–206

Henao A, Marshall WE (2019) The impact of ride hailing on parking (and vice versa). J Transp Land Use 12(1):127–147. https://doi.org/10.5198/jtlu.2019.1392

Kenney M, Zysman J (2016) The rise of the platform economy. Iss Sci Technol 32(3):61

Lee A, Mackenzie A, Smith GJ, Box P (2020) Mapping platform urbanism: charting the nuance of the platform pivot. Urban Plan 5(1):116–128

Leszczynski A (2020) Glitchy vignettes of platform urbanism. Environ Plan D: Soc Space 38(2):189–208

Odhiambo E, Kipkoech D, Hegazy M, Hegazy A, Manuel M, Schalekamp H, Klopp JM (2021) The potential for minibus electrification in three African cities. Volvo Research and Educational Foundations (VREF), Cairo, Nairobi and Cape Town

Papadopoulos S, Eck SV (2021) Disruptive technologies and taxi rides in South Africa: what is the 'Uber' uproar about? In: Global perspectives on legal challenges posed by ridesharing companies. Springer, Singapore pp 199–229

Partnership on Sustainable, Low Carbon Transport (SLoCaT) (2018) Transport and Climate Change Global Status Report 2018. http://slocat.net/tcc-gsr

Rayle L, Dai D, Chan N, Cervero R, Shaheen S (2016) Just a better taxi? A survey-based comparison of taxis, transit, and ridesourcing services in San Francisco. Transp Policy 45:168–178

United Nations Environment Programme (2020) Used vehicles and the environment. A global overview of used light duty vehicles: flow, scale and regulation. https://www.unep.org/resources/report/global-trade-used-vehicles-report

Young M, Farber S (2019) Ride-hailing platforms are shaping the future of mobility, but for whom. https://osf.io/pz7fk. Accessed 30 sep 2022

Ransford A. Acheampong is a Lecturer and Presidential Academic Fellow in Transport and Urban Futures at the Department of Planning and Environmental Management, University of Manchester, UK. The first strand of his research is focused on the nexus between transportation and spatial structure and their integrated planning in cities. The second strand of his research addresses questions around the diffusion and socio-spatial impact of new and emerging transport technologies and mobility solutions, such as app-based mobility, autonomous vehicles, shared-mobility and Mobility-as-a-Service.

Chapter 21
Opportunities and Challenges for Addressing Personal Security on Public Transport Through ICTs in South African Cities

Bianca Ryseck and Roger Behrens

Abstract While technologies providing basic passenger information on public transport stops, fares, routes, and timetables have proliferated in South Africa, information needs surrounding personal security have gone largely unmet. Personal security on public transport is a major concern in South African cities, where crime levels are some of the highest globally. Despite municipal governments' responsibility for promoting security in public transport, and for making relevant information on public transport accessible, fragmented authority for ensuring passenger security makes a coordinated approach through traditional methods like policing, that rely on external collaboration, difficult. Information and communications technologies offer the opportunity to bridge this fragmentation to gather a more nuanced understanding of security concerns, provide passengers with valuable information, and implement targeted strategies to reduce incidents onboard, en route, and at stations/stops. This chapter discusses the opportunities for, and challenges of, addressing personal passenger security through the use of information and communications technology in South African cities given potential barriers to user adoption. Local and international precedents tapping into technologies offer diverse approaches to the multifaceted security challenges South African public transport users face on a daily basis. Given the growth in smartphone adoption, communications technologies are a promising avenue to bridging the data gap needed to support data-driven decisions, but barriers to technological use prevail such as mobile data access and institutional distrust. Overcoming challenges to user adoption can pave the way for technologies to aid in providing essential passenger security information despite the hurdles siloed operators and authorities present.

Keywords Passenger security · Technology · Information · Public transport · South Africa

B. Ryseck (✉) · R. Behrens
University of Cape Town, Rondebosch, Cape Town 7700, South Africa
e-mail: biancaryseck@gmail.com

R. Behrens
e-mail: roger.behrens@uct.ac.za

R. A. Acheampong et al. (eds.), *Transport and Mobility Futures in Urban Africa*, The Urban Book Series, https://doi.org/10.1007/978-3-031-17327-1_21

21.1 Introduction

Personal security on public transport is a major concern for users in most cities of the world. This is especially true in South African cities, where urban crime levels are some of the highest globally. Security is a major factor in mode choice and potentially also a barrier to public transport use. For those public transport users who have a train or a minibus taxi available to them, a concern for security is one of the three main reasons for not using the train or taxis (SSA 2013).

While technologies providing basic information on public transport stops, fares, routes, and timetables to passengers have proliferated in South Africa, information needs surrounding personal safety have gone largely unanswered. There is a disjuncture between the persistent security concerns facing public transport users, the prevalence of Information and Communications Technologies (ICTs), and the lack of research into the applicability of these ICTs in security strategies in public transport situations. This chapter discusses the opportunities for, and challenges of, addressing personal passenger security within the context of ICTs in South African cities. Even though focused on the South African context, this discussion has relevance for cities experiencing similar problems elsewhere.

This chapter opens with a review of the international and South African literature at the nexus between public transport passenger security and ICTs. It then describes public transport security challenges in several South African cities, followed by a discussion of how these concerns have been addressed through policies at a metropolitan and national level, and subsequently translated into traditional interventions locally. The chapter then turns to an exploration of the ways ICTs have been leveraged globally within the transport sector specifically and applied more broadly to address the personal security gaps left by traditional approaches. The chapter then ends with a discussion of the challenges of implementing and adapting these precedents in South African cities from the perspective of potential barriers to user adoption.

21.2 Previous Research into Addressing Personal Security Concerns in Public Transport Environments

Safety and security are often bundled together when discussing public transport, where together these mean that "personal security is an objective freedom from security and safety risks combined with a subjective freedom from fear and uncertainty" (Beecroft and Pangbourne 2015a). In public transport, passenger security can be tied to three different situations: at the station or stop, onboard, and walking between transport points (Kruger and Landman 2007). Security of public transport users has been widely studied in terms of individual perception (e.g. Delbosc and Currie 2012; Nordfjærn and Rundmo 2018) and gendered-dimensions around sexual harassment and assault (e.g. Vanderschuren et al. 2019), which has been disaggregated based on demographics of riders and environmental characteristics (e.g. Masoumi and Fastenmeier 2016) and how carrying a mobile phone can act as a type of social presence and reduce perceived risk (Reichow and Friemel 2020). Perceived personal security is heightened in travel scenarios (Lemieux and Felson 2012; Newton and Ceccato 2015) and influences passengers' mode and operator choices as well as travel frequency (Sam and Abane 2017). Collective efficacy or the ability of people to enforce social cohesion within their community, associated with the spatial context is a strong predictor of crime in travel situations – more crime tends to happen in areas where there is low trust between neighbours and in expectations that neighbours will intervene to protect individuals from crime (Gerell 2018).

When addressing the security concerns of public transport users, the "whole journey approach" method incorporates a mix of different strategies to target specific challenges in different situational contexts throughout the public transport user journey (Smith and Cornish 2006). Traditionally, efforts to curb real and perceived public transport security threats have relied on methods such as improving environmental factors (e.g. lighting) or surveillance (e.g. CCTV and policing). However, this has been met with mixed results. Despite efforts by transport authorities to install CCTV or improve lighting to alleviate real threats and perceived fears of crime, such efforts are not necessarily met with the same levels of enthusiasm and confidence (Gardner et al. 2017; McCarthy et al. 2016). Conversely, passengers value measures like visible policing, public awareness campaigns, and means of reporting incidents (Gardner et al. 2017). However, a lack of data and accurate reporting impede a nuanced understanding of security incidents like harassment and effects on travel behaviour, thereby undermining targeted methods to improve passenger security (Ibid.).

Both public transport providers and users can leverage ICTs to address security concerns. ICTs can be used to gather data on crime hotspots disaggregated by crime type and demographics that decision-makers can use to plan targeted interventions within resource constraints, to provide information to users to plan safer public transport journeys more confidently, and as a tool to monitor and access safer en route journeys. Overall, research into the potential and on-ground implementation of such ICTs is limited. While research has looked at the potential role of ICTs in security applications, this has primarily focused on counterterrorism or ticket enforcement (e.g. Bennetts and Charles 2016; Beecroft 2019). Though security information can reassure and give confidence to public transport users, few technologies have emerged to provide such information (Beecroft and Pangboure 2015b). That said, the limited available options for ICTs are not necessarily restricted by user willingness to use mobile security applications, as McCarthy et al. (2016) found that respondents were willing to use security apps and even share their location data to report anti-social behaviour.

Despite the widespread security concerns that public transport users have in South Africa, research into a nuanced understanding of these concerns is limited, and even more absent on the different approaches taken or could be taken to address these challenges. A study on crime and prevention in three major South African cities surveyed respondents to investigate the types and locations of crimes committed against public transport users and made recommendations rooted in physical interventions, increased security presence, and effective communication (Page 2001). Kruger and Landman (2007) explored crime reduction through a situational crime prevention approach whereby environmental interventions were stressed as a solution in the local South African public transport context. Since this body of literature was published, the mobility landscape has changed and so have the opportunities that have come with the growing popularity of personal ICTs and increased connectivity. However, these technological changes have not been addressed in literature as potential avenues for reducing security threats, nor have ICTs from the public or private sector grappled head-on with local public transport security challenges.

21.3 Current Crime and Security Measures

21.3.1 South African Public Transport Security Context

A third of South Africans rely on public transport as their main mode of travel, with public transport use in 2013 accounting for half of all motorised trips in South Africa (SSA 2013). The available public transport modes in South African cities include minibus taxis (MBT), buses, and trains, accounting for 69.2%, 17.1%, and 13.7% respectively of all public transport trips nationally, though in urban areas where passenger trains operate these numbers vary considerably (Ibid.). MBTs are privately owned, typically 16-seater vehicles that operate semi-flexible routes. Owners, who do not operate their own vehicles, will hire their taxis out to drivers. Conventional buses and Bus Rapid Transit (BRT) systems both operate in urban areas and are mostly initiatives of the local metropolitan governments (e.g. Cape Town's MyCiTi, Tshwane's A Re Yeng, Johannesburg's Rea Vaya, and Durban's Go!Durban), though Golden Arrow Bus Services (GABS) in the Western Cape contracts with the provincial government. In the cities, Metrorail, which is run by PRASA (the Passenger Rail Agency of South Africa) at the national level, is the predominant rail service, though Johannesburg and Pretoria are also served by a provincial public–private partnership, the Gautrain. In recent years, Metrorail has been plagued by vandalism and infrastructure theft, that has led to severe reductions in service across all its metropolitan service areas.

Household travel statistics on public transport users' security perceptions onboard, at stations, and en route to public transport, from the most recently available survey data on security concerns, provide a glimpse into the magnitude of security concerns across South African public transport users. The most recent National Household Travel Survey from 2013 reveals that 53.8% of all Metrorail users are dissatisfied or very dissatisfied with security on the walk to and from stations, and 45.7% are unsatisfied with onboard security (see Fig. 21.1). A third of all users are unsatisfied with security at stations. Despite low satisfaction ratings across a range of service characteristics, it should be noted that the continuing popularity of Metrorail lies in

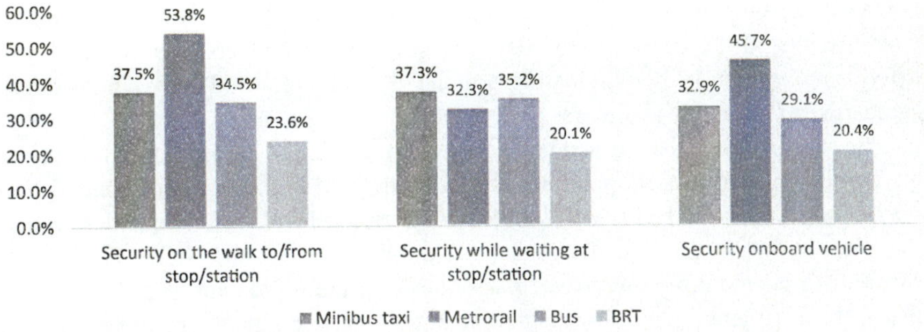

Fig. 21.1 Security concerns according to heads of households by mode (SSA 2013)

its relative affordability compared to alternative modes, leaving its captive users with little choice of other safer forms of mobility (Vanderschuren et al. 2019). MBT users are marginally less dissatisfied with security, with a third each dissatisfied with security en route, onboard, and at ranks. Similar dissatisfaction levels are also reported across the traditional bus services, while BRT users are generally less dissatisfied. Across all modes, perceived security risk in terms of waiting, on board, and walking to the station varies by area (Ibid.). Fear of violent crime, harassment, and sexual assault are security concerns across all three travel scenarios (Eagle and Kwele 2021; Luke and Heyns 2019; Mabaso 2019). Women and girls are particularly subjected to forms of gender-based violence and will often avoid boarding vehicles if no other women are onboard and avoid using public transport at night (Mabaso 2019). Safety from accidents is an additional concern passengers have onboard vehicles—43.3% of MBT users are dissatisfied with safety from accidents, while 29.2, 27.5, and 18% are dissatisfied with safety from accidents onboard Metrorail, conventional bus services, and the BRT respectively (SSA 2013).

21.3.2 Governing Security: Institutional Measures to Provide Safe + Secure Public Transport

Nationally, policies have tended to limit the scope of discussion around security issues on public transport to mechanical and road engineering aspects. The *1996 White Paper on National Transport Policy*, which provides the foundation for many subsequent transport policies, lays out the aim to improve the safety and security of transporting people, with a stated emphasis placed on road safety (DoT 1996). This is reiterated in the revised *White Paper on National Transport Policy* in 2017, along with the stress on road and rail safety (DoT 2017). Beyond the *National Development Plan 2030*, where personal security is emphasised in the context of communities and their environments (primarily enforced through policing) and thereby indirectly public transport environments, security directly within the public transport context is not explicitly mentioned as a priority area in national policies (National Planning Commission 2011). In effect, the national government sees its role in improving safety as a regulatory body, creating policies that oversee road and vehicle safety standards for transport operators.

The responsibility for passenger security is fragmented across different tiers of government and transport operators, as loosely defined in the *National Land Transport Act of 2009* (NLTA) (DoT 2009). At a national level, when considering any measures and strategic objectives related to public transport, the NLTA requires the Minister to promote the security of passengers. At the provincial level, in the case that there is violence, unrest or instability in the public transport sector or between operators in an area that risks the security of passengers, residents or others, the Member of the Executive Council, in consultation with relevant planning authorities, can suspend operations on the public transport routes or ranks concerned (NLTA Section 91[2]).

Though it is the municipal government's responsibility for promoting security in public transport, as well as making relevant information on public transport accessible (NLTA Section 11 [c] [xii] and [xiii]), current strategies to implement such measures are limited in scope. While the City of Cape Town's *Comprehensive Integrated Transport Plan 2018–2023* mentions safety as an access priority across all income groups on public transport, plans to address safety are limited to rail and then primarily to infrastructure safety such as addressing vandalism and infrastructure theft (CoCT 2018). The City of Johannesburg stressed the importance of road safety and transport security data in monitoring and measuring transport safety and security and identifying hotspots in its 2013 *Strategic Integrated Transport Plan Framework* (CoJDT 2013). The city of Durban's *Integrated Transport Plan Update 2010–2015* mentions that a policy for the onboard and terminus safety and security of public transport passengers has been developed, but no strategy for the implementation of this policy is in place (eTA 2010). Improving safety and security on public transport is consistently mentioned across all plans as a key objective and yet is overwhelmingly limited to targeting road safety strategies rather than passenger security. Without clear implementation strategies to apply measures cohesively across the public transport network as a whole, the responsibility to provide safe and secure transport becomes further muddled on an individual operator level.

21.3.3 Operator Responsibilities: Minibus Taxis

On the whole, responsibilities and approaches to passenger security around MBT use are splintered across different public and private stakeholders, fracturing the possibility of a holistic response to passenger security. Regulatory bodies at the provincial level issue operating licenses which dictate the routes the license holder can operate on, though national bodies can devolve this power to the municipalities. To apply for a license, the MBT operator must be part of an MBT association, which requires paying a membership joining fee and weekly financial contributions (Barrett 2003).

The operator can then hire out their vehicles to drivers, predominately men, to provide the passenger services, however, there is currently no clear obligation on the operators to ensure that the drivers are licensed or have safe driving records. Nor are there clear consequences for operators when the security of their passengers is compromised. To some extent, however, local municipal laws can regulate onboard security. For example, the *City of Cape Town's Traffic By-law* (2011) prohibits certain actions on a public transport vehicle, including compromising the security of passengers and gives power to authorised city officers to remove people who violate the by-law in public transport facilities.

At the rank, though queue marshals are delegated a level of responsibility over passenger security, this is often not fulfilled. Queue marshals are responsible for loading passengers into the vehicles and ensuring vehicles do not exceed legal seating capacity (although vehicles are often overloaded). They are responsible for taking the complaints of passengers. They may also be instructed by the taxi owners to confiscate the keys of a driver suspected of drinking on the job, but in practice are nervous to do so, opting to not load the vehicle instead (Barrett 2003). In this case, the driver can go ahead and gather passengers on the route.

Furthermore, drivers can remain relatively anonymous to their passengers, reducing their accountability. Sonke Gender Justice found that the most commonly reported form of sexual assault was rape by minibus taxi drivers, onboard or at ranks (Mabaso 2019). In response to ongoing sexual assault, a women-only taxi service, with only women drivers, was launched in 2021 in Cape Town, but likely will face pushback from the local industry who see it as a threat to their market, as has happened in the past with other such initiatives (Ludidi 2020).

21.3.4 Operator Responsibilities: Passenger Rail

All rail systems are subject to the Railway Safety Regulator, an agency established in terms of the *National Railway Safety Regulator Act* of 2002 that primarily regulates the operational and infrastructural safety of railway systems, but makes no explicit mention of the need to ensure passenger security onboard or at stations (DoT 2002). This ambiguity leads to diverse approaches to passenger safety and security.

PRASA, the state-owned rail agency, focuses on infrastructure safety as opposed to personal security, and its limited resources have restricted its ability to respond to passenger security concerns. Following the fatal stabbing of a Metrorail passenger, a landmark 2004 case, *Rail Commuters Action Group v Transnet Ltd*, found that PRASA bears responsibility for the security of passengers from crime on trains and is obliged to ensure that reasonable measures are instated for the security of passengers (Rail Commuters Action Group v Transnet Ltd 2004). In a 2015 case, *Mashongwa v PRASA,* involving a man who was attacked and thrown out of an open door on a moving train, the court found that PRASA had been negligent in not ensuring that the doors were closed prior to the train leaving the station (Mashongwa v PRASA 2015). However, the court also found that requiring PRASA to post a security guard in each carriage was unreasonable because it was seen to exceed reasonable measures. Beyond limited security personnel spot-checking carriages, CCTV and some personnel in major terminal stations, intermediary stations are usually unstaffed, carriages are unmonitored and pedestrian subways underneath the railway between station platforms are prone to crime. The most recent *PRASA Corporate Plan of 2020–22* has taken a targeted approach, tasking the national police rail unit with ensuring passenger security in high-risk areas and relying in part on collaborating with national police to gather and share data to identify these high-risk areas (PRASA 2019).

The Gautrain, a public–private partnership in Gauteng province, is perhaps the most digitised system in South Africa, employing multiple technologies to ensure security. The Gautrain Management Agency enacts security measures in the station precinct and vehicle parking areas, as well as in carriages. However, their security jurisdiction is limited to their property, i.e. trains, stations and parking, while the areas surrounding their property are under metropolitan jurisdiction, meaning that the rail service relies on the metro and national police to patrol adjacent areas that passengers use to access Gautrain services. Given this, Gautrain trains and stations are equipped with CCTV, some of which have face recognition. The drivers are in direct communication with an operation centre, a police emergency contact centre and the Gauteng Provincial Disaster Management Centre.

21.3.5 Operator Responsibilities: Commuter Buses

Like the rail systems, security measures employed on passenger buses vary based on the operator's initiative and contractual agreements with regulatory authorities. Generally, BRTs in South Africa are equipped with CCTV and work with the City's security personnel as well as national police to respond to emergencies. Staff are trained in customer service and security, though staff may not be present at intermediary stops along routes. Enclosed stations are staffed with customer service staff and, in some instances, security staff. The intermediary stops, that are either demarcated with a shelter or a signpost, are not lit but rely on secondary light sources like streetlamps. Conventional bus systems have generally less security measures in place. GABS passengers are continuous targets for onboard armed robbery, more so on some routes, and have lamented that the lack of CCTV and onboard security presence is concerning (iOL 2021). In response, GABS has maintained that, while they are investigating implementing some security measures, it is the responsibility of the national police to curb robberies.

21.3.6 Towards ICTs for Approaching Passenger Security

As shown with this brief overview of the different transport stakeholders, the responsibility for ensuring the security of passengers is fragmented across different authorities and operators, making a coordinated approach through traditional methods like policing, that rely on external collaboration, difficult. Resource constraints further complicate security measures, limiting enhanced approaches to providers with larger operating budgets such as the more expensive Gautrain services or those with greater subsidy allocations. These limitations could be eased through increased access to data on incidents to inform areas where resources could be targeted. Furthermore, there have been no attempts at communicating security information to passengers to help inform their journey choices and bolster confidence in the public transport system, despite the mandates in the NLTA that state municipal governments have to implement such communication systems. ICTs, as will be illustrated in the subsequent section, offer these stakeholders the opportunity to gather a more nuanced understanding of security concerns, provide passengers with valuable information, and implement targeted strategies to reduce the rate of security incidences onboard, at stations/stops, and en route.

21.4 Survey of Existing ICT Methods to Address Personal Security

The ubiquity of personal security concerns in public transport environments across global cities is not reflected in the widespread availability of ICT tools to address these concerns. ICTs that target the security concerns of public transport passengers are largely limited to Indian cities, while more generic personal security technologies are available more widely. The following are a collection of ways (see Table 21.1) that ICTs have been leveraged in South Africa, and globally, to prevent or react to personal security challenges either in more general, everyday situations, while on the move, or specifically in public transport environments. While many studies have assessed perceptions towards current public transport security (e.g. Sam and Abane 2017) and towards CCTV, security personnel and other traditional forms of security interventions (e.g. McCarthy et al. 2016), similar research has yet to be conducted around the impact of ICT interventions on perceived and real security challenges. Some research has investigated whether these features increase perceptions of security and willingness to use an app's services, especially e-hailing apps, but not whether these features directly mitigate real security risks in public transport scenarios (e.g. Acheampong 2021; Giddy 2019). Though difficult and sensitive to measure, evaluation of these initiatives is still needed to provide evidence of effectiveness (e.g. incidents reductions, improved perceptions of risk) to inform policy development around the use of ICTs to address security challenges in public transport.

Table 21.1 Selection of personal security ICTs available globally on the playstore and iOS app store

Product name	Implementing agency	Regions	Platform	Installs-Android	Specific information types provided	Other features/services	Cost
bSafe	Private	All	iPhone, Android	500,000+	Share tracked locations with trusted contacts Send exact pick-up points Send SOS signal to trusted contacts Record/stream emergency video Receive fake phone call to exit social situation	In 'free' mode, user can only act as a trusted contact	Subscription needed to access all features

(continued)

Table 21.1 (continued)

Product name	Implementing agency	Regions	Platform	Installs-Android	Specific information types provided	Other features/services	Cost
Disha SOS	Public	India	Android	500,000 +	Track location and share with emergency contacts Locate nearby police stations and other emergency services Police contact numbers Call police and receive calls from control room in case of emergency	–	Free
CitizenCOP	NGO	India	iPhone, Android	500,000+	Incident reporting SOS help with user's real-time location Local police phone directory	Coordinating with various police departments	Free

(continued)

Table 21.1 (continued)

Product name	Implementing agency	Regions	Platform	Installs-Android	Specific information types provided	Other features/services	Cost
Namola	Private	South Africa	iPhone, Android, Huawei	100,000+	Share tracked locations with trusted contacts Smart alerts notifying if trusted contact arrived safely Connect to police or other emergency services in an emergency SOS button to receive call from *Namola* team	–	Subscription to access emergency services or private response
Shake2safety	Private	India	Android	100,000+	Shake phone or pressing power button 4 times to send SMS to emergency contacts Saves 4 s audio recording during SOS event	Works on a locked screen and also when there is no Internet connection	Free

(continued)

Table 21.1 (continued)

Product name	Implementing agency	Regions	Platform	Installs-Android	Specific information types provided	Other features/services	Cost
Women safety	Private	All	Android	100,000+	Send email to emergency contacts along with location Photos, video and audio uploaded to the Women Safety server Link to these is emailed to emergency contacts Three buttons to state seriousness of the situation	–	Free

(continued)

Table 21.1 (continued)

Product name	Implementing agency	Regions	Platform	Installs-Android	Specific information types provided	Other features/services	Cost
Himmat Plus	Public	Delhi, India	Android	50,000+	Send an SOS alert which shares the user's location and video-audio feed with the Delhi police control room Verify drivers are registered with Delhi police	–	Free
MySafetipin	NGO	16 cities across N. & S. America, Africa, and Asia	iPhone, Android	50,000+	Share tracked locations with trusted contact Safe pick-up points Safe route options Get safety score of an area	User can contribute a 'safety audit' which creates a safety score based on 9 parameters	Free

(continued)

Table 21.1 (continued)

Product name	Implementing agency	Regions	Platform	Installs-Android	Specific information types provided	Other features/services	Cost
Raksha	Private	India	Android	10,000+	Notify emergency contacts via email and SMS Locate nearby police stations Share location with trusted contacts	If the app is not active, the user can send alerts by pressing the volume key for three seconds	Free
MBTA See Say	Public (MBTA)	Massachusetts Bay area	iPhone, Android	5,000+	Authorities alert users in emergencies Riders can comment on alerts Delay and service disruption notifications User reports	SeeSay is an SDK provided by Elerts company and is used in US transit systems. It can be added onto existing apps	Free

(continued)

Table 21.1 (continued)

Product name	Implementing agency	Regions	Platform	Installs-Android	Specific information types provided	Other features/services	Cost
Go Safe Nirbhaya	Private	India	Android	1,000+	Users upload a photo of the driver, their vehicle, and the registration number online with the time stamp and location	–	Upgrade option to keep photos for a year

21.4.1 Awareness Campaigns

Awareness campaigns can be powerful tools towards driving social change with low technological barriers to entry that do not require costly computer expertise. NGOs launched several campaigns in reaction to MBT-related violence in South Africa, leveraging radios, a common device in taxis, and social media to generate widespread awareness of Gender-Based Violence (GBV). In 2016, Sonke Gender Justice (2018) launched the 'Safe Ride Campaign' in response to the challenges women and girls faced every day at the hands of male MBT personnel. The campaign was formed in partnership with the South African National Taxi Council which enabled Sonke to gain access to the MBT ranks to engage directly with drivers and queue marshals. Sonke focused on community radio stations to reach their target audience with a five-part radio drama meant to incentivise dialogue around GBV and individual roles in prevention. The episodes were sent to radio stations together with guiding discussion questions for radio hosts to use to engage their listeners and guests in the studio in a debate around GBV in taxis. In reaction to a spate of taxi-related rape crimes in Johannesburg and Soweto between March and April 2017, Soul City launched the 'Safe Taxis Now Campaign' (Ibid.). The campaign relied heavily on social media channels, like Twitter, to raise awareness and generate engagement around focus areas to address GBV.

21.4.2 Crowdsourced Incident Data

Crowdsourced data can become a resource for contextualised intervention planning and also a resource for users in information-scare regions. Initially launched in Indian cities, the *Safetipin* app has since expanded to 16 cities including Johannesburg and Durban to provide women with a means to contribute to security audits of the various areas in their cities. While the app is applicable in non-public transport situations, the app has several functionalities that specifically target safer mobility. Users can share their tracked location with a trusted person who will be notified in the event that the user is either in an unsafe area (as recognised by the collated security audits), has been stationary for long, or diverted from their pre-selected route. Users can see nearby places (e.g. hospital, restaurant) they can safely wait for their pick-up in the case that they feel unsafe in their current waiting location. In theory, users can also plan the safest route from one point to another, though this relies on having enough up-to-date user contributions. Local governments have used the data from these mapping exercises to improve street lighting around low-income areas adjacent to metro stations in Delhi and monitor the change in security and improve lighting and CCTV installations along bike routes in Bogota to improve security for women cyclists after dark (Safetipin and Cities Alliance 2016).

Somewhat similarly, the Flone Initiative, an organisation that promotes gender-sensitive public transport spaces for women across Africa, launched an app and website in Kenya called *Report It! Stop It!* with the aim of gathering geolocated user-report data of harassment cases. This data is then given to the government to inform where hotspots are in order to focus attention and resources. Other users can also benefit from access to the incident reports to inform their own knowledge of crime spots in the city.

21.4.3 Security Alert Features

A plethora of applications have been designed to connect app users with emergency contacts to respond to immediate personal security threats. These include *Himmat* and *Namola* amongst others. While some of these apps are country-specific, others are available in any context as they do not connect the user to a municipal or other governmental security units. *Himmat* is a free app for women travelling alone late at night launched by the Delhi police in 2015 following the brutal gang rape and murder of a young woman travelling on a bus at night. The app directly connects users with the police in case of an emergency. The app was designed for Android phones after a survey revealed that the intended user-base primarily has access to Android devices. The app allows users to send alerts multiple ways without actively being in the app including a shake to alert and power button alert features. In the event that an alert is activated, the app captures audio and video recordings that are transmitted to the police control room. *Namola* is a security app available in South Africa for iPhone, Android, and Huawei phones. The app offers to connect users with a private response team in the case of an emergency, which may be appealing over public police response in terms of response time, and/or response effectiveness.

21.4.4 Trip Tracking

Trip tracking is one of the most widespread ICT security-related applications (e.g. *bSafe, Raksha, Uber*) and is often paired with other security features like alerts. Trip tracking can reduce users' perceived fears by actively allowing trusted contacts to monitor the users' movements and respond to potentially suspicious situations. However, a drawback to such a feature is privacy concerns around sharing location information with applications and third parties beyond trusted contacts.

21.4.5 Driver Verification

Driver verification features have emerged in response to fears of heightened exposure to crime at the hands of drivers in single-passenger systems, like tuk tuks and e-hailing vehicles. To improve trust in riding with strangers, driver verification enables passengers to match the identification details on the app to verify the driver is the same person. However, the initial verification process itself might inherently be flawed—companies do not necessarily disclose how they verify the identity of a driver and whether their identities are checked against official state records (Acheampong 2021). Thus, those with an official criminal record may slip through the verification process.

In response to a string of incidents including the violent rape of female passengers, *Uber* launched a series of security features in their app. In 2016, Uber added real-time ID checking to their services in India and the United States, which was subsequently added a year later in South Africa, in order to ensure that non-Uber registered drivers were not making use of the app under a registered driver's name. This feature asks drivers to send selfies to Uber, enabling Uber to verify that the selfie is a match with the driver listed on their database and in turn give the driver access to their account. However, in a series of high-profile rape cases committed by a registered Uber driver, the victims did check the vehicle registration number before getting into the vehicle. It was only after these cases that the company also added an in-app emergency button that allows users to directly connect with the police, except in the case of South Africa where activating the button connects users to a private security company.

Another method of driver verification that bypasses the need for official identity document checks, is to hold the driver accountable through traceability measures. In an effort to prevent onboard incidents perpetrated by drivers in vehicles ranging from tuk tuks to buses, *Go Safe Nirbhaya* was developed in India. With the driver's permission, users can upload a photo of the driver, their vehicle, and the registration number online with a time stamp and location. If the driver refuses, the user is then cautioned against boarding the vehicle. The idea is that by having the driver's identifying information uploaded online, the driver is unlikely to commit a crime against the user.

21.4.6 Driver Rating

Driver rating systems, such as those typically employed by e-hailing apps, serve as preventative security measures for future users of services. In the case of apps like *Uber* and *Bolt*, riders and drivers rate each other and have the opportunity to provide feedback or report a serious incident (e.g. dangerous driving) directly to the service company. If a rating dips below a certain acceptable threshold as defined by the service company, the app user may be excluded from making future use of the services. Rating systems theoretically positively affect a driver's driving behaviour given that they are incentivised to mitigate reckless behaviour to prompt high ratings from passengers, which in turn increases the likelihood that they will attract further customers (Acheampong 2021).

21.5 Discussion

While these precedents offer diverse approaches to the multifaceted security challenges South African public transport users face on a daily basis, several potential barriers to user adoption may limit the widespread uptake of these security technologies. The majority of these precedents rely on smartphones to deliver security solutions, and some have the additional challenge of gaining a critical mass of users to create value for users through an accumulation of user-generated data.

Mobile phone penetration in cities is likely high enough within the public transport user group to position smartphones as an effective method of communicating security-related information, though it is difficult to pinpoint exact statistics on mobile phone uptake in South Africa. There are several ways of measuring mobile phone penetration (e.g. SIM card penetration, unique subscriber penetration, mobile access penetration) that each leads to different answers. For example, the PEW Research Centre (2018) reported that 51% of South African adults owned a smartphone in 2017, while in the same year Deloitte (2017) reported that 93% of the population has access. Urban and rural mobile phone penetration rates may vary within a country as well, with urban rates skewed to a higher proportion of smartphone-owning residents.

The ubiquity of smartphone use in urban centres is supported by widespread network coverage. As of 2019, all urban areas have full 2G coverage, and more than 98% 3G and LTE coverage (ICASA 2020). Though the necessary infrastructure is abundant, data affordability and local mobile phone storage space present potential challenges to the uptake of mobile security applications. A single gigabyte of mobile data costs people living in Africa on average 7.1% of their monthly income in 2019, and 3.1% in 2020 (A4AI 2020). As opposed to regions where mobile data is readily available and relatively inexpensive compared to income, in areas where data is expensive or limited, users will restrict and optimise their data use (Mathur et al. 2015). In South African cities, readily available fibre and broadband tend to be restricted to higher income areas, leaving Internet connectivity options in low-income areas limited to mobile data. The lack of widely available public Wi-Fi exacerbates the dependency on mobile data in urban townships (Phokeer et al. 2016). Mobile data contract plans are restricted to those who can provide proof of a stable monthly income and bank statements, which makes such plans inaccessible to informal workers. Because of this, low-income mobile users in particular rely on expensive prepaid data plans or out-of-bundle data.

These high data costs relative to income mean that data use is more likely to be constrained to mobile applications users believe are valuable to them, as opposed to experimentally downloading and using an application without demonstrated value. This presents a challenge, particularly for mobile applications whose value is generated through critical mass use of the app. Initially such apps, like the *Safetipin* app present little benefit for users as a place to actively acquire information on personal security. Without any real incentive to download the application and subsequently contribute data points, available information is limited. Furthermore, South African cities are less densely populated than other cities on the *Safetipin* app, such as Cairo and Mumbai. This makes it more challenging to acquire the local active user-base necessary to accumulate widespread up-to-date security reports.

Outside of technological barriers, localised circumstances and transport operating structures can pose challenges to the implementation of similar intervention types. For example, driver rating apps would need to take into account that the specific vehicles MBT drivers drive and the routes they run are not necessarily the same from day to day. This complicates technological solutions like Uber's identity verification and driver rating where a driver's identity is bound to a vehicle license plate number. However, apps like *Himmat Plus* that verify drivers independently of the vehicle would be able to authenticate licenses with online verification tools that look for specific markers that indicate authenticity. Additionally, drivers may be resistant to the accountability that comes with rating apps as evidenced by opposition to tracking devices on their vehicles.

The paramount for user uptake of information technologies is trust in the institutions that provide them, particularly those that rely on user-tracked locations, and the parties these technologies rely on to act on security incidents (e.g. Nguyen-Phuoc et al. 2021). In South Africa, the low confidence in national police services is potentially problematic for alert-type apps that require users to trust that the local police will respond to an emergency alert. Burkhard et al. (2013) studied whether people's trust in the government as an information source affected their intentions to use the government's online public transport information service. They found that trust in information provided digitally correlates with trust in the public transport service reliability. In South Africa where service unreliability is common, this finding may bear implications for the effectiveness of relaying security information in tandem with information on other public transport service attributes.

In South Africa, it is no surprise that security forums have largely taken the form of community-organised *WhatsApp* groups which are relatively data-light solutions on a pre-existing app that is already on many people's phones. Groups have taken off amongst workers organising shared transport home at off-peak times, hikers sharing their routes with others to make sure they are safe while on the mountain, and neighbourhood watch groups as a place to report incidents. While these groups have offered a rudimentary way to monitor security they come with several drawbacks. These groups are prone to 'noise'—without strong moderation on the administrator's part, irrelevant conversations clutter the channel making it more difficult to identify actual emergencies. Data on incident reports that could have been invaluable in informing preventative measures are siloed in different closed channels, and precise locations of these incidents are not guaranteed.

21.6 Conclusion

Without clear responsibilities assigned to transport stakeholders, security measures become implemented at the discretion of public transport operators and at the expense of passengers. Lower-income passengers in particular are disadvantaged as they cannot financially turn to alternative means of private, more secure mobility. Approaches are unimaginative, tending towards interventions that implementers assume are useful, rather than demonstrably effective on the ground. However, traditional methods like CCTV and personnel-heavy security checks are not financially viable if implemented across the full system and may be susceptible to vandalism. Data-driven decisions are needed to prioritise resources across high incident areas, but disparate data sources are difficult to tap into. In areas where distrust is high and confidence is low in local police forces, incidents may even go underreported, creating a misconstrued image of crime hotspots.

Given the growth in smartphone ownership in South Africa amongst public transport users, ICTs are a promising avenue to bridging the data gap and providing solutions directly to people without having to invest in costly fixed technologies that can be subject to theft and vandalism. Yet without demonstrable evidence of the effectiveness of ICTs to ease perceived and real security threats, traditional methods may continue to take precedence in policy and planning. Further research is needed to gauge the effectiveness of different ICT approaches in a context like South Africa, where barriers to technological use such as mobile data access are prevalent and institutional distrust is high.

The legislation is currently not clear on who bears responsibility for passengers' security onboard public transport, but the onus is on municipalities to provide information related to security as it affects public transport use. Furthermore, security measures cannot be siloed to individual operators as it currently is, where certain operators can afford to provide their passengers with more enhanced security measures than others whose main consumers are lower-income users captive to low fares. While implementing security measures across disparate operators may be difficult to coordinate, municipal governments could provide the infrastructure to make solutions possible, whether that be open-data portals, zero-rated services that make online resources accessible regardless of mobile data, or other methods to break down the barriers to consolidating information and to individual ICT access.

References

Acheampong R (2021) Societal impacts of smart, digital platform mobility services—an empirical study and policy implications of passenger safety and security in ride-hailing. Case Stud Transp Policy 9(1):302–314

Alliance for Affordable Internet (AAI) (2020) The Affordability Report 2020. Web Foundation. https://a4ai.org/affordability-report/

Barrett J (2003) Organizing in the informal economy: a case study of the minibus taxi industry in South Africa. International Labour Office

Beecroft M (2019) The future security of travel by public transport: a review of evidence. Res Transp Bus Manag 32:1–7

Beecroft M, Pangbourne K (2015a) Future prospects for personal security in travel by public transport. Transp Plan Technol 38(1):131–148

Beecroft M, Pangbourne K (2015b) Personal security in travel by public transport: the role of traveller information and associated technologies. IET Intel Transp Syst 9(2):167–174

Bennetts C, Charles M (2016) Between protection and pragmatism: passenger transport security and public value trade-offs. Int J Public Adm 39(1):26–39

Burkhard R, Schooley B, Horan T (2013) Public use of an online advanced traveler guidance information system: trust in the e-service and the agency. J Int Technol Inform Manag 22(2):1+

City of Cape Town: Traffic By-Law, C 53/05/11 (2011). https://resource.capetown.gov.za/documentcentre/Documents/Bylaws%20and%20policies/Traffic%20%20By-law%202011.pdf

City of Cape Town (CoCT) (2018) Comprehensive Integrated Transport Plan 2018–2023

Delbosc A, Currie G (2012) Modelling the causes and impacts of personal safety perceptions on public transport ridership. Transp Policy 24:302–309

Deloitte (2017) Global Mobile Consumer Survey 2017. The South African Cut. https://www2.deloitte.com/content/dam/Deloitte/za/Documents/technology-media-telecommunications/ZA-Deloitte-South-Africa-Mobile-Consumer-Survey-2017-Mobile_090718.pdf

Department of Transport (DoT) (1996) White Paper on Transport Policy. Republic of South Africa

Department of Transport (DoT) (2002) National Railway Safety Regulator Act. Republic of South Africa

Department of Transport (DoT) (2009) National Land Transport Act (NLTA), No. 5 of 2009. Republic of South Africa

City of Johannesburg Transport Department (CoJTD) (2013) Strategic Integrated Transport Plan Framework. City of Johannesburg

Department of Transport (DoT) (2017) Revised White Paper on National Transport Policy (Draft). Republic of South Africa

Eagle G, Kwele K (2021) "You Just Come to School, If You Made It, Its Grace": young black women's experiences of violence in utilizing public "Minibus Taxi" transport in Johannesburg, South Africa. J Interpers Viol 36(15–16):8034–8055

eThekwini Transport Authority (eTA) (2010) Integrated Transport Plan Update 2010–2015. eThekwini Municipality

Gardner N, Cui J, Coiacetto E (2017) Harassment on public transport and its impact on women's travel behaviour. Austral Planner 54(1):8–15

Gerell M (2018) Bus stops and violence, are risky places really risky? Eur J Crim Policy Res 24:351–371

Giddy J (2019) The influence of e-hailing apps on urban mobilities in South Africa. Afric Geogr Rev 38(3):227–239

ICASA (2020) The state of the ICT sector report in South Africa. https://www.icasa.org.za/uploads/files/State-of-the-ICT-Sector-Report-March-2020.pdf

iOL (2021) Golden Arrow commuter fears for life after robbery, upset at 'lack of protection'. iOL, January 26. https://www.iol.co.za/news/south-africa/western-cape/golden-arrow-commuter-fears-for-life-after-robbery-upset-at-lack-of-protection-fa52300a-0716-499e-9604-62da635fd608

Kruger T, Landman K (2007) Crime and public transport: designing a safer journey. In: Proceedings of the 26th annual South African transport conference, pp 112–126

Lemieux AM, Felson M (2012) Risk of violent crime victimization during major daily activities. Viol Vict 27(5):635–655

Ludidi V (2020) Teen owner of women-only taxi service back on the road despite taxi drivers' threats. iOL, September 16. https://www.iol.co.za/news/south-africa/western-cape/teen-owner-of-women-only-taxi-service-back-on-the-road-despite-taxi-drivers-threats-bbb3b42c-a70a-42e8-8112-866074a0e56f

Luke R, Heyns GJ (2019) An analysis of the quality of public transport in Johannesburg, South Africa using an adapted SERVQUAL model. Transp Res Procedia 48:3562–3576

Mabaso Z (2019) Women and girls' experiences of gender-based violence in Gauteng and the Western Cape Province. Sonke Gender Justice

Mashongwa v PRASA (2015) CCT 03/15

Masoumi H, Fastenmeier W (2016) Perceptions of security in public transport systems of Germany: prospects for future research. J Transp Secur 9:105–116

Mathur A, Schlotfeldt B, Chetty M (2015) A mixed methods study of mobile users' data usage practices in South Africa. In: Proceedings of the 2015 ACM international joint conference on pervasive and ubiquitous computing, pp 1209–1220

McCarthy O, Caulfield B, O'Mahony M (2016) How transport users perceive personal safety apps. Transp Res Part F 43:166–182

National Planning Commission (2011) National Development Plan: Vision for 2030

Newton A, Ceccato V (2015) Theoretical perspectives of safety and security in transit environments. In: Ceccato V, Newton A (eds) Safety and security in transit environments; an interdisciplinary approach. Palgrave Macmillan, pp 23–36

Nordfjærn R, Rundmo T (2018) Transport risk evaluations associated with past exposure to adverse security events in public transport. Transp Res Part F 53:14–23

Nguyen-Phuoc DQ, Oviedo-Trespalacios O, Vo NS, Le PT, Nguyen TV (2021) How does perceived risk affect passenger satisfaction and loyalty towards ride-sourcing services? Transp Res Part D 97:102921

Page O (2001) Crime and crime prevention on public transport—reducing crime on public transport in South Africa, appropriateness of crime prevention strategies. In: Proceedings of the 20th South African transport conference

Pew Research Centre (2018) Internet connectivity seen as having positive impact on life in sub-Saharan Africa. https://www.pewresearch.org/global/2018/10/09/internet-connectivity-seen-as-having-positive-impact-on-life-in-sub-saharan-africa/. Accessed 13 Nov 2020

Phokeer A, Johnson D, Densmore M, Feamster N (2016) A first look at mobile internet use in township communities in South Africa. In: Proceedings of the 7th annual symposium on computing for development, New York, vol 15, pp 1–10

PRASA (2019) Corporate Plan 2020–2022

Rail Commuters Action Group v Transnet Ltd, CCT 56/03 (2004)

Reichow F, Friemel T (2020) Mobile communication, social presence, and perceived security on public transport. Mob Media Commun 8(2):268–292

Safetipin & Cities Alliance (2016) Using data to build safer cities. https://safetipin.com/report/cities-alliance-project/

Sam E, Abane A (2017) Enhancing passenger safety and security in Ghana: appraising public transport operators' recent interventions. Manag Res Pract 9(3):62–75

Smith M, Cornish D (eds) (2006) Secure and Tranquil travel: preventing crime and disorder on public transport. UCL Jill Dando Institute of Crime Science

Sonke Gender Justice (2018) Public transport and safety symposium. https://www.saferspaces.org.za/uploads/files/Public-Transport-Safety-Symposium1.pdf

Statistics South Africa (SSA) (2013) National Household Travel Survey 2013. DataFirst. https://doi.org/10.25828/zs4k-hb60

Vanderschuren M, Phayane S, Gwynne-Evans A (2019) Perceptions of gender, mobility, and personal safety: South Africa moving forward. Transp Res Rec 2673(11):616–627

Bianca Ryseck is a Transport Studies Ph.D. candidate at the University of Cape Town where she is investigating the role of information technologies in enabling equitable access to the hybrid public transport system in Cape Town. She holds an M.Sc from the London School of Economics in City Design and Social Science where her passion for the relationship between social equity and urban development was born.

Roger Behrens is a Professor in the University of Cape Town's Department of Civil Engineering. He is Director of the Centre for Transport Studies. He graduated with a Master Degree in City and Regional Planning from UCT in 1991, and with a Ph.D. degree in 2002. His current research activities relate to: the regulation and improvement of informal public transport services, and their integration with scheduled formal services; analysis of the dynamics of changing travel behaviour, and the implications this has for the management of travel demand; and the identification of urban form preconditions for effective and viable public transport networks.

Chapter 22
Smart Mobility in Urban Africa: Geography of Diffusion, User Characteristics and Emerging Impacts of Digital Platform/app-Based Mobility Services

Ransford A. Acheampong

Abstract Smart, digital platforms or app-based mobility solutions are becoming common in global south cities. Over the last few years, major Transportation Network Companies (TNCs) have expanded their services into several African countries with unfolding disruptive consequences. The overarching aim of this chapter is to explore the geography of the diffusion of these new and emerging mobility services in Africa and their emerging socio-environmental impacts. Firstly, the chapter presents an inventory of and maps known app-based mobility services in Africa, to reveal, for the first time, the distribution of these ICT-mediated mobility solutions across the continent, and their key drivers. Secondly, drawing on a large sample survey conducted in two of Ghana's major urban centres (i.e., Accra and Kumasi), the chapter provides empirical insights into the emerging impacts of app-based on-demand mobility solutions focusing on internet-based ride-hailing. To this end, the following key questions are addressed: (a) who are the users (and non-users) of app-based ride-hailing and the reasons for doing so? (b) what are the associated travel behaviour impacts, in terms of mode substitution and ride-hailing trip characteristics? (c) What are the safety and security impacts from the perspective of passengers? Finally, the chapter reflects on the implications of app-based mobility services for creating sustainable transport and mobility futures in urban Africa.

Keywords Digital platform mobility · App-based mobility · Smart urban mobility · Ride-hailing · Travel behaviour · Sustainable futures

22.1 Introduction

Digital platforms or app-based mobility solutions are now a global phenomenon. Constituting one of the common expressions of the smart city development model

R. A. Acheampong (✉)
Department of Planning and Environmental Management, University of Manchester, Manchester, England
e-mail: Ransfordantwi.acheampong@manchester.ac.uk

R. A. Acheampong et al. (eds.), *Transport and Mobility Futures in Urban Africa*, The Urban Book Series, https://doi.org/10.1007/978-3-031-17327-1_22

and emerging platform urbanism, these new mobility solutions are leveraging advances in Information and Communication Technology (ICT) and associated digital technologies to transform the way we move in towns and cities.

Globally, the availability and widespread adoption of the internet and smartphones have enabled the delivery and adoption of ICT-mediated mobility solutions. Major Transportation Network Companies (TNCs), such as *Uber*, *Lyft* and *Bolt*, as well as other service providers, provide digital platforms deployed as smartphone applications with which users can meet their everyday mobility needs on-demand. Within the emerging paradigm of 'collaborative consumption' (Botsman and Rogers 2010) or 'sharing-economy' (see e.g., Sundararajan 2016) in which shared-use of resources and assets is promoted and encouraged as opposed to outright ownership, digital platforms enable users to access shared-mobility services such as car-sharing, bike-sharing and e-scooter-sharing (Shaheen and Cohen 2013). Other options, including ride-hailing (non-pooled), motorcycle-hailing and water taxis are also increasingly being offered through smartphone applications and other digital platforms.

ICT-mediated mobility solutions have emerged with disruptive consequences for cities and their transportation systems, triggering wider societal impacts that are not yet fully-understood (see e.g., Palm et al. 2021; Lucas 2019). They are radically transforming urban mobility and individual travel behaviours by providing flexible, on-demand alternatives to hitherto rigid conventional transport systems, such as taxis and public transport. A case in point is internet-based ride-hailing which is increasingly becoming a popular way of meeting various travel needs, including for social and recreational trips, as well as work journeys in cities globally (see e.g., Acheampong et al. 2020; Grahn et al. 2020; Lavieri and Bhat 2019; Pham et al. 2017). Evidence from the aforementioned studies shows that people prefer ride-hailing services because they are convenient to use; can be affordable compared with other options; and offer a much flexible alternative to individuals not wanting to use personal cars for trips, especially in busy urban centres where parking is limited.

In African cities, the disruptive consequences and societal implications of the on-going diffusion of ICT-mediated mobility solutions are even more profound. For example, minibuses and taxis (including cars, bicycles and motorcycles used as taxis) have long served the mobility needs of different groups of the growing urban populations of Africa (see e.g. Sietchiping et al. 2012; Ehebrecht et al. 2018). These conventional, mostly informal mobility systems have operated in very rigid ways, such as requiring passengers to board a taxi from a designated station/rank or walk to the nearest service route in order to hail one, such as what is known locally as 'dropping' in urban areas in Ghana. Thus, in most African cities, the arrival of TNCs, such as Uber and Bolt with their digital platforms often mark the introduction of flexible, on-demand, ICT-mediated mobility options.

While from the perspective of individual users TNCs and their ICT-mediated mobility solutions, such as ride-hailing are providing convenient, flexible and clean travel alternatives, from the point of view of other actors such as taxi businesses, the immediate impact has largely been negative. With existing transport governance regimes grappling with how to regulate the activities of TNCs and capture maximum public value (see e.g. Agyemang 2020; Papadopoulos and van Eck 2021), local taxi businesses are struggling to compete, with profound implications for employment and livelihoods of the thousands of workers in the taxi industry. In some cases, local operators have sought to compete and reclaim their market share by digitizing their services through the adoption of platforms similar to those used by the major TNCs. A case in point is the recent digitization of '*Boda Boda*' (motorcycle and bicycle taxis) in urban Uganda and '*Okadas*' in Nigeria (see e.g. Doherty 2020; Ibrahim and Bize 2018).

This chapter investigates the diffusion and emerging impacts of digital platform mobility solutions in urban areas across the African continent. The research and approach are outlined as follows: Firstly, an inventory of ICT-mediated or app-based mobility solutions present across the continent as of the time of conducting this research was derived through a desk study. The inventorization involved searching the world-wide-web for any documented evidence of the presence of app-based mobility solutions including the popular ones provided by TNCs such as Uber and Bolt and relatively smaller, less-known local providers/platforms. The internet-based search was done on country-by-country basis, recording for each country the names of the providers/platform and the cities where they are present. The output is mapped to show the geography of the diffusion of the digital platform mobility solutions across the continent. Secondly, the chapter draws on survey data conducted in Accra and Kumasi, Ghana two largest urban centres on the adoption and use patterns of internet-based ride-hailing. Using the survey data, the chapter explores key questions including: (a) who are the users (and non-users) of app-based ride-hailing and the reasons for doing so? (b) what are the associated travel behaviour impacts, in terms of mode substitution and ride-hailing trip characteristics? (c) What are the safety and security impacts from the perspective of passengers? And (d) what are the implications of platform mobility services for sustainable transport and mobility futures in African cities?

The rest of the chapter is organized as follows. Following this introduction section, the second section will focus on the geography of the diffusion of app-based mobility solutions in cities across Africa and the factors underlying the observed distribution. Next, the chapter will present insights from the survey-based empirical studies from Ghana on user characteristics and travel behaviour impacts of the digital platform mobility solutions. The penultimate section will highlight and discuss the key insights of the findings with a particular focus on their implications for sustainable transport and mobility futures, followed by conclusions and directions for future research in Sect. 22.4.

22.2 Geography of the Diffusion of Digital Platform Mobility Solutions

The geographical distribution of TNCs providing app-based mobility solutions across the continent of Africa is mapped in Fig. 22.1 while Fig. 22.2 provides a summary of the number of cities for each of the TNCs found through the desk study. As the visualization shows, digital platforms and their app-based mobility solutions are present across all of the continent's sub-regions. In total, the study found app-based mobility solutions in 95 major towns and cities in 29 countries across Africa.

There are two main categories of app-based mobility service providers. These are the major multinational TNCs (i.e., Bolt and Uber) and other local, relatively smaller app-based mobility service providers. Bolt (formerly Taxify), an Estonian-based TNC is by far the dominant app-based mobility provider on the continent of Africa. It operated in 75 out of the 95 cities represented in this study. Uber, the second largest TNC in terms of presence operated in 19 of the 95 countries represented.

The two major TNCs (Bolt and Uber) tend to be the sole app-based mobility providers in most of the cities where they operate. However, there are also instances where they compete in the same city. The study found that in 12 of the major cities represented, including Abuja (Nigeria), Accra (Ghana), Cape Town, Durban and Johannesburg (South Africa), Dar es salaam (Tanzania) and Kampala (Uganda), both Uber and Bolt were present. This means that Bolt operates exclusively in 59 of the cities compared to Uber that does so in only seven of the 95 cities represented in this study.

Furthermore, the desk study found that TNCs other than Bolt and/or Uber provided app-based mobility solutions in 17 of the cities represented in this study (Fig. 22.1b). The majority of this second category of TNCs were present in cities in East, North and Central Africa where none of the two major providers (i.e., Bolt and Uber) had any presence. One major exception is the *'Boda Boda'* app-based motorcycle-hailing service that competes with Uber in Kampala, Uganda.

The diffusion of app-based mobility solutions across Africa is largely an urban phenomenon. There are a number of reasons why this is the case. TNCs and their ICT-mediated mobility services are present in the most urbanized areas with higher population concentrations in the countries represented in this study. With the high population and economic activity concentrations, these locations are strategically chosen by TNCs to offer the critical mass of early adopters for the new mobility solutions. In addition, major urban areas tend to have the basic satellite-based mapped information that digital platforms require for drivers to be able to navigate in providing mobility services on-demand. Thus, usually, such services first become available in the capital city and subsequently spread to other large urban centres. For example, in Ghana, Uber started operating ride-hailing services around 2016, first in Accra, the capital and subsequently in other large urban areas, including Kumasi, the second largest city.

Aside from the large population concentrations as a key driver of the presence and diffusion of digital platform mobility solutions, the major cities represented in this study have the ICT infrastructure needed to support app-based mobility solutions. In recent years, the percentage of the African population using mobile phones and connected to mobile internet has been increasing steadily. Regarding mobile phone usage, a recent Pew Research Centre study on mobile phone ownership in six sub-Saharan African countries namely South Africa, Tanzania, Ghana, Kenya, Nigeria and Senegal found that large majorities of the population in these countries own

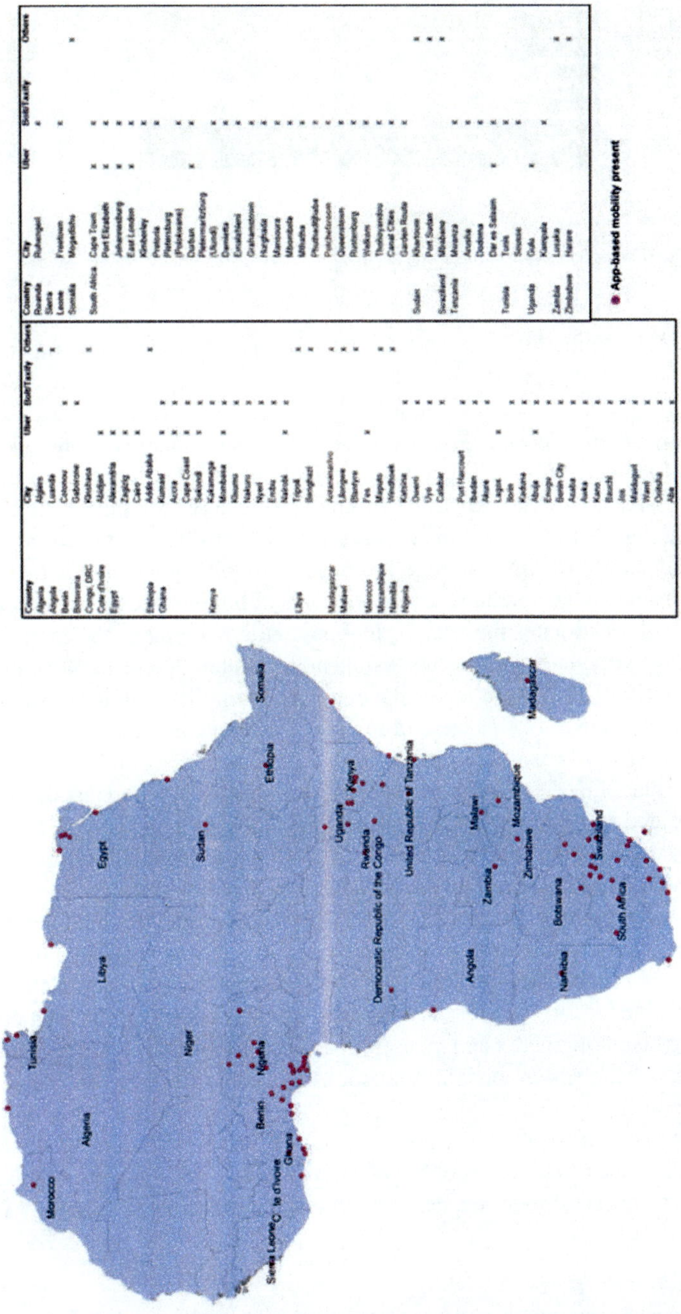

Fig. 22.1 Distribution of TNCs and app-based mobility services in Africa

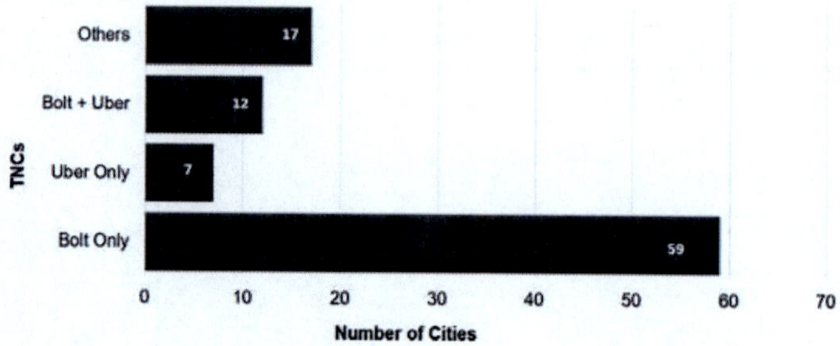

Fig. 22.2 Summary of TNCs presence across Africa

mobile phones. In South Africa, more than half (51%) owned a smartphone that can access the internet and apps while in countries such as Ghana, Senegal and Kenya, about one-third of the adults owned smartphones (Pew Research Centre 2018).

Furthermore, in Sub-Saharan Africa, it is estimated that some 120 million people, representing 13% percent of the continent's population were connected to mobile internet in 2014 (GSMA 2020). By 2019, this had increased to 270 million or 26% of the continent's population. In addition to actual mobile broadband connections, more of the populations are living within the footprint of mobile broadband networks in Africa, suggesting a huge potential for higher levels of internet access in the coming years. The GSMA estimates the usage gap—populations living in the footprint of a broadband network but who are not using mobile internet—to be 49% or 520 million as of 2019. The distribution of the app-based mobility solutions mapped in this study was found to mirror patterns of mobile broadband internet availability and access across Africa. For example, comparing the distribution of app-based mobility solutions with that of the Mobile Connectivity Index, this study found that the former were concentrated in areas that score higher on the latter.[1]

A third key driver of the adoption and diffusion of app-based mobility solutions is the unmet travel demand generated by a mix of historical policy and market failures in the transport and mobility domains. Major transportation problems, including urban congestion, unreliable public transport services and the existence of huge first-and-last-mile accessibility deficits typify many urban areas in Africa (see e.g. Sietchiping et al. 2012) and indeed those represented in this study where new app-based mobility solutions have taken off. TNCs and their app-based mobility solutions can therefore be seen as part of the wider portfolio of responses, mainly from the private sector, to meet the growing mismatch between travel demand and the supply of modern transport services in urban Africa. They are a manifestation of an unmet need and a market-driven, profit-oriented response by multinational TNCs to historical public sector failures to invest in modern transportation systems in rapidly urbanizing Africa.

[1] Mobile Connectivity Index combines a number of indicators including 3G coverage, mobile ownership and mobile app accessibility to map levels of ICT penetration and connectivity globally. More information can be found via this link: www.mobileconnectivityindex.com.

It is important to establish that the foregoing is not to argue that TNCs and their app-based mobility services are necessarily providing effective solutions to the prevailing urban mobility challenges in the cities they operate. Indeed, prior to the advent of TNCs, informal para-transit and various forms of traditional taxis, including car-based, motorcycles and bicycle taxis, provided and continue to provide the means by which the majority of the urban population in Africa move around and access opportunities in their cities. However, with the introduction of apps, these hitherto rigid conventional means of travel are now becoming more responsive to the travel needs of users, by offering flexible, convenient and affordable mobility on-demand. Even so, one could also argue that app-based mobility solutions are contributing to or even exacerbating some of the existing transport and mobility challenges in African cities. For example, instead of the more sustainable mobility alternatives such as bicycling, the large majority of the app-based mobility solutions available in the cities mapped in this study offer motorized options, with car-based mobility such as ride-hailing being the most common service provided.

There are also wider social equity issues associated with app-based mobility solutions. The ICT infrastructure and services required to enable people to access app-based mobility solutions are not available to and accessible by everyone. Regarding mobile internet connectivity, the GSMA (2020) estimate suggests that a large gender gap and rural–urban gap persist in mobile internet use in Africa, standing at 37% and 60%, respectively. Furthermore, mobile phone ownership in general and smartphone ownership in particular tend to be significantly higher among young, highly educated, high-income and male populations (Pew Research Centre 2018). Thus, within urban areas where these mobility services exist, the prevailing digital divide and its intersection with socio-demographic factors are major determinants of who can use these services. Indeed, TNCs are aware of these factors and configure their service coverages to respond to the existing potential demand. In Ghana, for example, previous research (see e.g. Acheampong et al. 2020) has shown that rid-hailing services provided by Uber and Bolt tend to be concentrated on University campuses and their immediate catchments and around the catchments of major commercial centres, such as malls with the goal of attracting the more tech-savvy and financially capable demographics in these areas. App-based mobility services also tend to serve major tourism districts in sub-Saharan Africa (see e.g. Park et al. 2021; Henama and Sifolo 2017).

The foregoing discussions suggest that, with the prevailing digital divide, app-based mobility solutions could not be meeting the mobility needs of the majority of Africa's poor and low-income urban populations unless fundamental gaps in access to ICT infrastructure and services are addressed. In the section that follows, issues around user characteristics, usage patterns and the associated travel behaviour and passenger safety and security in ride-hailing are discussed, drawing on two major survey-based studies conducted in Ghana.

22.3 Platform Mobility Services in Ghana: User Characteristics and Emerging Impacts

22.3.1 *Overview of the Survey Data*

In Ghana, internet-based ride-hailing was introduced in 2016 by Uber in Accra metropolitan area, the country's capital and largest conurbation. While Uber's main competitor is the major TNC Bolt, (formerly known as Taxify), there are a number of relatively smaller digital platform mobility service providers currently operating in the country's major urban centres. With the presence of TNCs offering new on-demand mobility services, a survey was conducted between May and August 2019 to explore the adoption and user characteristics of ride-hailing and the associated travel behaviour and safety and security impacts in Accra and Kumasi, Ghana's two major urban centres.

The original survey (see Acheampong et al. 2020; Acheampong 2021) covered a number of themes, including the respondents' socio-demographic characteristics; locality of residence; reasons for using internet-based ride-hailing and the associated mobility patterns; as well as their perception of safety and security in ride-hailing. The data collection involved online surveys, using various online and social media platforms, including direct emailing, Facebook, WhatsApp and LinkedIn. Field Assistants also conducted face-to-face surveys, using versions of the questionnaire hosted on Mobile Phones and Tablets, with the aim of targeting respondents who could not be reached online and/or self-complete the questionnaire. Ultimately a total of 1,188 valid responses from adults aged over 18 years were obtained. An estimated 65% of the respondents were attracted via the online platforms while the remaining 35% of responses were obtained through personal interviews. Of the total sample, a sub-sample of 548 individuals responded to the open-ended question below about their safety and security experiences and perceptions:

> Overall, do you consider internet-based ride-hailing services such as *Uber* and *Taxify/Bolt* as being safe and secure? Please state your view or experience regarding this…you may also identify if your view or experience applies specifically to any of the ride-hailing service providers in your city/metropolis. Please provide as much information as you possibly can.

An overview of the sample characteristics is presented in Table 22.1. In the sections that follow, broad insights from the survey data are presented with a particular focus on user characteristics, patterns of usage, travel mode substation behaviour in the context of ride-hailing and passenger's safety and security perceptions and experiences.

Table 22.1 Summary of characteristics of the survey respondents

City of residence	Accra: Kumasi	51%: 49%
Gender	Female: Male	48%:52%
Age-groups	18–24	18%
	25–39	53%
	40–54	26%
	55 +	3%
Coupling	Couple: Single: Divorced/Widowed	45%:53%:2%
Have children?	Yes: No	42%:58%
Children in school (KG-SHS)? (n = 497)	Yes: No	84%:16%
Education	Basic school (Primary + Junior High School)	16%
	Senior High School (SH)	12%
	Tertiary (undergraduate/Diploma)-currently enrolled	13%
	Tertiary (undergraduate/Diploma)-completed	34%
	Postgraduate (Master's degree or higher)	25%
Employment	Employed (Full-time)	50%
	Employed (Part-time)	18%
	Retired	1%
	Student (Full-time)	17%
	Unemployed	14%
Income (n = 812)	≤ GH¢1,050	55%
	GH¢1,051–2,000	19%
	GH¢2,001–3,000	10%
	GH¢3,001–4,000	5%
	GH¢4,001–5,000	4%
	≥ GH¢5,001	7%
Car Ownership/Use	I have a car	15%
	I DON'T have my own car, but there is at least one car in my household that I can use if it is available	22%
	None of the above applies to me (don't have a car)	63%
Driver's Licensure	Valid License: Learner, provisional: No	32%:5%:63%

22.3.2 Who is Using Ride-Hailing and Why?

Of the total sample, 59% had used ride-hailing, suggesting that nearly three out of five of the survey respondents were using this new form of mobility in the case study areas (Fig. 22.3a). For the seven-day period prior to completing the survey, 36, 30 and 14% of this sub-sample of respondents had used the app-based service for one, two and three days, respectively. About 5% of the respondents had used the available app-based mobility services every day of the week prior to taking the survey. The users included those who indicated being able to use the app to request a ride themselves (76%); those who had a ride for their reference trip booked by another person, but indicated that they were capable of doing so if they had to (21%) and those who were assisted by another person to use the app because they could not do so on their own (3%).

Furthermore, the survey data showed a generational, educational and income divide regarding who uses internet-based ride-hailing (Fig. 22.3b,c). A larger proportion of the users were 18–24 year-olds (25%) and 25–38 year-olds (66%) who were either currently in higher education and/or employed and earning relatively higher incomes. For example, among individuals with monthly earnings of ≤ GH¢1,050,

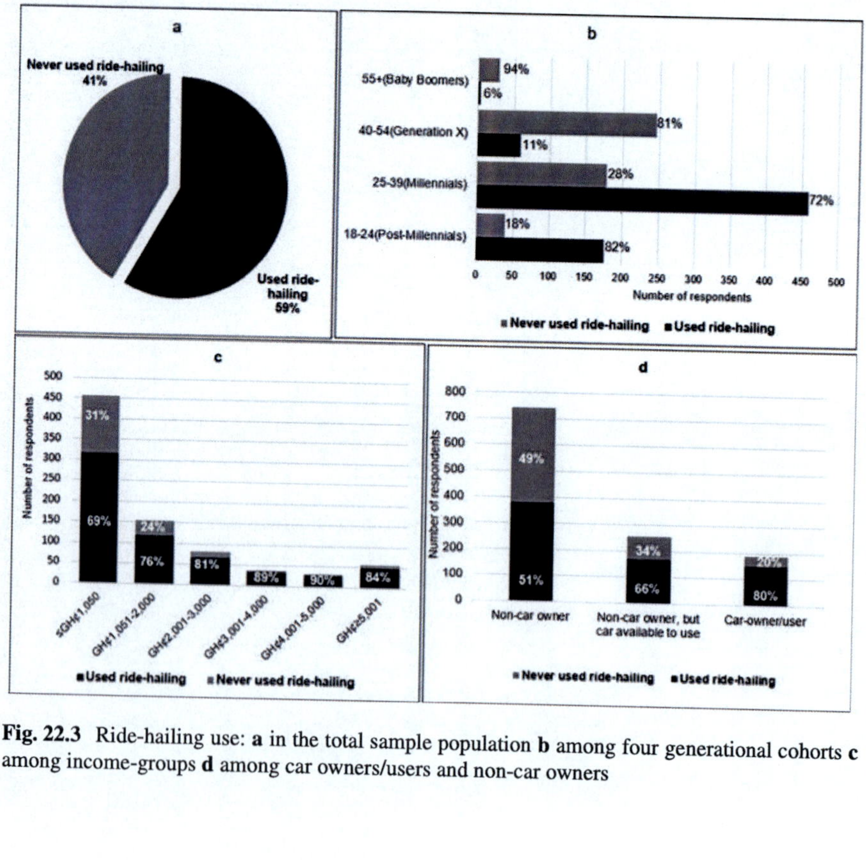

Fig. 22.3 Ride-hailing use: **a** in the total sample population **b** among four generational cohorts **c** among income-groups **d** among car owners/users and non-car owners

69% had used a ride-hailing service but among those with higher earnings (i.e. > GH¢1,050), at least 81% used a ride-hailing service. The proportion of ride-hailers was high for both car owners/users and non-car owners (Fig. 22.3d).

Individuals use app-based ride-hailing for their unique instrumental utilitarian and hedonic benefits. More than three quarters of all the reasons given by the respondents in the survey reflect these benefits, including ride-hailing being a convenient, cheaper, reliable, comfortable and quicker travel option, compared to other available public transport modes (Fig. 22.4). Indeed, the self-reported wait times between passengers requesting a ride and being picked up by the driver, demonstrate the time-saving benefits of using these on-demand mobility services. The majority of individual users reported having a vehicle available to them in less than five minutes (65%) and between 5 and 10 min (42%) of requesting one via an app. The waiting time for an additional 17% of all online vehicle requests were reported to have been met between 10 and 15 min. The respondents indicated additional reasons for using ride-hailing, including the need for privacy while travelling; difficulty in accessing public transport for the trips in question; and ride-hailing being perceived as 'classy' and 'hip' and 'trendy'.

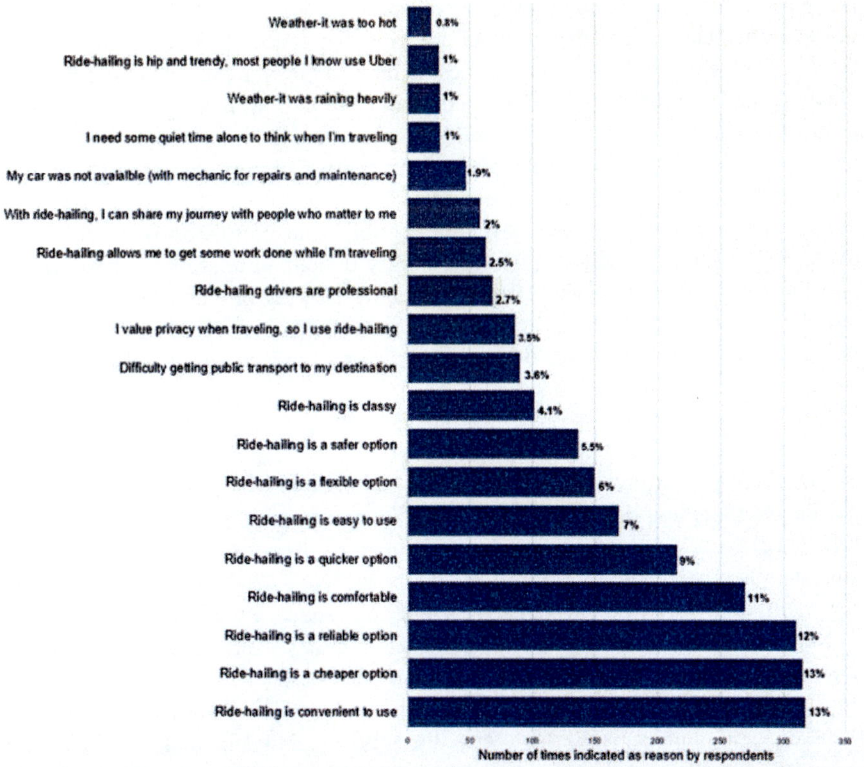

Fig. 22.4 Reasons for using a ride-hailing service for adult respondents' reference trip

22.3.3 What Are the Travel Behaviours Associated with Ride-Hailing?

The characteristics of the surveyed trips (Table 22.2) reveal a number of useful insights based on which the emerging travel behaviour impacts of ride-hailing can be assessed. Individuals ride-hailed to work and school (mainly university). Besides these core activities, 50% of all the ride-hailing trips could be classified as 'special occasion' journeys, such as booking the ride to attend social functions (e.g. weddings and parties); visiting friends and family and for religion-related journeys (e.g. church or mosque attendance). The remaining 9% of journeys involving ride-hailing were unknown as the respondents chose to not disclose the purpose of those trips. The self-reported in-vehicle travel times show that ride-hailing trips typically covered relatively shorter travel times and distances, with about 78% of the surveyed trips lasting between 10 and 30 min.

Furthermore, as the trip pattern analysis show (Fig. 22.5) a larger proportion of the ride-hailing journeys had a suburban location of origin and destination: For example, an estimated 70% and 64% of the ride-hailing trips originated from an outer-suburban locality in the Kumasi and Accra metropolitan areas, respectively. A significant proportion of the reference trips have the same zones of origins and destinations. This means that ride-hailing trips that started in a suburban location were more likely to be a suburban destinations. For example, 63% and 54% of the trips that originated in the outer-suburban zone in both metropolitan areas had destinations in the same zone. Typical suburban localities represented in the ride-hailing trip flow data included the public University Campuses in each of the study areas, and their relatively affluent catchment neighbourhoods. Most suburban areas are not adequately covered by conventional public transport (Trotro) and therefore experience high first-and-last-mile accessibility deficits in relation to public transport (Acheampong and Asabere 2022). Thus, by offering on-demand door-to-door mobility in these suburban locations, ride-hailing becomes a convenient and flexible alternative mode of transport, enabling users to overcome the existing public transport accessibility challenges.

The survey also explored vehicle occupancy and mode substitution behaviours in the presence of internet-based ride-hailing. The results show that a significant share of ride-hailing trips involved lone travel—single occupancy of the vehicle by the passenger, excluding the driver. This information is inferred from the self-reported vehicle occupancy at the trip origins and destinations. About 87% of the respondents indicated that there were zero passengers in the vehicle when it arrived to pick them, suggesting that they would have been the only passenger at the trip's origin.

Similarly, 85% of the respondents indicated that they were the only passenger in the vehicle at the trip's destination and that no other passengers boarded the vehicle during the entire journey of the ride-hailing trip in question. In order to understand the associated mode substitution effects, the survey respondents indicated which of the existing conventional modes they typically would have used for their reference journeys when the option to ride-hail was not available. The results show that in the

Table 22.2 Ride-hailing trip characteristics and travel behaviours

Trip purpose	Work and school-related	41%
	Special occasion	50%
	Undisclosed	9%
Journey start time	Early morning (between 5 and 7am)	7%
	Morning (between 7 and 9am)	19%
	Late morning (between 9 and 11am)	18%
	Early afternoon (between 12 noon and 2 pm)	18%
	Late afternoon (between 2 and 5 pm)	15%
	Evening (between 5 and 9 pm)	19%
	Late Evening (after 9 pm)	12%
Journey times (minutes)	About 10	12%
	10–15	16%
	15–20	23%
	20–25	13%
	25–30	14%
	30–60	18%
	60 +	4%
Vehicle occupancy (trip destination)	Zero	85%
	One person	8%
	Two persons	4%
	Three persons	2%
	Four persons	1%
Mode substitution	Public transport (Trotro)	36%
	Taxi (conventional)	51%
	Private car	10%
	Public transport (bus)	1%
	Walking	1%

absence of ride-hailing, 51%, 36% and 1% of the surveyed journeys would have been undertaken using conventional taxis, public transport (Trotro) and buses respectively (Table 22.2). An additional 10% of the journeys would have been completed using a private car in the absence of internet-based ride-hailing. Ride-hailing also replaced walking trips, although this was significantly lower, constituting only 1% of the reference trips.

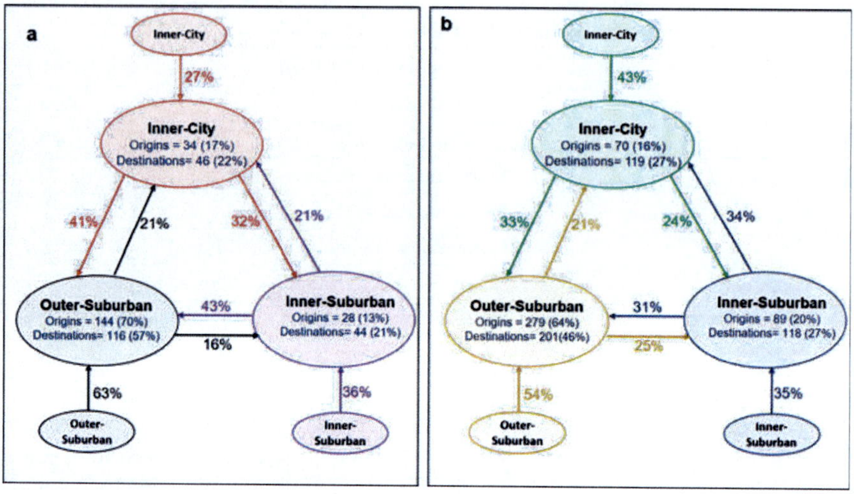

Fig. 22.5 Reference ride-hailing trips distributions among three broad zones in **a** Kumasi and **b** Accra

22.3.4 *What Are the Emerging Safety and Security Impacts?*

In addition to the travel behaviour changes explored in the previous sections, the presence of digital platform mobility services are triggering safety and security consequences in the study areas. While traditional traffic safety concerns and challenges persist with this new form of mobility, other safety and security challenges are emerging that are inherent in ride-hailing being ICT-mediated. As indicated in Sect. 22.3.1, the survey on safety and security elicited detailed qualitative responses from the respondents. This section outlines and discusses the key findings of the survey responses.[2]

The individual respondents perceived the issues of passenger safety and security regarding ride-hailing differently. Overall, the reported safety and security perceptions and experiences could be summarised along the continuum of responses that reflected internet-based ride-hailing being safe (48%, n = 236), somewhat safe and secure (24%, n = 132) and not at all safe and secure (25%, n = 137). As both individuals who have used ride-hailing before and those who had not as of the time of the survey expressed their views about the safety and security consequences of ride-hailing, the extent to which their responses differed are also examined.

The results show that among the former sub-group of respondents, more than half (54%) indicated that they felt safe, overall, in using internet-based ride-hailing. Within the latter sub-group of respondents, 41% perceived this new form of mobility as not being safe and secure. For this sub-group of respondents, it could be deduced from the information they provided that their safety and security perceptions have

---

[2] See Acheampong (2021) for a detailed analysis of the qualitative responses with the codes and emergent themes on which the discussion here is based.

been shaped largely by media reports and in some instances by the experiences of significant others such as friends and families.

A summary of the main factors influencing passenger safety and security that emerged from the qualitative data analysis is presented in Table 22.3.

Table 22.3 A summary of factors influencing passenger safety and security in ride-hailing based on the detailed qualitative survey responses

Factors	Influence on passenger safety and security (perceived and experienced)
Inbuilt app safety and security features: driver and vehicle identification and performance ratings	• Positive influence on safety and security as passengers see basic identification of the driver and vehicle via the ride-hailing app ahead of the trip. Higher safety and security perception by those who trust the apps' inbuilt identification system. Others believed that they asserted and ensured their safety through the instituted regulatory mechanism of reward and punishment, in the form of driver performance ratings • Some respondents, however, simply did not trust in the app's identification features in guaranteeing their safety and security for reasons including possible lapses in driver vetting and recruitment processes; susceptibility of apps to hacks; fears and anxiety about possible driver impersonations and identity swapping; and fake driver accounts
Inbuilt app safety and security features: trackability	• The apps' trackability function provided a number of safety and security benefits to passengers by making it possible to: share one's travel with significant others, such as family and friends to track in real time; alerting significant others to suspicious and potentially safety compromising behaviours during the trip; and track on-going trip on their own phones, so that any unexpected detours en-route by the driver could be detected and questioned
Inbuilt app safety and security features: traceability	• Traceability is both an 'in-trip' and 'after-trip' benefit that accrues from the app allowing passengers to be able to trace a vehicle and/or driver to recover personal belongings left onboard. Both traceability and trackability also improved safety through the notion of 'surveillance'— the belief that the driver is being monitored by someone or that someone is watching over both the driver and passenger

(continued)

Table 22.3 (continued)

Factors	Influence on passenger safety and security (perceived and experienced)
Possible exposure to malicious and criminal activities	• The risk of exposure to criminal activities while using ride-hailing services, such as robbery, kidnapping and sexual harassment were identified as causing major safety and security concerns. Most of the respondents had not been victims of such criminal activities, but their fears and concerns stemmed largely from news media reports on the subject of robbery and kidnaping around ride-hailing services
Privacy and lone travel	• On the one hand, some individuals seemed to derive a sense of safety and security from the privacy and lone travel in ride-hailing. On the other hand, they expressed fear that that lone travel and the privacy it affords could expose passengers to danger if they find themselves with a ride-hailing driver who has intentions of harming them
Emergency use	• Respondents also associated and experienced safety and security regarding internet-based ride-hailing in times of emergencies, such as ill-health. In a context where access to reliable ambulance services is extremely limited, ride-hailing services provide public value by becoming a rapid response alternative, to accessing health care, thereby contributing to saving lives
Driver behaviour	• Driver behaviours such as reckless driving and associated risk-taking behaviour, smartphone use while driving (e.g. for navigation) were identified by passengers as one of the major sources of their anxiety while using ride-hailing

It becomes clear from the summary that the platforms' inbuilt safety features had major implications for perceived and/or experienced safety and security in ride-hailing. Inbuilt features such as driver and vehicle identification and trackability and traceability functions of the ride-hailing apps contributed largely positively to passenger safety and security. At the same time, some of the responses reflected significant distrust in these inbuilt safety and security features, stemming mainly from lapses, including driver vetting and recruitment processes and the apps' susceptibility to hacks, leading to fake driver accounts, impersonation and identity swapping.

From the perspectives of the respondents, these app-related lapses were exposing passengers to malicious and criminal activities including armed robbery, kidnapping and sexual harassment. The overall view was that females in particular were more at risk of exposure to these safety and security compromising situations in using internet-based mobility services. There were genuine concerns and fears about individuals who dabble in 'Sakawa'—a local term for illegal practices which combine internet-based fraud with African traditional rituals—leveraging ride-hailing platforms and posing as drivers of requested rides to target, kidnap and/or rob their rider victims. While the survey response captured these perceptions from the perspective of passengers, drivers are also exposed to criminal and malicious activities that leverage ride-hailing apps and vehicles.[3]

Moreover, the analysis found that safety and security considerations were partly responsible for the large proportion of lone-travel journeys in ride-hailing in the case study areas. On the one hand, some passengers preferred to travel alone in order to avoid the possibility of being exposed to risks and dangers associated with having to travel with complete strangers in a vehicle, as has long been the case with conventional taxis. Indeed, most of the respondents were not open to the option of pooled-rides that would mean having to share the same vehicle with other passengers, especially if the other passengers were going to be complete strangers. On the other hand, there were concerns that the perceived safety and security benefits of lone travel is only guaranteed as long as drivers of requested rides are genuine. To the extent that ride-hailing apps were susceptible to hacking and fake driver accounts, the risks to passenger safety and security remained even for the preferred option of lone passenger journeys.

22.4 Discussion and Policy Implications

Following global trends, digital platforms or app-based mobility services can now be found in several African countries, as revealed by the inventory and mapping of app-based mobility solutions that were found through the web-based desk study (see Fig. 22.1). As the analysis has shown, app-based mobility solutions are an urban phenomenon, with the available services concentrated mainly in the largest cities, mainly the national capitals in all cases, as well as other major urban centres. With the distribution of app-based mobility solutions mirroring mobile broadband internet availability, African countries that have seen investment and growth in the latter have also become fertile grounds for TNCs and other service providers to offer app-based mobility services.

The reverse holds true for countries that score low on the Mobile Connectivity Index. Thus, in the future, as smartphone penetration and broadband connectivity expands across the continent, as the trend already shows, service providers would seek to expand their markets to more countries. App-based mobility solutions are therefore expected to become more widespread in major towns and cities across the

[3] The gruesome murder of an Uber driver is reported by one of the major media outlets in Ghana here: https://www.myjoyonline.com/decapitated-man-was-an-uber-driver-and-headmaster-feyiase-assemblyman-reveals/?param=.

African continent. Moreover, app-based mobility solutions have emerged partly as a response to the unmet mobility and accessibility needs in Africa's rapidly growing cities, created by a mismatch between the ever-growing travel demands on the one hand, and the supply of affordable and efficient public transport services on the other hand.

The geography of diffusion observed at the country level is also a reflection of the digital divide within individual countries, whereby ICT diffusion favours the most urbanized areas, such as capital cities and other major towns and cities. Moreover, within these urban areas, we know that smartphone ownership and mobile broadband internet availability required to access the available app-based mobility services differ considerably among social groups. Indeed, as studies in some African countries have shown (see e.g., Pew Research Centre 2018) smartphones and mobile broad internet are accessible to younger, highly educated and/or individuals with relatively higher earnings. Access to the mobility services provided by digital platforms is therefore skewed toward this demographic who can afford smartphones and available mobile broadband internet. Indeed, the empirical evidence presented in this chapter support this assertion. Thus, within urban areas where these mobility services exist, the digital divide is a major determinant of who can have access to these new app-based mobility services. This also implies that the larger share of the urban population who are poor or earn low wages face significant barriers in using app-based mobility solutions in the areas where they are present.

These findings have implications for the wider equity impacts of app-based mobility solutions now and in the long-term. We know that the prevailing mobility challenges in urban Africa, to which TNCs are partly responding with their app-based mobility solutions, disproportionately affect low-income and poor households and individuals. This is because, while groups that are more affluent increasingly own personal vehicles and can also afford the car-based alternatives offered by digital platforms, the rest of the population must rely on public transport that is under-funded, inefficient, unsafe and offering low quality levels of service. Thus, app-based mobility solutions have not emerged to serve the mobility needs of the larger mass of the urban populations whose mobility needs are already not being adequately met. Instead, they appear, at least in the early years of their diffusion, to add to the pool of alternatives that is accessible to a rather narrow sub-group of the urban population—the affluent or the smartphone-owning, tech-savvy younger population.

This is expected, as like many other technical innovations, the diffusion of app-based mobility solutions, requires a niche market of a critical mass of early adopters, who tend to fit the aforementioned demographics. That said, given the nature and extent of mobility and accessibility challenges in Africa's major cities and urban areas, app-based mobility solutions would only become beneficial when they are able to serve the populations who are most affected by the prevailing urban transport and accessibility challenges. In practice, this would mean app-based mobility solutions expanding their service coverage geographically; providing ways by which both individuals who have access to broadband internet and those who do not can access available services on-demand; and becoming affordable.

A combination of market competition among service providers and public sector interventions and a regulatory response would be crucial to making app-based mobility solutions accessible to different social groups in urban Africa. While in many countries the landscape of app-based mobility solutions is dominated by major multinational TNCs such as Uber and Bolt, increasingly, new and relatively smaller players are emerging to rival the major players. In the process, app-based mobility services are expanding to a much wider population in some contexts. The recent adoption of digital platforms by *Boda Boda* motorcycle taxis in Uganda is a case in point. Prior to this, Uber, the major TNC operating in the country's capital, Kampala, catered to a rather narrow catchment population, targeting affluent neighbourhoods and central areas of the city. By adopting digital platforms similar to those used by Uber, *Boda Boda* informal transport service providers are now rivalling Uber in offering on-demand motorcycle taxi services. In the process, app-based mobility services have not only become available to a much wider population in the city through *Boda Boda*, but Uber is also responding by expanding their service coverage to areas which were hitherto not covered as a way of increasing their market share of users of these new mobility solutions.

While the emergence of competition among app-based mobility providers is inevitable, free market forces alone may not be enough in bringing about the changes and outcomes that we are starting to see in cities such as Kampala. Instead, governments should bring their outdated urban transport regulatory regimes in line with the emergence of digital platforms and the associated mobility services. Governments could also provide critical support by investing in the necessary digital infrastructure and platforms as a way of enabling existing informal public transport service providers to transition to and benefit from the emerging platform economy. Doing so would not only protect employment and livelihoods, but also yield wider public value by enabling the expansion of and access to app-based mobility solutions to meet the mobility needs of the urban population.

The presence of app-based mobility solutions also raises a number of implications for travel behaviour and the sustainability of urban transport and mobility in African cities. The available app-based mobility services identified through this study are dominantly car-based and/or other motorized forms of transport, such as app-based motorcycle taxis that are becoming common in countries such as Nigeria, Uganda and Kenya.

Furthermore, as the empirical findings from urban Ghana on the travel behaviours associated with ride-hailing found, they are characterised by low vehicle occupancy with a significant share of usage being for lone travel; and ride-hailing tends to be used for relatively short journeys in both urban and suburban localities. What is more, these single occupancy car-based journeys offered by ride-hailing services are replacing conventional alternatives such as taxis and public transport that are shared by several passengers. For example, of all the ride-hailing trips surveyed in Ghana's two major cities, Accra and Kumasi, 51 and 36% of them would in the past, before app-based mobility became available, have been undertaken by traditional taxis and public transport (Trotro) respectively. About 10% of the journeys would have been undertaken in a private car prior to ride-hailing becoming available. These findings raise questions about the extent to which the presence of ride-hailing is really filling a gap in public transport supply. The substitution behaviours suggest that ride-hailing service providers are duplicating and even possibly undermining existing public transport provisions.

The above findings also suggest that the presence of ride-hailing is triggering travel behaviour changes that are not all positive or beneficial, especially from a sustainability point of view for reasons outlined as follows: Being ICT-mediated and offering mobility on-demand, app-based solutions such as ride-hailing are enabling access to car-based transport in ways that were not possible a few years ago in urban Africa. Indeed, as the evidence from the survey-based studies from Ghana has shown, while ride-hailing's impact on personal vehicle use is relatively lower, they appear to be replacing more public transport and conventional taxi trips, with most of the journeys involving a single passenger. In most cases, lone travel is mainly because service providers do not offer pooled or shared alternatives. Thus, the car-based options that app-based mobility solutions are enabling access to could be contributing to the already rising levels of motorization and car use in Africa's major cities. Moreover, at the time of this study, none of the TNCs offering car-based ride-hailing services across the continent offered cleaner energy alternatives such as electric vehicles. This also implies motorized forms of app-based mobility solutions could be exerting further negative impacts in terms of energy consumption and contributing to congestion as well as the emission of $C0_2$ and other pollutants.

Within the existing car-based on-demand mobility systems, providing and promoting ride-share and ride-pool options to users could contribute to reducing the overall impacts of pervasive car-based on-demand mobility services. More crucial to creating sustainable mobility futures in urban Africa is the urgent need for non-motorized alternatives to on-demand, app-based mobility alternatives. Now, car-based alternatives dominate the landscape of digital platform mobility services in Africa, but non-motorized alternatives are also becoming available, although this is happening on a much smaller scale in a few cities. A case in point is GURARIDE,[4] a new app-based shared-mobility service offering non-motorized alternatives including bicycles, e-bicycles and e-scooters in Kigali, Rwanda's capital. The diffusion of these non-motorized alternatives would be contingent on sustained investment in infrastructure such as protected bicycle lanes and overcoming the general negative public perceptions and attitudes towards bicycling and other forms of micro-mobility.

Finally, as this chapter has briefly indicated, the emergence of digital platform mobility solutions now raises new safety and security concerns, some of which go beyond the traditional considerations of traffic safety, such as preventing road accidents and fatalities. By nature, app-based mobility services mediate interactions between drivers and passengers via their digital platforms. User data is also generated and exchanged in the process. We already know that vulnerabilities in digital platforms could be exploited by persons with criminal intent, thereby exposing users to safety- and security- comprising situations with potentially serious consequences.

[4] https://cleantechnica.com/2020/09/14/rwandas-guraride-looks-to-jumpstart-africas-bike-share-micromobility-industry/.

Thus, as digital platforms offering mobility services become pervasive, measures to ensure public safety and security would also become critical.

22.5 Conclusion

Globally, new transport technologies and ICT-mediated mobility services are having significant impacts on urban transport and mobility. Africa is no exception in the diffusion and adoption of these ICT-enabled mobility solutions. In Africa, as this chapter has shown, digital platform/app-based mobility is an urban phenomenon, necessitated in part, by the ever-growing, yet unmet travel demand in general, and the need for flexible, convenient and fast on-demand mobility options in particular.

The evidence presented in this chapter has shown that while app-based mobility services present several benefits, they are also exerting wider socio-economic and environmental impacts that are not well understood yet. Their user population are mostly younger and/or relatively affluent, own smartphones and have access to mobile broadband internet. This implies that app-based mobility services are not necessarily enabling the larger mass of urban poor and low-income populations to meet their everyday mobility needs. Currently, the landscape of digital platform/app-based mobility services in urban Africa is dominated by car-based demand-responsive offerings, such as ride-hailing, that are used as substitutes for conventional public transport and taxis. Shorter travel distances in urban and suburban areas, as well as lone journeys also typify usage of these car-based on-demand mobility services.

Following current trends, it is inevitable that digital platform/app-based mobility solutions will become pervasive across urban Africa in the coming years. Responsive policy-making and governance will become ever more critical for capturing public value from them, by ensuring that their presence do not widen existing urban transport inequalities, increase motorization levels and car-dependence, and act as barriers to creating sustainable urban transport and mobility futures.

As a fairly recent phenomenon, our understanding of the impacts of app-based mobility solutions, especially in African cities is rather limited. There is therefore the need for research to deepen our understanding of how these new mobility solutions are changing travel behaviours and shaping accessibility outcomes; their implications for employment and livelihoods; the associated social equity and environmental sustainability impacts and emerging governance responses and models.

References

Acheampong RA (2021) Societal impacts of smart, digital platform mobility services—an empirical study and policy implications of passenger safety and security in ride-hailing. Case Stud Transp Policy 9(1):302–314

Acheampong RA, Asabere SB (2022) Urban expansion and differential accessibility by car and public transport in the Greater Kumasi city-region, Ghana—a geospatial modelling approach. J Transp Geogr 98:103257. https://doi.org/10.1016/j.jtrangeo.2021.103257
Acheampong RA, Siiba A, Okyere DK, Tuffour JP (2020) Mobility-on-demand: an empirical study of internet-based ride-hailing adoption factors, travel characteristics and mode substitution effects. Transp Res Part C Emerg Technol 115:102638
Agyemang E (2020) "Uber is here to stay": exploring the policy implications of the Uber-Local Taxis turf war in Accra, Ghana. Case Stud Transp Policy 8(1)
Botsman R, Rogers R (2010) What's mine is yours. The rise of collaborative consumption. arper Business, New York
Doherty J (2020) Motorcycle taxis, personhood, and the moral landscape of mobility. Geoforum. https://doi.org/10.1016/j.geoforum.2020.04.003
Ehebrecht D, Heinrichs D, Lenz B (2018) Motorcycle-taxis in sub-Saharan Africa: current knowledge, implications for the debate on "informal" transport and research needs. J Transp Geogr 69:242–256
Grahn R, Harper CD, Hendrickson C, Qian Z, Matthews HS (2020) Socioeconomic and usage characteristics of transportation network company (TNC) riders. Transportation 47(6):3047–3067
GSMA (2020) Mobile internet connectivity sub-saharan Africa factsheet. https://www.gsma.com/r/wp-content/uploads/2020/09/Mobile-Internet-Connectivity-SSA-Fact-Sheet.pdf Accessed April 10 2022
Henama US, Sifolo PPS (2017) Uber: the south africa experience. Afric J Hosp Tour Leis 6(2):1–10
Ibrahim B, Bize A (2018) Waiting together: the motorcycle taxi stand as Nairobi infrastructure. Africa Today 65(2):73–92. https://doi.org/10.2979/africatoday.65.2.06
Lavieri PS, Bhat CR (2019) Investigating objective and subjective factors influencing the adoption, frequency, and characteristics of ride-hailing trips. Transp Res Part C Emerg Technol 105:100–125
Lucas K (2019) A new evolution for transport-related social exclusion research? J Transp Geogr 81:102529
Papadopoulos S, van Eck S (2021) Disruptive technologies and taxi rides in South Africa: what is the 'Uber'Uproar about?. In: Global perspectives on legal challenges posed by ridesharing companies. Springer, Singapore, pp 199–229
Palm M, Farber S, Shalaby A, Young M (2021) Equity analysis and new mobility technologies: toward meaningful interventions. J Plan Lit 36(1):31–45
Park SY, Kim J, Pan B (2021) the influence of uber on the tourism industry in Sub-Saharan Africa. J Travel Res 60(7):1598–1611
Pew Research Center (2018) Internet connectivity seen as having positive impact on life in Sub-Saharan Africa. https://www.pewresearch.org/global/2018/10/09/majorities-in-sub-saharan-africa-own-mobile-phones-but-smartphone-adoption-is-modest/ Accessed April 10, 2022
Pham A, Dacosta I, Jacot-Guillarmod B, Huguenin K, Hajar T, Tramèr F, Hubaux JP (2017) Privateride: a privacy-enhanced ride-hailing service. Proc Privacy Enhanc Technol 2017(2):38–56
Shaheen SA, Cohen AP (2013) Carsharing and personal vehicle services: worldwide market developments and emerging trends. Int J Sustain Transp 7(1):5–34
Sietchiping R, Permezel MJ, Ngomsi C (2012) Transport and mobility in sub-Saharan African cities: an overview of practices, lessons and options for improvements. Cities 29(3):183–189
Sundararajan A (2016) The sharing economy: the end of employment and the rise of crowd-based capitalism. MIT Press, Cambridge

Ransford A. Acheampong is a Lecturer and Presidential Academic Fellow in Transport and Urban Futures at the Department of Planning and Environmental Management, University of Manchester, UK. The first strand of his research is focused on the nexus between transportation and spatial structure and their integrated planning in cities. The second strand of his research addresses questions around the diffusion and socio-spatial impact of new and emerging transport technologies and mobility solutions, such as app-based mobility, autonomous vehicles, shared-mobility and Mobility-as-a-Service.

Chapter 23
e-Quantum Leap: Planning for Electric Minibus Taxis in Sub-Saharan Africa's Paratransit System

Marthinus J. Booysen, Chris J. Abraham, Innocent Ndibatya, and Arnold J. Rix

Abstract Minibus taxis are ubiquitous in the developing cities of the Global South. This versatile and somewhat chaotic public transport system is now faced with the need to move to renewable energy. However, the looming roll-out of electric vehicles poses a threat to the already fragile electrical grids of African cities. This chapter evaluates the energy requirements of decarbonising paratransit and evaluates two types of data, passenger-based and vehicle-based, from research in South Africa that has modelled these taxis. Using these two data capture methods, we assess the energy requirements and charging opportunities for electric minibus paratransit in three African cities and compare the results of the two methods to assess their suitability for planning minibus taxi electrification. We found that vehicles need approximately 300 kWh of energy per day, with stationary charging times up to approximately 8 h per day. The results demonstrate the need for reliable tracking data when assessing vehicular energy requirements, which directly impact infrastructure requirements. We conclude by highlighting the remaining challenges on the road to decarbonisation.

Keywords Electric vehicle · Paratransit · Minibus taxi · Passenger data · Tracking data

---

The methodology, data and other parts of the chapter have been extended from Abraham et al. (2021), Booysen et al. (2022), Rix et al. (2022).

M. J. Booysen (✉) · C. J. Abraham · I. Ndibatya · A. J. Rix
Department of E&E Engineering, Stellenbosch University, Stellenbosch, South Africa
e-mail: mjbooysen@sun.ac.za

A. J. Rix
e-mail: rix@sun.ac.za

M. J. Booysen
Department of Industrial Engineering, Stellenbosch University, Stellenbosch, South Africa

I. Ndibatya
Alabama Transportation Institute, University of Alabama, Tuscaloosa, AL, USA

R. A. Acheampong et al. (eds.), *Transport and Mobility Futures in Urban Africa*, The Urban Book Series, https://doi.org/10.1007/978-3-031-17327-1_23

23.1 Introduction

Paratransit is sub-Saharan Africa's main public transport system, being the most used mode of motorised transport and creating countless jobs (Behrens et al. 2015a; Ehebrecht et al. 2018). The modes are various, including even motorcycle or bicycle taxis, but most passengers use minibus taxis (Behrens et al. 2017). Powered by internal combustion engines, these vehicles contribute to greenhouse gas emissions and air pollution.

Africa's paratransit differs in many ways from that of developed countries. In the latter, paratransit is usually an on-demand, flexible transport service which provides facilities for transporting elderly or disabled people (Askari et al. 2021; Behrens et al. 2017). But in Africa, it is an informal public transport service, which has organically evolved from customer demand, in the absence/collapse of formal mass transport services. African paratransit is usually run by private operators who autonomously control the cost, scheduling, routes, and quality of their service (Ndibatya and Booysen 2021; Neumann and Joubert 2016). In sub-Saharan Africa, examples of paratransit include the minibus taxis of Johannesburg, Lagos, Kampala and Nairobi or Kampala's motorcycle taxis ("boda bodas") and Nairobi's tricycle taxis ("tuktuks") (Booysen et al. 2013; Diaz Olvera et al. 2019; Mutiso and Behrens 2011). Paratransit accounts for approximately 70%, 90%, 91%, and 98% of the road-based public trips in Johannesburg, Lagos, Kampala, and Dar es Salaam, respectively (Behrens et al. 2015b; Evans et al. 2018), and around 80% of these are by minibus taxi (Dorothy et al. 2016; Evans et al. 2018; KCCA 2016).

The public transport industry in sub-Saharan Africa underwent a substantial change towards the end of the twentieth century. Firstly, public transport was shifted from state ownership to private ownership due to the IMF and World Bank's structural adjustment policies, which reduced the proportion of funds which they allocated to state-owned entities, many of which eventually collapsed. (Kumar and Barrett 2008; Cervero and Golub 2007; Kumar 2011). As a result of this, a public transport vacuum was created. Gradually, low-capacity vehicles were introduced by the private sector to fill this void. (Behrens et al. 2015a; Diaz Olvera et al. 2019; Jennings and Behrens 2017; Mutiso and Behrens 2011). These vehicles have been of various kinds, from the five-seater Ford Anglia of the 1970s, to the 10-seater Peugeot 204 of the 1980s, to the current 16-seater Toyota HiAce. The fourth and fifth generations (H100; 1989 and H200; 2004) of the Japanese Toyota HiAce have dominated the paratransit market in Africa since the 1990s, trading under several names depending on the country of assembly, such as Toyota Quantum in South Africa and Toyota Venture in Thailand.

The first big step forward—or "Quantum leap"—came when minibus taxis began to dominate the mobility lifestyle of the urban poor in sub-Saharan Africa. These taxis are unlikely to be phased out anytime soon, given their ubiquity and schedule flexibility, and the socio-cultural lifestyles of the urban poor. However, the environmental cost of running them is worrying. It has triggered discussions about the possibilities of transitioning to electric minibus taxis (eMBTs) as part of the global electrification and sustainability agenda. In this chapter, "minibus taxi" refers to an internal combustion engine vehicle unless "electric minibus taxi" or "eMBT" is specified.

The development of low-carbon transport in cities is part of the global agenda to deal with possible climate change (Collett and Hirmer 2021; Collett et al. 2021; Odhiambo et al. 2021). The IPCC (Intergovernmental Panel on Climate Change) estimated in 2014 that the transport sector generates 23% of the global energy related greenhouse emissions (Sims et al. 2014). In sub-Saharan African cities, the deteriorating air quality resulting from ambient air pollution and a high concentration of particulate matter ($PM_{2.5}$) is partly attributed to vehicular emissions (Lozano Gracia et al. 2021; Raje et al. 2018; Singh et al. 2020).

Akumu estimates the cost of air pollution in African cities to be as high as 2.7% of the GDP (Akumu 2014). In the UN General Assembly's post-2015 development agenda, three of the seventeen Sustainable Development Goals, one, eleven and thirteen, are clean energy, sustainable cities and climate action (Zinkernagel et al. 2018). Consequently, electrification is promoted as a low-carbon transport strategy to reduce combustion emissions. The transition from internal combustion engine (ICE) vehicles to electric-powered vehicles (EVs) is gradually increasing in developing countries. Some vehicle manufacturers are planning to phase out ICE vehicles in favour of EVs. A few isolated pilot EV projects have cropped up in sub-Saharan Africa, mainly focusing on micro-mobility (two- and three-wheelers, such as motorcycles and tricycles), but also on buses and private cars (Black et al. 2018). At the time of writing, there was no known ICE to EV transition initiative targeting the paratransit industry, let alone the minibus taxis that are responsible for the vast majority of the public transport trips in the region.

To date, the literature on paratransit in sub-Saharan Africa focuses on sector governance (Goodfellow 2017) and regulation and reforms (Jennings and Behrens 2017; Lucas et al. 2019), and seldom on operations (Ndibatya and Booysen 2020), mobility characteristics (Ndibatya and Booysen 2021), and future prospects of electric mobility integration (Galuszka et al. 2021).

Studies from China (Du et al. 2017) and Europe (Weiss et al. 2020) found that EVs are three times more efficient than ICE-powered cars and twice as efficient as hybrid cars (a vehicle powered by a combination of electricity and an internal combustion engine). This efficiency is partly achieved because the good braking systems and elimination of idling losses save energy for the actual EV movements (Weiss et al. 2020). Although debate continues on the economic and environmental trade-offs associated with EVs (Li et al. 2016), evidence of sustainable EV deployment models also exists (Rnstvik 2013). Some researchers argue that deploying EVs shifts gasoline usage to coal-fired power generation, exacerbating CO_2 emissions (Buresh et al. 2020; Li et al. 2016). However, EV proponents counter-argue that this effect can be reduced by a charging strategy that optimises the use of renewable energy sources such as solar photovoltaics (Buresh et al. 2020; Schucking et al. 2017).

Two recent papers emphasise the need to decarbonise and electrify paratransit, but they also highlight the difficulties of achieving a more sustainable transport sector (Collett and Hirmer 2021; Odhiambo et al. 2021). To evaluate the readiness of three African cities to transition to electric minibus taxis, Odhiambo et al. (Odhiambo et al.

2021) interviewed key stakeholders and concluded that the electrification of the most common mode of transport, namely paratransit, needs to receive increased attention, government support, and cross-sector collaboration for timeous uptake.

Collett and Hirmer (Collett and Hirmer 2021) evaluated the readiness of paratransit in sub-Saharan Africa to transition to electric vehicles. They mention many benefits of the shift but identify the lack of data as the main impediment to making the transition a reality (Collett and Hirmer 2021). They say that

> for a sustainable transition to EVs, additional low-cost, clean generation will be required in many regions. Vital questions such as: "what is the potential demand?", "where and when will this demand occur?" and "how will consumers pay?" need to be answered. EV infrastructure planning, business model design, investment decisions, and policy making alike rely on the availability of adequate and reliable data to answer these questions. Such data are currently not well documented or aggregated in low-income countries, resulting in a data gap.

The words to focus on in this quotation are "adequate" and "reliable"; qualities that are often not applicable to data in developing counties. The literature reviewed above makes it clear that the road to electrification for transport in sub-Saharan Africa is far from smooth.

23.2 Method

23.2.1 Two Main Data Collection Methods Currently in Use

We assess the energy requirements of the paratransit system from the perspectives of two mobility data collection methods typically used: using handheld mobile devices and using GPS devices fixed in taxis.

The first method, using handheld mobile devices (often mobile phones) is commonly used by transport engineers in resource-constrained environments. This method involves field workers boarding the minibus taxis as passengers and tracing the routes for the duration of the trip. We show that although this type of data is readily available and often used, its usefulness in energy analyses is limited, for several reasons:

- It does not adequately reflect the mobility patterns from a vehicle perspective. It gives information on a particular route's energy requirements, which is less useful. However, with route frequency data, the route energy requirements with route frequency data can be extrapolated to obtain the aggregated energy needs for a transport system, which may be useful at a grid level.
- It is susceptible to errors introduced by the field workers' behaviour. They often start work late and go home early. They may be contracted to track for only a specified time of day for legal, oversight or convenience reasons.
- Although field workers may stick to their route and wait with passengers at stops, thereby capturing passenger waiting time, they are unlikely to capture the vehicle's

waiting time. The vehicle stop time, rather than the passenger waiting time (see (Ndibatya and Booysen 2021)), is crucial for energy analysis because the vehicle stops provide information about charging opportunities. An example of this is the first dataset evaluated below, which captures typical route information (origin, destination and temporal frequencies), but fails to specify how long the vehicle waits at formal and informal stops.

The second method which is slightly more invasive and has a higher setup cost, uses vehicle-fixed GPS trackers. The trackers continuously log and transmit their location and velocity to remote data collection centres through the communications network (usually cellular). This data is then made available from a cloud server either as timestamped GPS traces or as processed timestamped trip information that captures the origin and destination. This method is much more useful for energy analysis than the first method as it adequately and reliably captures the vehicle's moving and stopping patterns.

This chapter builds a foundation for evaluating the eventual impact of eMBT transition on sub-Saharan African cities' electric grids, localised pollution, carbon footprint and taxi owners' profitability. We look specifically at the energy requirements of these vehicles, and the potential for installing charging stations at the many formal and informal stops. We explore the energy requirements of electric minibus taxis in an urban context, which includes paratransit between and within towns and metros.

We use two publicly available datasets. The first, called the "Digital Matatu" dataset, was collected with mobile phones in Kampala, Uganda, and Nairobi, Kenya, and stored in the GTFS (general transit feed specification) format (DigitalTransport4Africa 2021; Williams et al. 2015). The Kampala dataset was actually collected by Transport for Cairo, and not by the Digital Matatu team. Although some useful information can be extracted from this dataset, we highlight the shortcomings and remaining challenges of using data captured by passengers on-board or at the roadside. The second consists of data from over a year of GPS vehicle tracking in Kampala and in Stellenbosch, South Africa. We analysed these datasets to assess the energy demands of minibus taxis and to explore the charging opportunities at the formal and informal stops. For both the datasets we used a micro-transport simulator.

23.2.2 *Data*

23.2.2.1 Passenger-Based Data

The first dataset we used, downloaded from DigitalTransport4Africa, is for the cities of Kampala and Nairobi (DigitalTransport4Africa 2021). This dataset captures the minibus taxis' schedules, routes and stops locations. Importantly, it does so from a transport system rather than a vehicle perspective. The focus is therefore on the route frequency, route length, origin location and destination location. Although this

Table 23.1 Properties of the first dataset, captured by passenger data collectors

City	Collectors	Weeks	Routes	Trips	Area (Km^2)	References
Kampala	Not specified	10	4,400	92,262	2,478	Transport for Cairo (2020)
Nairobi	5	24	264	35,640	3,097	Williamsetal. (2015)

may seem a fully descriptive dataset, the temporal coverage is dependent on the data collection method, as we will show. The use of fieldworkers meant that the degree of scientific rigour and reliability varied considerably.

Table 23.1 shows the number of collectors, weeks used to capture the data, taxi routes captured, trips captured, and surface area covered. If nothing else, a lot of data was captured. Kampala and Nairobi have similar population sizes –4.0 million and 4.7 million. The minibus taxi numbers reported for the two cities were 25,000 and 15,000. The difference is because Kampala's paratransit is 63% minibus taxis (as opposed to other forms, such as motorcycles), whereas Nairobi's is only 43% minibus taxis (other forms being more common here) (Bruun et al. 2016; Graeff 2009; Spooner et al. 2020; Watundu 2015).

23.2.2.2 Vehicle-Based Data

The second dataset was collected using tracking data from a fleet of eight minibus taxis in Kampala and nine minibus taxis in Stellenbosch. GPS trackers were installed in the vehicles, allowing full temporal coverage of the spatial location information. The dataset properties are listed in Table 23.2.

The Kampala part of this dataset consisted of GPS tracking data obtained over ten months from eight minibus taxis operating on routes in Kampala. The study area is defined by a box with coordinates (0.170202, 32.181182) and (0.794505, 32.852554) and is shown with a heatmap of GPS traces in Fig. 23.1a. It consisted of timestamped geographical coordinates, speed and direction, logged at one-minute intervals. We removed irrelevant data from the dataset, such as public holidays and weekends, and days when a minibus taxi left the area of study. We were left with an average of 150 days of data per taxi.

Table 23.2 Properties of the second dataset, captured by vehicle-based GPS trackers

City	Taxis	Weeks (After cleaning)	Daily distance (km)	Area (km^2)	References
Kampala	8	21	224	5,194	Booysen et al. (2022)
Stellenbosch	9	29	228	1,442	Abraham et al. (2021)

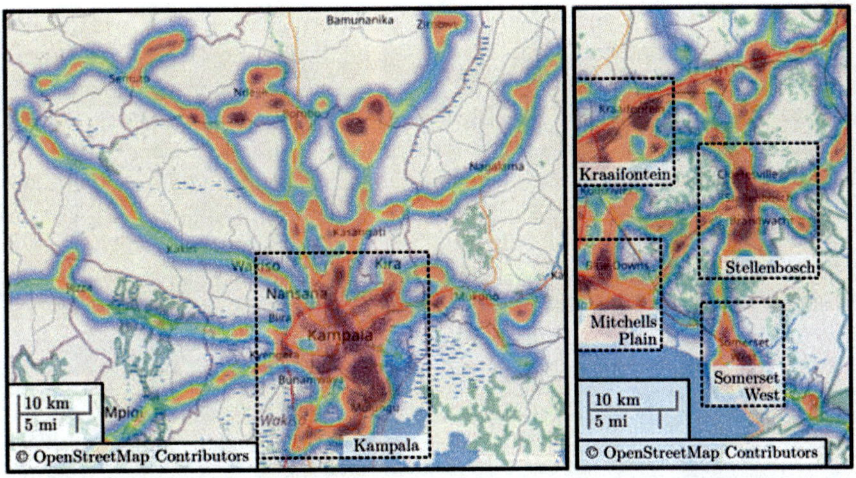

(a) Kampala (b) Stellenbosch

Fig. 23.1 Heatmaps of the two cities' GPS tracking data

The Stellenbosch part of this dataset consisted of GPS tracking data obtained over two years from nine minibus taxis operating on bi-directional routes connecting Stellenbosch with the northern suburbs of Cape Town and the town of Somerset West. The study area is defined by a box with coordinates (34.229224, 18.656884) and (−33.786222, 18.969438) as shown in Fig. 23.1b. The data was obtained from a local fleet management service provider, Mix Telematics. It consisted of timestamped geographical coordinates, speed and direction, logged at one-minute intervals. After cleaning the data, we were left with an average of 201 days' worth of data per minibus taxi.

23.2.3 Minibus Taxi Mobility Modelling from Passenger-Based Data

For each of the routes defined in the manually collected data, the software extracted the route's sequence of stops. After this, for each of the routes, it generated simulated *route plans*. A route plan generates a path (a sequence of *roads* connecting the stops), which the simulated vehicle must follow to go from one stop to the next. The route plan also specifies the time the vehicle must depart from each stop.

We identified a road network by downloading the OpenStreetMaps (OSM) raw data for the country in which each city lies (Geofabrik 2021). We cropped the OSM file to define an appropriate section of the network for our study, creating a boundary box defined by longitude and latitude coordinates (Markus Weber 2013). The software searched this chosen road network to find the paths that the simulated taxi should take along its route. We converted the cropped OSM file to a road network file which was readable by the mobility simulator, SUMO (Lopez et al. 2018).

The algorithm used for solving the paths was SUMO's implementation of the Dijkstra algorithm (Lopez et al. 2018). This algorithm chooses between various optimisation objectives when solving for a path: time, distance, energy usage, etc. We chose to use distance, as it is the computationally cheapest optimisation (Lewis 2020). However, for more realistic paths the *time* objective should be used instead. With the *distance* objective, the algorithm might choose short paths that go along roads with low speed limits and possible congestion. For example, it might choose a path through the city rather than along the highway. With the *time* objective, the algorithm might choose the highway option, which is longer in distance but shorter in time. This would be more realistic because the taxi driver would prefer the quicker option.

For each route, the route plan was generated by traversing through the route's sequence of stops. For each stop, the nearest road on the road network from the stop's coordinates is found, and the shortest path from the road of the previous stop to the current road is calculated. This path is appended to the route plan being built. The route plan also specifies that the simulated vehicle should stop on the current road until the departure time of the current stop.

From the generated routes, we computed extra information that affects an electric vehicle's energy usage, such as the total distance, the road inclination, and the road curvature.

23.2.4 Minibus Taxi Mobility Modelling from Vehicle-Based Data

Much more processing was required for the GPS-tracker data since the raw data does not identify any routes or indicate which datapoints constitute a stop. We, therefore, did spatio-temporal clustering to identify the stop locations and times. After this, we generated the route plans by computing the paths between stops in preparation for electric minibus taxi modelling.

We ran the EV models independently in a micro-transport simulator (SUMO) (Kurczveil et al. 2014; Lopez et al. 2018) and then recorded and analysed the vehicle's energy requirements for each simulated route.

We did three stages of spatial GPS data analysis to generate the mobility simulation data from measured GPS traces, using custom scripts written in Python. The first stage was a visual inspection of the raw GPS data on a map to identify the main areas where the minibus taxis operated. The second stage was spatial clustering of the geo-locations to identify regions with dense points of interest, and then categorising these points and identifying GPS-related errors. We also removed errors introduced by divergent driver behaviour, such as taking a trip to a beach or other unusual behaviour. The third stage was using the identified stops and SUMO's ancillary

Table 23.3 EV model parameters

Parameter	Value	Unit
Height	2.3	m
Width	1.9	m
Front surface area	4	m^2
Weight	2,900	kg
Constant power intake	100	W
Propulsion efficiency	0.8	–
Recuperation efficiency	0.5	–
Roll drag coefficient	0.01	–
Radial drag coefficient	0.5	–

routing function to generate the paths between consecutive stop events to use in a SUMO simulation with the built-in EV model. More information on the method can be found in Abraham et al. (2021), Booysen et al. (2022).

23.2.5 The eMBT Model Simulation Setup

To measure the temporal variation of power and energy usage, and the relationship between power consumption and eMBT speed, we set up a simulation model using a custom SUMO electric vehicle simulation module (SUMO EV). This model's parameters were specifically designed to be similar to those of the prevailing model of minibus taxi presently used in South Africa, the Toyota Quantum. Accordingly, the weight and front surface area of the eMBT model were taken from an existing Quantum. The rest of the parameters, summarised in Table 23.3, were approximated based on recommendations by Fridlund and Wilen (2020).

This eMBT model was applied to the route plans generated in Sects. 23.2.3 and 23.2.4. The simulation program initialised the eMBT model for each date that was simulated. For every second of simulation time, the simulator logged the energy consumption and speed of the eMBT as it progressed along its route plan.

23.3 Results and Discussion

23.3.1 Passenger-Based Data

The passenger-based energy results are shown in Fig. 23.2. Figure 23.2a shows the aggregated results for Kampala, in which all the individual routes in the dataset and their related frequencies were aggregated to estimate energy requirements for the paratransit system (minibus taxis only). The figure shows the power requirements throughout the day and the aggregated energy demand as the day progresses. Importantly, this only includes the routes covered in the dataset, and no extrapolation was done to include other routes. The figure shows a clear peak in the morning between 6 and 9 am, followed by a plateau up to 6 pm, after which demand quickly decreases. The energy profile (power and aggregated energy) is shown for a 43 km route in Fig. 23.2c.

Figure 23.2b. shows the aggregated results for Nairobi, again with all the routes and their corresponding frequencies aggregated for a system-level energy representation. The system profile shows the expected morning peak from 6 to 10 am, a lull between 10 and 3 am, and a longer peak from 3 to 9 pm, after which the power quickly decreases.

The aggregated power profile is substantially lower than that of Kampala, which peaks at 280 MW, while Nairobi peaks at a mere 90 MW. This may be partly because Nairobi has fewer minibus taxis than Kampala, as noted earlier. However, a simple proportional calculation shows that the difference may also be partly because of

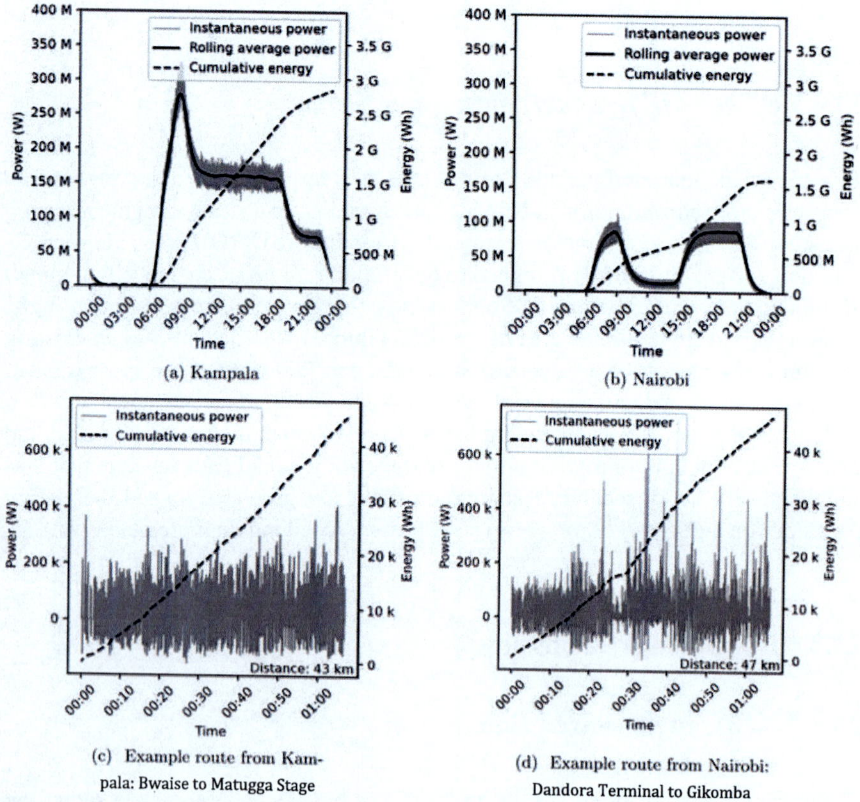

(a) Kampala

(b) Nairobi

(c) Example route from Kampala: Bwaise to Matugga Stage

(d) Example route from Nairobi: Dandora Terminal to Gikomba

Fig. 23.2 Daily power and energy profiles of minibus paratransit systems in Kampala and Nairobi, from passenger-based data

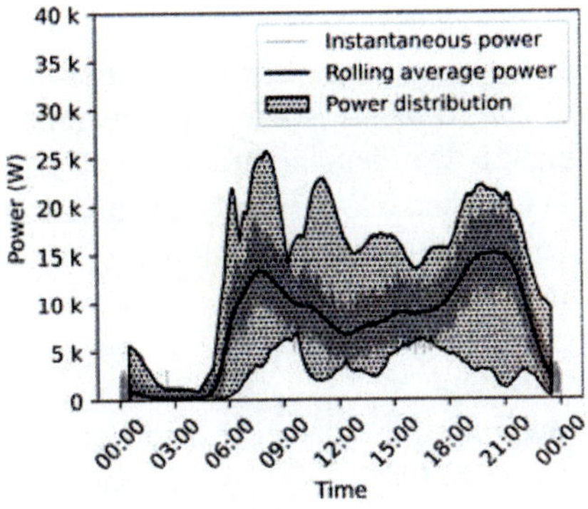

(a) Kampala

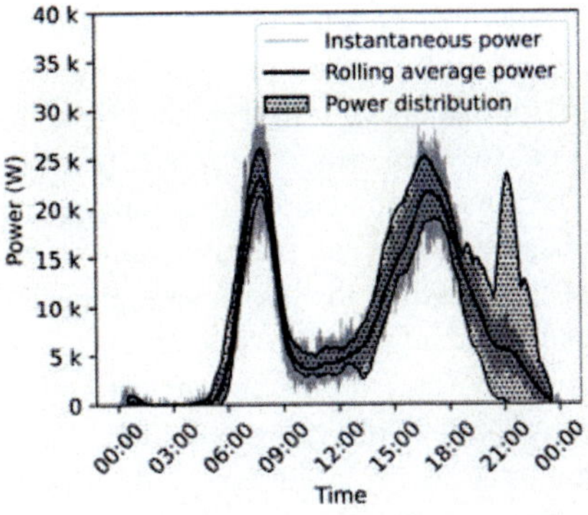

(b) Stellenbosch

Fig. 23.3 Summary of electrical demand for all the simulated eMBTs' daily power (instantaneous and rolling average) sampled per second, expressed per taxi

under-representation in the passenger-based acquisition of routes. It is a drawback that we have no way of knowing the taxis' destinations unless they had passengers collecting data.

23.3.2 Vehicle-Based Data

The minibus taxi mobility modelling provided spatial clustering of the identified stop events while generating paths used to determine the power requirements of the individual eMBTs and the fleet.

23.3.2.1 Energy Demand

The output of the eMBT simulation is shown in Fig. 23.3 for Kampala and Stellenbosch. For both the cities, the temporal power profiles seem to match the expected hours of peak traffic. Both profiles indicate that the energy requirements of the eMBTs would be highest during the peak traffic hours and would be lowest during midday and midnight. This hints that there may be a possibility of charging from grid power during midnight hours and charging from solar power during midday hours. The instantaneous power demand profiles of both the cities are only computed from taxi mobility data collected during weekdays (working days). The power demand shown in these profiles is the *net* power that was drawn from the battery of the vehicle.

The mean instantaneous power demand profile of Kampala is shown in Fig. 23.3a. In the morning, from around 7 am to 9 am, the taxis demand a high amount of power from their batteries, with a peak of around 15 kW on average. At around midday, from around 11 am to 4 pm, there is a period of relative inactivity. This causes the average power demand of the taxis to be reduced to around 10 kW. This demonstrates a possibility for the taxis to charge from solar energy during this time. After this period, the mean power demand gradually increases until the evening. From around 7 pm to 9 pm, we find another pronounced peak power demand of around 17 kW. The power demand declines until midnight, to less than 3 kW. Finally, the taxis are observed to be completely inactive from midnight until 5 am. This allows a period of 5 h in which the taxis can replenish their batteries, using grid power. The typical behaviour of these taxis over this 24-h period corresponds to a qualitative study done in Kampala by Spooner et al. (2020, p.29), who found that the working day of some taxi drivers was around 15 h. By integrating the power demand profile, the total energy required for the average taxi was 220 kWh per day. Similarly, by integrating the speed profile, the average distance was 224 km. This resulted in an energy efficiency of 0.98 kWh/km.

The mean instantaneous power demand profile of Stellenbosch is shown in Fig. 23.3b. During the morning hours, from around 6 am to 9 am, the taxis demand a high amount of power, with a peak of around 23 kW on average. At around midday, from around 9 am to 1 pm, there is a period of substantially reduced activity. This causes the average power demand of the taxis to be reduced to around 5 kW. This further demonstrates the potential for solar charging. After this period, the mean power demand gradually increases to a slightly less pronounced peak of 21 kW in the evening, from around 5 pm to 6 pm. The power demand declines until 9 pm, to less than 6 kW. Finally, the taxis are observed to be completely inactive from 11 pm until 5 am. The power demand profile of all the taxis which were observed in Stellenbosch was very similar. This is shown by the power distribution envelope in Fig. 23.3b. The envelope was computed by shading the area between the minimum and the maximum of the power profiles of all the taxis in the fleet. The envelope shows that the deviation between the taxis is very minimal, except for the peak, just after 9 pm, in the upper bound of the envelope. This was because of the behaviour of some taxis to start long-distance weekend trips on Friday evenings (Akpa Ebot Eno et al. 2016; Ebot Eno Akpa and Booysen 2014). The total energy required for the average taxi was 212 kWh per day, travelled over an average distance of 228 km. This resulted in an energy efficiency of 0.93 kWh/km.

The Kampala profile shows more variance in the power distribution and has less accentuated evening and afternoon peaks than the midday lull. This Kampala profile from vehicle tracking is substantially different from the passenger-based profile captured in Fig. 23.2. The profile generated from manually tracked data falls away shortly after 6 pm. We believe this is due to field workers going home after the day's work, which erroneously creates the impression of a system with diminished activity from 6 pm onward. Relying on the data captured by field workers would therefore lead to a substantial error in energy estimations, exemplifying the crucial need for vehicle-based data collection.

The daily energy usage of each of the Kampala taxis is shown as boxplots in Fig. 23.4a. The median energy usage ranged from a minimum of 108 kWh to 335 kWh, and the mean of this range was 220 kWh. As computed earlier, the mean distance travelled by the taxis was 224 km. Hence, the average energy efficiency was

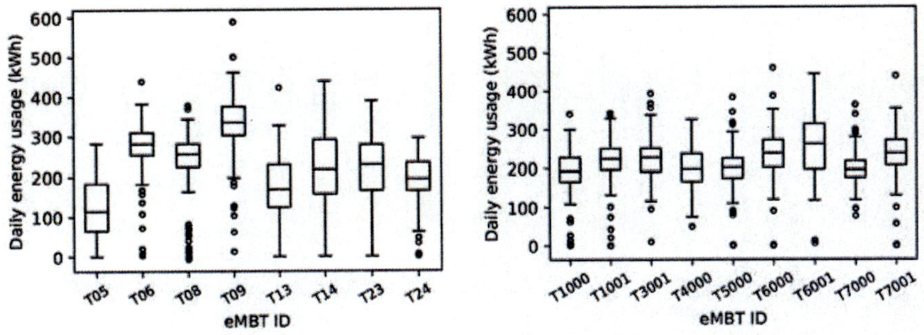

(a) Kampala (b) Stellenbosch

Fig. 23.4 Daily energy usage for each simulated eMBT

0.98 kWh/km. According to the energy usage distribution, one of the taxis (T09) used more than 375 kWh per day, 25% of the time, and required up to 491 kWh on its worst days. The situation is not as alarming for the other eight taxis, which use less than 314 kWh, 75% of the time.

The daily energy distribution of the Stellenbosch taxis is shown in Fig. 23.4b. The median energy usage of each of the nine taxis is more similar than in the Kampala dataset. It ranged from a minimum of 189 kWh to 252 kWh, resulting in a mean energy usage of 215 kWh. According to the energy usage distribution, each taxi is guaranteed to use less than 303 kWh per day, on 75% of its days. Eight of the nine taxis, on all days, used less than 420 kWh, while the remaining taxi used up to 490 kWh.

The results show that a maximum usable battery capacity of approximately 500 kWh would be sufficient for urban travel, if charging is limited to the stationary period before the day's first trip, and for 75% of the time a 303 kWh battery would be sufficient. This indicates that the battery capacities currently available in equivalent vehicles are insufficient. Charging during stationary times would be required to reduce the battery size requirements, which is a key driver for EV costs. Although the (in)efficiency numbers are high (a mean efficiency of 0.98 kWh/km), it sits in-between the reported energy efficiencies of electric varieties of light vehicles and large buses (Gao et al. 2017). Moreover, the stop-start nature of this mode of transport will have an adverse impact on efficiency.

23.3.3 Charging

With the large number of minibus taxis and the large energy requirements and fast charging, electrification of the fleet could place a burden on both local electrical infrastructure and the generation plants. Therefore, the opportunities for augmenting charging from solar photovoltaics are evaluated here. This assessment was done assuming no battery storage, to limit the infrastructure investment cost. Therefore,

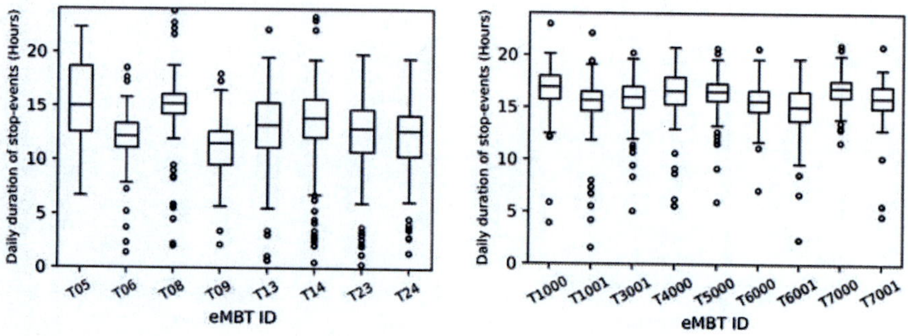

Fig. 23.5 Daily durations of minibus taxi stop events with duration thresholds (20 min to 8 h)

we had to determine the time during which vehicles are stationary and the extent to which those times overlaps with insolation and PV generation. To do this, we analysed the stationary times by identifying their start times and the durations of these stop events. From these stop times and durations, we could also determine the charging rate if the vehicles were to be charged from the electrical grid only. We discarded stops shorter than 20 min, as they would be considered too short to achieve meaningful charging and could be drop-and-go operations rather than proper stops.

Figure 23.5 shows the distribution across days of stop events with the minimum and maximum stop duration thresholds applied. Figure 23.5a shows that the Kampala minibus taxis' cumulative stop durations vary considerably, with the median duration per day ranging from a minimum of 8 h for taxi T05 to a maximum of 12 h for taxi T08. Figure 23.5b shows that there is also a substantial variation in stop duration times between the taxis in Stellenbosch, with the median stop duration per day ranging from a minimum of 8 h for taxi T3001, and a maximum of 11 h for taxi T6000.

The charger's charging rate needs to cater for the faster end of the charging spectrum. Accordingly, we used a representative vehicle that requires a relatively high energy, namely the average of the 75th percentile of energy required from Fig. 23.4. The charger rate is also affected by the duration since a shorter stop will require a faster charging time. Accordingly, we use the average of the 25th percentile of the stop duration times in Fig. 23.5. For Kampala, these were 273 kWh and 6.7 h, which necessitates a charging rate of 40.6 kW. To simplify the calculation, we assumed a constant charge rate, with which a taxi would be recharged from 40.6 kW for 6.7 h on most days. For Stellenbosch, these were 247 kWh and 7.7 h, which results in a charging rate of 32 kW. Again, this assumes a constant charge rate that is drawn from only the existing electrical grid.

23.4 Discussion and Implications: Using Data to Plan for Decarbonisation

The traditional domain of civil engineers is infrastructure. Transport engineers typically obtain information by employing people in two outdated ways, either to stand next to the road to record inflows and outflows or to get into vehicles to act as human carriers of tracking devices. These methods are particularly attractive in developing countries where operational traffic monitoring infrastructure is sparse and cheap labour is abundant. But, as mentioned in Sect. 23.1, this shortcut has many pitfalls.

Developments in vehicle tracking technology have changed the game. Although the setup cost is more, tracking devices have the substantial advantage that they are not susceptible to human behavioural problems. For example, they don't wake up late, get tired and need eating breaks, and they do not sleep.

To illustrate the difference, in Fig. 23.6 we show Kampala's energy profile from the two vantage points. The overlay shows the passenger-based energy profile, which was down-scaled by the number of taxis in Kampala (25,000 according to (Spooner et al. 2020)), and our vehicle-based energy profile. The differences are stark. First, the taxis started moving between 4 and 5 am, before the fieldworkers managed to get on- board. Second, the passenger-based profile grinds to a halt just before supper time. But we know that the minibus taxis in our samples happily chug away until after 9 pm.

Clearly, if we make assumptions about the power profile and energy requirements of electric minibus taxis from passenger-based data, we will miss the mark by a substantial margin. This adds weight to the statement by Collett and Hirmer that we quoted earlier stressing the need for adequate and reliable data (Collett and Hirmer 2021).

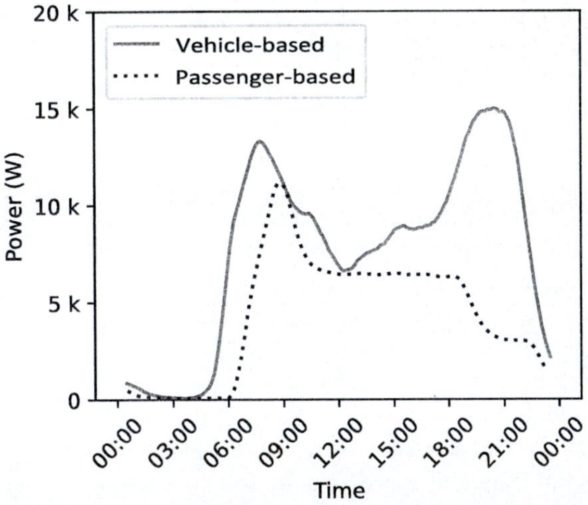

Fig. 23.6 Comparison of Kampala per-vehicle power profiles derived from passenger-based and vehicle-based data

23.5 Conclusion

The threat of transport emissions that exacerbate climate change is causing an energy revolution in electric vehicles in developed countries. Due to the availability of vehicles, market forces and manufacturer policies from those countries, this revolution will eventually reach developing countries in sub-Saharan Africa with its informal and chaotic paratransit systems and fragile electricity networks. Paratransit has many unique characteristics, challenges and opportunities that will shape the eventual adoption of electric minibus taxis. This chapter focused on paratransit in the sub-Saharan region, which is responsible for the transportation of an overwhelming number of the region's commuters.

We evaluated the electrical requirements of these minibus taxis with their unique mobility characteristics. With their spontaneous and flexible stopping behaviour, for durations determined by passenger demand, the stationary charging potential is unknown. We also assessed two ways of modelling electrical energy requirements, passenger-based and vehicle-based.

We found that passenger-based data may be useful for determining the aggregate energy load of a whole city or a single route. However, these results could be wholly wrong if the passenger-based tracking is not reliable. Vehicle-based tracking provides a reliable means of determining the energy requirements of a vehicle, and with sufficient adoption it could be used to determine system demand too.

Our results showed that the electricity demand of the taxis was similar, with a nominal 250 kWh required per median day if no additional charging capacity is provided. This increased to 420 kWh when we included all days, except for one taxi, which required 490 kWh. The median stops per day ranged from 7.7 h to 10.6 h, suggesting considerable potential for charging.

The taxis with shorter stopping periods, and hence the lower potential for charging, will need more energy because they are more mobile. Nevertheless, a nominal 41 kW charger will suffice when only charging only from the grid—if the grid is fully operational, which should not be taken for granted.

This chapter presents an informative and instructive view of how paratransit could be electrified, but challenges remain.

The first is the lack of sufficient energy generating capacity to power the paratransit charging stations. The blackouts and load-shedding experienced in sub-Saharan Africa may hamper the transition from combustion engine powered minibus taxis to electric minibus taxis. Research should therefore explore the possibilities of harnessing alternative sources of energy such as wind and solar photovoltaic to power the electric minibus taxi charging stations.

Separately, the electricity distribution infrastructure around minibus taxi ranks and informal stops cannot be assumed to be sufficient for charging numerous vehicles concurrently. It is expected that these vehicles will require charging after the morning peaks. For example, the relatively small town of Stellenbosch in South Africa has a population of approximately 150,000 people, and more than 100 minibus taxis. If these taxis were to initiate a fast charge at a nominal 60 kW, it would introduce a new load of 6000 kW, which is equivalent to introducing approximately 1,000 households or 120 schools. Such demands will require careful assessment of the local medium voltage (MV) and low voltage (LV) distribution network, the MV to LV transformers (which are typically limited to 30 to 50 kW). To overcome these physical limitations, a scheduled charging system will have to be employed, which will have to be closely tied to vehicle battery capacity, charging rates, and the times at which demand is expected.

Second, although our simulation included micro-simulation of traffic scenarios, it would be prudent to expand the traffic models to include different scenarios for different cities and different modes of transport, such as motorbike taxis. Third, our eMBT model was extended from an existing EV model in SUMO, a transport simulator. Although the parameters were adjusted to reflect a minibus taxi, validation and potential further development of such a vehicle would probably provide more accurate results. Finally, the battery storage requirements for both the vehicle and roadside infrastructure have not been explored. These will have a large impact on business models and the viability of the transition.

Apart from these two main challenges, practical challenges need to be addressed too. For example, the operations at a taxi rank are not set up for charging at present. Vehicles fall into queues, according to destination, and move to the front as vehicles depart. High-powered charging infrastructure tends to be stationary and does not allow for movement. Other practical considerations include electricity theft, which is rife in the region, theft of charging equipment, and payment solutions for electricity that support a large number of unbanked operators.

Although these challenges will address urban weekly transport, a major remaining challenge that needs to be addressed is that of long-distance transport, which the same vehicles currently support over weekends. Since these vehicles must cover more than 4000 km in 48 to 60 h, recharging needs to happen in minutes, which is the case with pumped fuel. More work needs to be done on battery swapping and exploring mobile energy-as-a-service, which seems to be the only viable electric option.

Finally, the elephant in the room is the taxi associations to which the owners belong. This sector is notoriously violent, territorial, and cut-throat. It will take more than technical know-how, good data, and environmental incentives to convince the sector to decarbonise.

Acknowledgements We acknowledge MTN South Africa (Contract S003061), Eskom (Tertiary Education Support Programme) and the National Research Foundation of South Africa (Reference number: MND200609529659) for funding this research through the MTN Mobile Intelligence Lab.

References

Abraham CJ, Rix AJ, Ndibatya I, Booysen MJ (2021) Ray of hope for sub-Saharan Africa's paratransit: solar charging of urban electric minibus taxis in South Africa. Energy Sustain Dev 64:118–127. https://doi.org/10.1016/j.esd.2021.08.003

Ajay Mahaputra K, Foster V, Barrett F (2008) Stuck in traffic: urban transport in Africa (tech. rep.). Washington, DC: World Bank

Akpa Ebot Eno NA, Booysen MJ, Sinclair M (2016) A comparative evaluation of the impact of average speed enforcement (ASE) on passenger and minibus taxi vehicle drivers on the R61 in South Africa. J South African Inst Civil Eng 58:2–10

Akumu J (2014) Improving air quality in African cities. UN Environment, Nairobi

Askari S, Peiravian F, Tilahun N, Yousefi Baseri M (2021) Determinants of users' perceived taxi service quality in the context of a developing country. Transp Lett 13(2):125–137. https://doi.org/10.1080/19427867.2020.1714844

Behrens R, McCormick D, Mfinanga D (2015a) Paratransit in African cities: operations, regulation and reform. https://doi.org/10.4324/9781315849515

Behrens R, McCormick D, Mfinanga D (2015b) In: Behrens R, McCormick D, Mfinanga D (eds) Paratransit in African cities: operations, regulation and reform, 1st ed. Routledge. https://doi.org/10.4324/9781315849515

Behrens R, McCormick D, Orero R, Ommeh M (2017) Improving paratransit service: Lessons from inter-city matatu cooperatives in Kenya. Transport Policy 53:79–88. https://doi.org/10.1016/j.tranpol.2016.09.003

Black A, Barnes J, Makundi B, Ritter T (2018) Electric two-wheelers in Africa? Markets, production and policy conference on green transformation and competitive advantage evidence from developing countries

Booysen MJ, Abraham CJ, Rix AJ, Ndibatya I (2022) Walking on sunshine: pairing electric vehicles with solar energy for sustainable informal public transport in uganda. Energy Res Soc Sci 85:102403. https://doi.org/10.1016/j.erss.2021.102403

Booysen MJ, Andersen SJ, Zeeman AS (2013) Informal public transport in Sub-Saharan Africa as a vessel for novel intelligent transport systems. 16th International IEEE conference on intelligent transportation systems (ITSC 2013), 767–772. https://doi.org/10.1109/ITSC.2013.6728324

Bruun E, Mistro RD, Venter Y, Mfinanga D (2016) The state of public transport systems in three Sub-Saharan African cities. In Behrens R, McCormick D, Mfinanga D (eds), Paratransit in African cities: operations, regulation, and reform (1st ed.). Routledge. https://doi.org/10.4324/9781315849515

Buresh KM, Apperley MD, Booysen MJ (2020) Three shades of green: perspectives on at-work charging of electric vehicles using photovoltaic carports. Energy Sustain Dev 57:132–140. https://doi.org/10.1016/j.esd.2020.05.007

Cervero R, Golub A (2007) Informal transport: a global perspective. Transp Policy 14(6):445–457. https://doi.org/10.1016/J.TRANPOL.2007.04.011

Collett KA, Hirmer SA (2021) Data needed to decarbonize paratransit in Sub-Saharan Africa. Nature Sustain 4(7):562–564. https://doi.org/10.1038/s41893-021-00721-7

Collett KA, Hirmer SA, Dalkmann H, Crozier C, Mulugetta Y, McCulloch MD (2021) Can electric vehicles be good for Sub-Saharan Africa? Energ Strat Rev 38:100722. https://doi.org/10.1016/j.esr.2021.100722

Diaz Olvera L, Plat D, Pochet P (2019) Looking for the obvious: motorcycle taxi services in SubSaharan African cities. J Transp Geogr, 102476.https://doi.org/10.1016/J.JTRANGEO.2019.102476

DigitalTransport4Africa (2021) Data [Online; accessed 17. Oct. 2021]

Dorothy M, Herrie S, David M (2016) The nature of paratransit operations. In: Roger B, Dorothy M, David M (eds), Paratransit in African cities: operations, regulation and reform, 1st ed. Routledge, pp 59–78

Du J, Ouyang M, Chen J (2017) Prospects for Chinese electric vehicle technologies in 2016–2020: ambition and rationality. Energy 120:584–596. https://doi.org/10.1016/j.energy.2016.11.114

Ebot Eno Akpa NA, Booysen MJ (2014) Minibus driving behaviour on the cape town to mthatha route. In: Proceedings of the 33rd Southern African transport conference (SATC), pp 761–770

Ehebrecht D, Heinrichs D, Lenz B (2018) Motorcycle-taxis in sub-Saharan Africa: current knowledge, implications for the debate on "informal" transport and research needs. J Transp Geogr 69:242–256. https://doi.org/10.1016/j.jtrangeo.2018.05.006

Evans J, O'Brien J, Ch Ng B (2018) Towards a geography of informal transport: mobility, infrastructure and urban sustainability from the back of a motorbike. Trans Inst Br Geogr 43(4):674–688. https://doi.org/10.1111/tran.12239

Fridlund J, Wilen O (2020) Parameter guidelines for electric vehicle route planning

Galuszka J, Martin E, Nkurunziza A, Achieng' Oginga J, Senyagwa J, Teko E, Lah O (2021) East Africa's policy and stakeholder integration of informal operators in electric mobility transitions—Kigali, Nairobi, Kisumu and Dar es Salaam. Sustainability 13(4). https://doi.org/10.3390/su13041703

Gao Z, Lin Z, LaClair TJ, Liu C, Li J-M, Birky AK, Ward J (2017) Battery capacity and recharging needs for electric buses in city transit service. Energy 122:588–600. https://doi.org/10.1016/j.energy.2017.01.101

Geofabrik (2021) Geofabrik download server. Accessed 24 Nov 2021

Goodfellow T (2017) Double capture' and de-democratisation: interest group politics and Uganda's 'Transport Mafia'. J Dev Stud 53(10):1568–1583. https://doi.org/10.1080/00220388.2016.1214722

Graeff J (2009) The organization and future of the matatu industry in Nairobi. Earth Institute, Columbia University, New York, Kenya. Paper prepared by Center for Sustainable Urban development, pp 1–3

Jennings G, Behrens R (2017) The case for investing in paratransit: strategies for regulation and reform

KCCA (2016) Multimodal urban transport master plan for greater Kampala metropolitan area (tech. rep.). Kampala Capital City Authority (KCCA). Kampala

Kumar AM (2011) Understanding the emerging role of motorcycles in African cities: a political economy perspective (tech. rep. No. 13). Washington, DC: World Bank. Washington, DC

Kurczveil T, L'opez PA, Schnieder E (2014) Implementation of an energy model and a charging' infrastructure in SUMO. In: Behrisch M, Krajzewicz D, Weber M (eds) Simulation of urban mobility. Springer, Berlin, Heidelberg, pp 33–43

Lewis R (2020) Algorithms for finding shortest paths in networks with vertex transfer penalties. Algorithms 13(11). https://doi.org/10.3390/a13110269

Li Y, Davis C, Lukszo Z, Weijnen M (2016) Electric vehicle charging in China's power system: energy, economic and environmental trade-offs and policy implications. Appl Energy 173:535–554. https://doi.org/10.1016/j.apenergy.2016.04.040

Lopez PA, Behrisch M, Bieker-Walz L, Erdmann J, Fl¨otter¨od Y-P, Hilbrich R, Lu¨cken L, Rummel J, Wagner P, Wießner E (2018) Microscopic traffic simulation using SUMO. In: 2019 IEEE intelligent transportation systems conference (ITSC), 2575–2582

Lozano Gracia N, Bainomugisha E, Soppelsa ME, Okure D (2021) Characterization of ambient air quality in selected urban areas in Uganda: a low-cost approach (tech. rep. No. 9512). The World Bank

Lucas K, Jennings G, Zuidgeest M, Behrens R, Venter C, Mitullah W, Thornton B (2019) Transport social exclusion in Five African Cities (tech. rep.). International network for transport and accessibility in low income communities. Leeds, UK

Markus Weber (2013) Osmconvert. (Software)

Mutiso W, Behrens R (2011) Boda Boda' bicycle taxis and their role in urban transport systems: case studies of Kisumu and Nakura, Kenya. In: 30th annual Southern African transport conference 11–14 July 2011 "Africa on the Move", CSIR International Convention Centre, 430–444

Ndibatya I, Booysen MJ (2020) Minibus taxis in Kampala's paratransit system: operations, economics and efficiency. J Transp Geogr 88. https://doi.org/10.1016/j.jtrangeo.2020.102853

Ndibatya I, Booysen MJ (2021) Characterizing the movement patterns of minibus taxis in Kampala's paratransit system. J Transp Geogr 92:103001. https://doi.org/10.1016/j.jtrangeo.2021.103001

Neumann A, Joubert JW (2016) The "Minibus" contribution. In: Horni A, Nagel K, Axhausen KW (eds) The multi-agent transport simulation matsim. Ubiquity Press. https://doi.org/10.5334/baw.17

Odhiambo E, Kipkoech D, Mohamed, Manuel M, Schalekamp H, Klopp AH, Jacquiline (2021) The potential for minibus electrification in three African cities: Cairo, Nairobi, and Cape Town. Volvo Res Educ Found. https://doi.org/10.13140/RG.2.2.26552.26882

Raje F, Tight M, Pope FD (2018) Traffic pollution: a search for solutions for a city like Nairobi. Cities 82:100–107. https://doi.org/10.1016/j.cities.2018.05.008

Rix AJ, Abraham CJ, Booysen MJ (2022) Computing our way to electric commuting in Africa: the data roadblock. Preprint currently under review at iScience. Available on engrXiv

Rnstvik HN (2013) Norway's electric vehicle deployment success. a historical review including plans for fast charging stations covering all of the country—by 2015. 2013 World Electric Vehicle Symposium and Exhibition (EVS27), pp 1–10

Schucking M, Jochem P, Fichtner W, Wollersheim O, Stella K (2017) Charging strategies for economic operations of electric vehicles in commercial applications. Transp Res Part D: Transp Environ 51:173–189. https://doi.org/10.1016/j.trd.2016.11.032

Sims R, Schaeffer R, Creutzig F, Cruz-Nu´n˜ez X, D'Agosto M, Dimitriu D, Meza MF, Fulton L, Kobayashi S, Lah O, McKinnon A, Newman P, Ouyang M, Schauer J, Sperling D, Tiwari G (2014). Transport. In: Edenhofer O, Pichs-Madruga R, Sokona Y, Farahani E, Kadner S, Seyboth K, Adler A, Baum I, Brunner S, Eickemeier P, Kriemann B, Savolainen J, Schl¨omer S, von Stechow C, Zwickel T, Minx J (eds) Climate change 2014: mitigation of climate change. contribution of working group iii to the fifth assessment report of the intergovernmental panel on climate change. Cambridge University Press, p 603

Singh A, Avis WR, Pope FD (2020) Visibility as a proxy for air quality in East Africa. Environ Res Lett 15(8):84002. https://doi.org/10.1088/1748-9326/ab8b12

Spooner D, Mwanika J, Manga E (2020) Kampala bus rapid transit: understanding Kampala's Paratransit market structure

Transport for Cairo (2020) Kampala and parantransit street usage study. Accessed 7 Nov 2021

Watundu S (2015). Urban household road travel demand and transport mode choice: the case of Kampala, Uganda

Weiss M, Cloos KC, Helmers E (2020) Energy efficiency trade-offs in small to large electric vehicles. Environ Sci Eur 32(1):46. https://doi.org/10.1186/s12302-020-00307-8

Williams S, White A, Waiganjo P, Orwa D, Klopp J (2015) The digital matatu project: using cell phones to create an open source data for Nairobi's semi-formal bus system. J Transp Geogr 49:39–51. https://doi.org/10.1016/j.jtrangeo.2015.10.005

Zinkernagel R, Evans J, Neij L (2018) Applying the SDGs to Cities: business as usual or a new sustainability 10(9). https://doi.org/10.3390/su10093201

M. J. Booysen Thinus is Professor of Electrical and Electronic Engineering at Stellenbosch University and Research Chair in the Internet of Things. His focus is on Smart Energy, Water and Vehicles in Sub-Saharan Africa. He is founding director of Bridgiot and GreenX, and Director of the MTN Mobile Intelligence Lab. He is a Member of the IET and senior member of the IEEE. He is Chartered at the Engineering Council (UK), and a Professional Engineer with the Engineering Council (ZA). He has industry experience in the aerospace and automotive industries with companies that include SunSpace, Rolls-Royce, Boeing, BMW, and Jaguar Land Rover.

Chris J. Abraham Chris first fell in love with Africa back when he rode his bicycle on the clay streets of rural Eastern Cape at age three. It saddens him to see Africa's inheritance of forests, minerals and wildlife pillaged and destroyed for short-term gain. Armed with a Bachelor's Degree in Mechatronics Engineering (CumLaude), Chris hopes to combat this problem through his research in renewable energy and green transport. With 30,000 views on his first YouTube lecture, he hopes to further educate and enable millions more to tackle these problems and to build the continent, rather than consume it.

Innocent Ndibatya Innocent Ndibatya (Ph.D.) is a Postdoctoral Researcher with the Alabama Transportation Institute, University of Alabama (USA). He holds a Ph.D. in Electrical & Electronics Engineering from Stellenbosch University (South Africa) and a MS.c. in Computer Science from Makerere University (Uganda). Innocent's research is in machine learning, artificial intelligence, agent-based modelling (ABM) and their applications to transportation systems decarbonisation. His lifelong career objective is to contribute towards carbon-neutral mobility by reducing mobility-related greenhouse gas emissions and mitigating climate change.

Arnold J. Rix With more than ten years of experience in the renewable energy industry Dr. Rix currently holds the Scatec Solar Chair in Photovoltaics Systems position at Stellenbosch University. The research group he is leading has a strong focus on Solar PV energy although other renewable generation technologies and batteries are included in his research. His path to renewable energy started after he completed his Ph.D. at the University of Stellenbosch in 2011 on the topic of electrical machine design for electric vehicles while working on the Joule electric vehicle project with Optimal Energy.

Part VI
Creating New Futures

Chapter 24
Conclusions: What Futures for Transport and Mobility in African Cities?

Ransford A. Acheampong, Karen Lucas, Michael Poku-Boansi, and Chinebuli Uzondu

As the chapter contributions by the various authors in this book have demonstrated, transport and mobility in Africa are not simply about shuttling people and goods from one point to another. Instead, they are fundamental to creating inclusive and sustainable cities. Some of the major challenges that cities in Africa face, including rising inequalities of access to economic and social opportunities, social exclusion, environmental deterioration and associated public health risks and climate change, stem directly from their current transportation situations. The interconnectedness of transport and mobility with critical social, environmental and economic systems means that addressing challenges in this sector of the urban realm would have a direct impact on tackling the continent's complex urban development challenges now and in the future. In recognition of this, we have aimed, through the whole set of issues and perspectives presented by the various authors in this book, drawing on multiple case studies from across the continent, to advance our understanding of the transport and mobility situations of African cities today. The rich empirical insights presented together serve to provide the canvas on which we can begin to imagine, explore and

R. A. Acheampong (✉)
Department of Planning and Environmental Management, University of Manchester, Manchester, UK
e-mail: ransfordantwi.acheampong@manchester.ac.uk

K. Lucas
Department of Geography, University of Manchester, Manchester, UK
e-mail: karen.lucas@manchester.ac.uk

M. Poku-Boansi
Department of Planning, Kwame Nkrumah University of Science and Technology, Kumasi, Ghana
e-mail: mpoku-boansi.cap@knust.edu.gh

C. Uzondu
Department of Transport Management Technology, Federal University of Technology, Owerri, Nigeria
e-mail: Chinebuli.uzondu@futo.edu.ng

R. A. Acheampong et al. (eds.), *Transport and Mobility Futures in Urban Africa*,
The Urban Book Series, https://doi.org/10.1007/978-3-031-17327-1_24

interrogate the kinds of futures cities in Africa should aspire to for their transport systems and the pathways toward these futures.

We have partly demonstrated that at a fundamental level, the urban transportation and mobility challenges facing cities in Africa today are deeply rooted in rapid levels of historical urbanization and the resulting urban spatial forms. As some of the case studies presented at the beginning of this book, as well as the wider scholarly literature have evidenced, the spatial forms of major urban centres across the continent have been radically transformed, largely through unplanned, unfettered outward expansion of human settlements. In the process, urban spatial structures have emerged in which individuals and households are increasingly having to live farther away from the already limited opportunities that their cities provide. The accessibility challenges resulting from the emergent spatial mismatch between where people live and where opportunities, such as jobs, healthcare and education are located manifest in two inter-linked and reinforcing ways. Firstly, the limited spatial coverage of public transport infrastructure and services, combined with the ever growing suburban and peri-urban populations means that in a typical African city, first-mile access to public transport—e.g. walking access to a bus stop or paratransit service route—is increasingly becoming a challenge for residents. Secondly, access by public transport to opportunities and valued destinations is worsening through the interplay of a number of factors, including the existing first-mile public transport accessibility deficits; poor integration between transport modes; congestion and long journey times; and rising transport fares that are increasingly becoming unaffordable, especially for the urban poor, vulnerable and those with low-incomes who depend on public transport to meet their everyday mobility needs.

What the foregoing implies is that without effective management of the rapid and uncontrolled expansion of cities that is already underway, a future where individuals and households have comparatively better and equitable access to the opportunities their cities provide, would not be realizable in Africa. We know that the on-going rapid urban expansion of cities across the continent is largely driven by individuals and households desperately seeking to meet their housing needs and increasingly doing so in peri-urban areas where land is relatively cheaper. Consequently, the emergent urban development patterns tend to be a patchwork of largely unplanned, low-density residential neighbourhoods with limited or no public transport connectivity to their established urban cores. Public transport service provisioning in such sprawling, low-density residential areas is extremely expensive and would be much more so in the future for city governments. Therefore, urban planning in African cities must rethink, evolve and implement bespoke development models and approaches that address the urban structural conditions underlying the accessibility challenges that residents face. To this end, there is a need to bring together the currently siloed sectors of land use and transportation planning. Instead of low-density sprawling development, cities must adopt and pursue an accessibility-oriented paradigm to urban development that prioritises higher density developments with good connections to reliable public transport, and offer people the opportunity to live closer to jobs, shops and other amenities.

As African cities continue to develop and transform over the next 10–20 years, one perspective is that the extreme challenges of continued material poverty and social deprivation of residents will persist. The African continent is likely to follow similar social development trends as almost low and middle-income countries worldwide. The continent's cities will display wide disparities in the wealth and wellbeing of their populations with many people still dependent on the informal economy for their survival and/or housed in slum settlements with limited or no access to basic utilities and commodities. Inadequate transport resources are simply one more denial of basic rights that the African urban poor will face on a daily basis in their struggle for survival.

Therefore, achieving transport justice and bringing an end to transport poverty are imperatives that African cities cannot ignore. Yet, the future trajectory toward the realization of these goals within African cities is quite bleak despite most countries' lip service to the United Nations Sustainable Development Goals for transport, which promise to: *provide access to safe, affordable, accessible and sustainable transport systems for all, improving road safety, notably by expanding public transport, with special attention to the needs of those in vulnerable situations, women, children, persons with disabilities and older persons by 2030'*. To meet the mobility and accessibility needs of poor and vulnerable citizens, as well as to address the severe environmental problems of urban transport systems, policymakers, planners and service providers would need to make a radical shift away from their currently favoured car-dominated spatial and infrastructure plans. Instead, the priority for future investment must be wholly targeted towards providing safe, healthy and liveable streets and multi-functional and inclusive activity spaces.

Moreover, new and emerging transport technologies and ICT-mediated mobility solutions that are increasingly becoming pervasive across cities in Africa may offer opportunities for people to meet their mobility needs. Technology will also have a role to play in decarbonising transport in light of the inescapable climate change imperative to reduce our carbon footprints. However, as the emerging evidence suggests, technologies alone are unlikely to resolve the significant mobility and accessibility barriers that urban poor and low-income households face in African cities. Thus, rather than invest large sums of money in road-base transport systems, smart technologies for automation and flagship public transport systems that mimic western models, which serve only a small minority of the overall population, what is needed is whole-system integrated urban design that reduces the need for people to travel and promotes the use of low-tech sustainable modes.

There is some evidence that the tide of opinion is slowly turning and that a more socially and environmentally sustainable transport agenda is beginning to gain traction on the global stage with the advent of a new UN Environment Programme-led African Walking and Cycling Network. Equally, some leading cities such as Nairobi are beginning to address the mobility needs of low-income settlements from the bottom up, as part of their wider slum uplift programmes. Local grassroots movements are also successfully promoting low-tech, community-led local transport initiatives. Such grassroots, citizen-led initiatives and experimentations have more potential in generating the experience and learning that are necessary for African cities to respond effectively to their unique transport and mobility challenges.

Major reforms are inevitable if the current transportation challenges in African cities are to be resolved. However, transport remains a highly politically charged and contested policy domain in African cities, with various actors who seem to recognise the need for transformative reforms, but have visions, motivations, interests and expectations that are often conflicting and seemingly irreconcilable. The stakes are often high for city mayors, politicians and public sector agencies pushing policy and regulatory reforms on the one hand and operators/service providers, drivers and other intermediary organizations on the other hand, who perceive such reforms as a threat to their livelihoods and survival. It will take genuine engagement with relevant stakeholders such as policymakers, regulators, incumbent paratransit operators and allied organizations, as well environments supportive of co-creation and implementation of transport policies and interventions, to bring about the needed reforms.

Understanding and recognising the enormity of the transport and mobility challenges that African cities face today is critical to designing and implementing the portfolio of interventions required to create desired futures. Today, urban areas across the continent exhibit similar transport and mobility situations and challenges, but the underlying causes are diverse and stem from the interplay of complex spatial, socio-cultural, political and economic forces that are unique to the individual countries and cities. There is therefore not a single transport and mobility future and/or pathway toward that future for African cities. Instead, the possibilities are endless for the creation of socially inclusive and environmentally sustainable transport futures, if African cities would strive to imagine, co-design, experiment and scale-up models and practices that work best within their unique contexts. Interventions must be conceived and targeted to respond to the needs of the continent's largely poor and vulnerable urban populations. Failure to ensure this will constitute major failings of African cities to create the enabling conditions critical to lift their populations out of extreme poverty and reducing inequalities.

Ransford A. Acheampong is a Lecturer and Presidential Academic Fellow in Transport and Urban Futures at the Department of Planning and Environmental Management, University of Manchester, UK. The first strand of his research is focused on the nexus between transportation and spatial structure and their integrated planning in cities. The second strand of his research addresses questions around the diffusion and socio-spatial impact of new and emerging transport technologies and mobility solutions, such as app-based mobility, autonomous vehicles, shared-mobility and Mobility-as-a-Service.

Karen Lucas is Professor of Human Geography at the University of Manchester where she also leads the Transport and Mobility hub for the Manchester Urban Institute. She has extensively researched issues of transport poverty in African cities and, since 2007 has led the International Network for Transport and Accessibility in Low Income Communities (INTALInC).

Michael Poku-Boansi is a Professor of Planning at the Department of Planning, and the Vice Dean, School of Graduate Studies, Kwame Nkrumah University of Science and Technology, Kumasi, Ghana. Michael has over fifteen (15) years' of transportation and land use planning, resettlement planning, climate change and resilience planning experience spanning both academic and private sector appointments. Professor Poku-Boansi is a Fellow of the Rockefeller Foundation Professional Fellowship Programme and a Commonwealth Academic Fellow. He has authored/co-authored several peer-reviewed research articles published in reputable international journals and numerous project reports. Prof. Michael Poku-Boansi is also a Fellow of the Ghana Institute of Planners (FGIP).

Chinebuli Uzondu is a Lecturer and Researcher at the Department of Transport Management Technology, Federal University of Technology Owerri Nigeria. She has previously worked as a Post-doctoral researcher with the Human Factors and Safety Group at the Institute for Transport Studies, University of Leeds, United Kingdom where she obtained her Ph.D. Her research seeks to understand road user behaviour through projects funded by various research organisations and the Nigeria and UK governments. This involves using both quantitative and qualitative research methods, to inform the design of transport infrastructure and/or technology interventions, and to evaluate the effectiveness of these interventions.